This book describes the stochastic method for ocean wave analysis. This method provides a route to predicting the characteristics of random ocean waves – information vital for the design and safe operation of ships and ocean structures.

Assuming a basic knowledge of probability theory, the book begins with a chapter describing the essential elements of wind-generated random seas from the stochastic point of view. The following three chapters introduce spectral analysis techniques, probabilistic predictions of wave amplitudes, wave height and periodicity. A further four chapters discuss sea severity, extreme sea state, directional wave energy spreading in random seas and special wave events such as wave breaking and group phenomena. Finally, the stochastic properties of non-Gaussian waves are presented. Useful appendices and an extensive reference list are included. Examples of practical applications of the theories presented can be found throughout the text.

This book will be suitable as a text for graduate students of naval, ocean and coastal engineering. It will also serve as a useful reference for research scientists and engineers working in this field.

T0275616

CAMBRIDGE OCEAN TECHNOLOGY SERIES: 6

General editors: I. Dyer, R. Eatock Taylor, J. N. Newman, W. G. Price

OCEAN WAVES

Cambridge Ocean Technology Series

OCEAN WAVES

The Stochastic Approach

Michel K. Ochi
University of Florida

CAMBRIDGE UNIVERSITY PRESS
Cambridge, New York, Melbourne, Madrid, Cape Town, Singapore, São Paulo

Cambridge University Press
The Edinburgh Building, Cambridge CB2 2RU, UK

Published in the United States of America by Cambridge University Press, New York

www.cambridge.org
Information on this title: www.cambridge.org/9780521563789

First published 1998
This digitally printed first paperback version 2005

A catalogue record for this publication is available from the British Library

Library of Congress Cataloguing in Publication data

Ochi, Michel K.
    Ocean waves / Michel K. Ochi.
        p.   cm. – (Cambridge ocean technology series ; 6)
    ISBN 0 521 56378 X (hc)
    1. Ocean waves.   I. Title.   II. Series.
GC211.2.026   1998
    551.47′02–dc21   97-16355 CIP

ISBN-13 978-0-521-56378-9 hardback
ISBN-10 0-521-56378-X hardback

ISBN-13 978-0-521-01767-1 paperback
ISBN-10 0-521-01767-X paperback

CONTENTS

PREFACE

This book is intended to provide uniform and concise information necessary to comprehend stochastic analyss and probabilistic prediction of wind-generated ocean waves.

Description and assessment of wind-generated ocean waves provide information vital for the design and operation of marine systems such as ships and ocean and coastal structures. Wind-generated seas continuously vary over a wide range of severity depending on geographical location, season, presence of tropical cyclones, etc. Furthermore, the wave profile in a given sea state is extremely irregular in time and space – any sense of regularity is totally absent, and thereby properties of waves cannot be readily defined on a wave-by-wave basis.

Characterization of the stochastic properties of ocean waves was first presented in the early 1950s; Neumann (1953), Pierson (1952, 1955), St Denis and Pierson (1953) introduced the stochastic approach for analysis of random seas, and Longuet-Higgins (1952) demonstrated the probabilistic estimation of random wave height. The four decades following the introduction of the stochastic prediction approach have seen phenomenal advances in the probabilistic analysis and prediction methodologies of random seas.

For the design of marine systems, information on the real world is required. Recent advances in technology permit the use of the probabilistic approach to estimate the responses of marine systems in a seaway, including extreme values, with reasonable accuracy. Such technology lends itself to application of the probabilistic approach as an integrated part of modern design technology in naval, ocean and coastal engineering.

In view of the growing need for more comprehensive advances in prediction methodologies and for application of the probabilistic approach in naval, ocean and coastal engineering, this book is designed as a text book at the graduate level and as a reference book for researchers and designers. The intent is to provide a thorough understanding of the modern concept of stochastic analysis and probabilistic prediction of wind-generated random seas. Specific efforts are made in this work to explain the basic principles supporting current prediction techniques and to provide practical applications of prediction methods.

Readers are expected to be familiar with basic probability theory and fundamental stochastic processes. For the readers' convenience, however, definitions, theorems and relevant formulae on probability and stochastic process theory used in the text are summarized in the appendixes without proof or derivation.

I am grateful to the College of Engineering, University of Florida, for granting me sabbatical leave to prepare this book. Significant progress was achieved toward its completion during this period of time. I would like to acknowledge the encouragement and support received from Professor Eatock Taylor of the University of Oxford during this undertaking. Thanks are also due to Professor Isobe of the Tokyo University who provided valuable suggestions on the section addressing directional wave spectra.

I am indebted to many learned scholars and researchers who directly or indirectly inspired me to study the stochastic analysis and probabilistic prediction of ocean waves. I thank those who sponsored my research which ultimately culminated in this book; in particular, Dr Silva, Office of Naval Research. Appreciation is extended to my graduate students; in particular, Drs C.H. Tsai, D.W.C. Wang, I.I. Sahinoglou, K. Ahn and Lieut. D.J. Robillard, US Navy, who through their dedicated project support had a significant influence on the final product. Finally, I would like to acknowledge the contribution of my wife, Margaret, who read the complete manuscript and provided valuable assistance with the editorial work.

1 DESCRIPTION OF RANDOM SEAS

1.1 STOCHASTIC CONCEPT AS APPLIED TO OCEAN WAVES

1.1.1 Introduction

The profile of wind-generated waves observed in the ocean changes randomly with time; it is non-repeatable in time and space. In reality, both wave height (peak-to-trough excursions) and wave period vary randomly from one cycle to another. It is often observed that waves break when the wave steepness exceeds a certain limit. Furthermore, during the process of the wind-generated waves traveling from one location to another after a storm, waves of shorter length gradually lose their energy resulting in the wave profile becoming less irregular (this situation is called swell) than that observed during a storm.

A more distinct difference in the wave profile can be observed when the water depth becomes shallow. As an example, Figure 1.1 shows portions of wave profiles recorded in severe seas; one in deep water in the North Atlantic, the other in a nearshore area of water depth 2.1 m. As seen in Figure 1(a), positive and negative sides of the wave profile in deep water are, by and large, similar, while for waves in shallow water (Figure 1(b)), peaks are much sharper than troughs, and the order of magnitude of the peaks is different from that of the troughs.

As seen in the examples shown in Figure 1.1, evaluation of the properties of random waves is almost impossible on a wave-by-wave basis in the time domain. However, if we consider the randomly changing waves as a

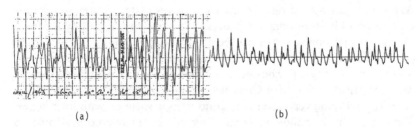

(a) (b)

Fig. 1.1. Wave profiles in severe seas. (a) deep water; (b) shallow water.

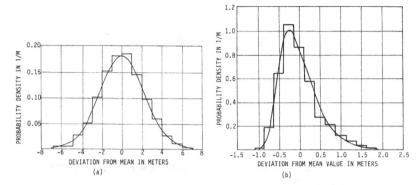

Fig. 1.2. Comparison between histogram of wave profile and theoretical
probability density function: (a) deep water; (b) shallow water.

stochastic process, then it is possible to evaluate the statistical properties
of waves through the frequency and probability domains.

In the stochastic process approach, waves in deep water are categorized as a Gaussian random process for which the probability distribution
of displacement from the mean value (wave profile) obeys the normal
probability law. On the other hand, waves in areas where the water depth
affects the wave properties are categorized as a non-Gaussian random
process. Examples of comparisons between histograms of the wave
profile constructed from data obtained in severe seas and the theoretical
probability density functions are shown in Figure 1.2. Figure 1.2(a)
shows a comparison for waves in deep water in which the theoretical
probability density function is a normal probability distribution with a
variance evaluated from the data. In Figure 1.2(b) the comparison pertains to waves obtained in shallow water, and the theoretical probability
density function given in the figure is as presented in Section 9.2.3. As
seen in these examples, wave profiles in both deep and shallow water are
well represented by theoretical probability distributions, and this permits
us to predict various statistical properties with reasonable accuracy.

1.1.2 Ocean waves as a Gaussian random process

As stated in the preceding section, ocean waves in deep water are considered to be a Gaussian random process. This was first found by Rudnick
(1951) through analysis of measured data obtained in the Pacific Ocean. In
general, the Gaussian property of ocean waves depends on sea severity and
water depth. It may safely be said that if the water is sufficiently deep, waves
may be considered to be a Gaussian random process irrespective of sea
severity, including very severe sea conditions associated with hurricanes.
Waves in relatively shallow water areas may also be considered as a Gaussian
random process if the sea severity is very mild. This will be discussed in

detail in Section 5.2.3 in connection with hurricane-generated seas and in Chapter 9 where wave properties in shallow water areas are presented.

The question arises as to the rationale for ocean waves being a Gaussian random process. This may be explained based on the central limit theorem in probability theory as follows.

Let η be the wave profile at a fixed time t. Here, η is a random variable defined in the sample space $(-\infty,\infty)$. We may assume that η is the sum of a large number of components X. That is,

$$\eta = X_1 + X_2 + \ldots + X_n \tag{1.1}$$

where the X_i are statistically independent random variables having the same probability distribution, although the form of the probability distribution is unknown. Let the mean value of X_i be zero and its variance (the second moment) be σ^2. Since the X_i are statistically independent, the probability distribution of η (which is unknown at this stage) has zero mean and variance $n\sigma^2$, where n is large.

We may standardize the random variable η and write the new random variable Z as follows:

$$Z = \eta/(\sqrt{n}\sigma) = \sum_{j=1}^{n} X_j/(\sqrt{n}\sigma) \tag{1.2}$$

Let the characteristic function of X be $\phi_x(t)$, though the form of $\phi_x(t)$ is unknown. Then, by using the properties of the characteristic function, the characteristic function of the standardized random variable $X/(\sqrt{n}\sigma)$ can be written from Eq. (1.2) as $\phi_x(t/\sqrt{n}\sigma)$. Hence, the characteristic function of Z becomes

$$\phi_z(t) = \{\phi_x(t/\sqrt{n}\sigma)\}^n \tag{1.3}$$

On the other hand, the characteristic function can be expanded in general as follows:

$$\phi_x(t) = 1 + it\,E[x] - \frac{t^2}{2}E[x^2] + \ldots \tag{1.4}$$

Since $E[x]$ and $E[x^2]$ of the standardized random variable are 0 and 1, respectively, $\phi_x(t/\sqrt{n}\sigma)$ may be written as

$$\phi_x(t/\sqrt{n}\sigma) = 1 - (t^2/2n) + o(t^2/n) \tag{1.5}$$

and thereby we have

$$\phi_z(t) = \{1 - (t^2/2n) + o(t^2/n)\}^n \tag{1.6}$$

By letting $n \to \infty$, Eq. (1.6) yields

$$\phi_z(t) = \exp\{-t^2/2\} \tag{1.7}$$

This is the characteristic function of the standardized normal distribution. This implies that the random variable Z obeys the normal distribution with zero mean and unit variance. Therefore, η has a normal distribution with zero mean and variance $n\sigma^2$, and hence it may be said that the waves are a Gaussian random process.

Following the concept of the central limit theorem, Pierson *et al.* (1958) show an explanatory sketch (given in Figure 1.3) indicating ocean waves consisting of an infinite number of sinusoidal waves having the same height with different frequencies and directional angles.

1.1.3 Random seas

The random fluctuation of the sea surface is generally attributed to energy transfer from wind to the sea. The interaction between air and sea which

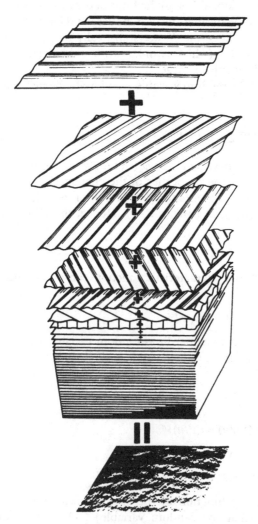

Fig. 1.3. Structure of random sea (Pierson *et al.*, 1958).

leads to wave formation is an extremely interesting phenomenon. However, the generation and growth mechanism of waves by wind is beyond the scope of this text, and it may suffice to list such well-known references on this subject as Jaffreys (1924, 1925), Miles (1957, 1959a, 1959b, 1960, 1962) and Phillips (1966, 1967). Further detailed discussions on this subject may be found in Sverdrup and Munk (1947), Pierson (1952, 1955), Kinsman (1965) and Donelan (1990), among others.

The potential and kinematic energies of random waves are represented by the wave spectral density function, often simply called the wave spectrum, and it plays a significant role in evaluating the statistical properties of random waves. The mathematical definition and a detailed discussion of the wave spectral density function will be given in Chapter 2. Only a conceptual description of the wave spectrum, necessary for the stochastic presentation of ocean waves, is given here as outlined below.

First, an example of the wave spectral density function, $S(\omega)$, computed from data obtained in the North Atlantic (Moskowitz et al. 1963) is shown in Figure 1.4. The spectral density function illustrates the magnitude of the time average of wave energy as a function of wave frequency, and the area under the density function represents the degree of sea severity. The most commonly used definition of sea severity is the significant wave height, denoted by H_s (defined in Section 3.6) which is equal to four times the square-root of the area under the spectral density function. The significant wave height of the spectrum shown in Figure 1.4 is 8.9 m. As seen in the figure, the spectrum peaks at the frequency $\omega=0.43$ radians

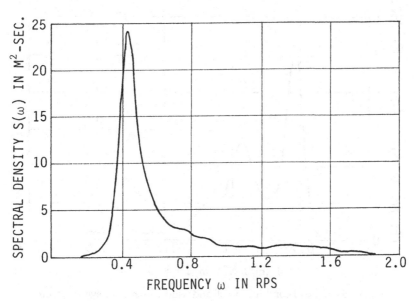

Fig. 1.4. Example of wind-generated spectrum (significant wave height 8.9 m).

per second (rps) in this example. This implies that waves of length $2\pi g/\omega^2 = 333$ m have the largest energy in this sea. It should not be interpreted from Figure 1.4 that waves of 330 m in length are the most dominant, frequently observed waves in this sea. The frequency of occurrence of waves having a specific length must be evaluated from the probability density function of wave period.

As stated earlier, the source of irregularity of waves observed in a sea is usually the local wind. Sometimes, however, another wave system, called *swell*, runs across or mixes with wind-generated local waves. Swell is defined as waves which have traveled out of their generating area. During the course of traveling, shorter waves are overtaken by larger waves resulting in a train of more regular long waves moving in its own direction. Fairly large waves observed at sea with minor or even no wind may be categorized as swell.

When swell mixes with the local wind-generated waves, it is not easily identified in the wave record; however, it can be clearly identified in the wave spectrum. Figure 1.5 shows a wave spectrum of significant wave height of 4.9 m obtained from data taken in the North Atlantic. The sea is much less severe than that shown in Figure 1.4. The spectrum has double peaks, one of which occurs at the same frequency, 0.43 rps, as observed in the spectrum shown in Figure 1.4. In this case, however, it is understood that a swell of length approximately 333 m is crossing or mixing with the local wind-generated seas.

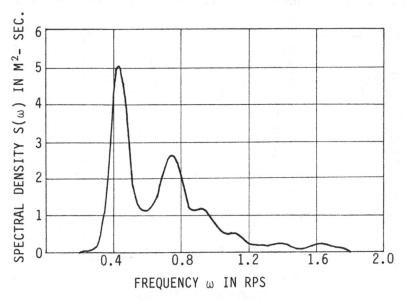

Fig. 1.5. Example of combined swell and wind-generated wave spectrum (significant wave height 4.9 m).

It is noted that, in extremely severe seas, the major portion of the wave energy is usually concentrated around a low frequency while the remaining energy spreads over a wide range of higher frequencies. The very large low frequency energy observed in this case, however, is not associated with swell. An example of a wave spectrum for extremely severe seas (significant wave height of 16.1 m) is shown in Figure 1.6. This spectrum is one of very few wave spectra obtained from records taken in extremely severe seas.

Since waves in the ocean are not necessarily moving in the same direction as the wind, the wave energy represented by the spectrum must consist of energies spreading in various directions. Therefore, we may consider a directional spectral density function, denoted by $S(\omega,\theta)$, and let the time average of wave energy at any frequency interval $\Delta\omega$ and for any directional angle interval $\Delta\theta$ of the random sea be equal to $(1/2)\,\rho g a_j^2$, where a_j is a positive random variable. That is, by ignoring the factor ρg, we may write

$$S(\omega,\theta)\Delta\omega\Delta\theta=(1/2)\,a_j^2 \qquad (1.8)$$

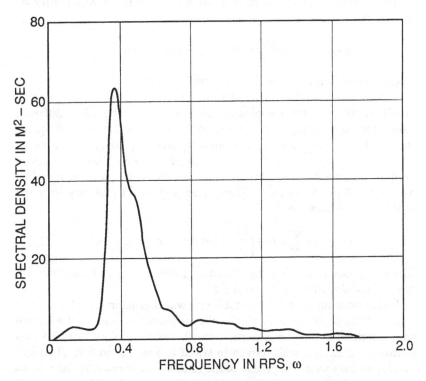

Fig. 1.6. Wave spectrum evaluated from data obtained in extremely severe seas (significant wave height 16.1 m).

The time average of the total energy of waves coming from various directions and including all frequencies is given by

$$\sum_{\Delta\omega}\sum_{\Delta\theta}\frac{1}{2}a_j^2 = \int_{-\pi}^{\pi}\int_0^{\infty} S(\omega,\theta)\,d\omega\,d\theta \qquad (1.9)$$

This is the basis of the stochastic description of waves in random seas.

1.2 MATHEMATICAL PRESENTATION OF RANDOM WAVES

For the mathematical presentation of waves in random seas, knowledge of the basic stochastic theory presented in Chapter 2 is required. Hence, the stochastic description of random waves will be discussed in Chapter 2. However, a mathematical presentation of random wave profiles based on Eq. (1.8) is presented in this section.

Let us consider a progressive wave in deep water using the coordinate system (X,Y,Z) fixed in space, and let θ be the angle taken in a counter-clockwise direction with respect to the X-axis. Then, the profile of simple harmonic waves traveling at an angle θ with the X-axis may be written as

$$\eta(x,y,t) = a\cos\left\{\frac{\omega^2}{g}(x\cos\theta + y\sin\theta) - \omega t + \epsilon\right\} \qquad (1.10)$$

where a=amplitude, ω=frequency in rps, ϵ=phase.

Next, let us consider the incidental wave profile at time t to be of an infinite number of sinusoidal wave components composed of different amplitudes, a_j, coming from divergent directions θ_j, with various frequencies ω_j. Here, a_j, θ_j and ω_j are random variables covering the range $0 < a_j < \infty$, $-\pi < \theta_j < \pi$ and $0 < \omega_j < \infty$, respectively. The phase ϵ is also a random variable distributed uniformly over the range $-\pi < \epsilon_j < \pi$, and its magnitude depends on the frequency and angle. Thus, we may write the profile of random waves as

$$\eta(x,y,t) = \sum_j a_j\cos\left\{\frac{\omega_j^2}{g}(x\cos\theta_j + y\sin\theta_j) - \omega_j t + \epsilon_j\right\} \qquad (1.11)$$

The amplitude a_j satisfies the condition given in Eq. (1.8) for any frequency and directional interval $\Delta\omega\,\Delta\theta$.

By considering the number of discrete waves given in Eq. (1.11) to be extremely large, while maintaining the frequency as well as directional angle components to be extremely small, the summation may be presented as an integral with respect to ω and θ. Then, from Eqs. (1.8) and (1.11), we may write the profile of the random sea surface by the following stochastic integral representation (Pierson 1955, St. Denis and Pierson 1953):

$$\eta(x,y,t) = \int_{-\pi}^{\pi} \int_{0}^{\infty} \cos\left\{\frac{\omega^2}{g}(x\cos\theta + y\sin\theta) - \omega t + \epsilon(\omega,\theta)\right\}$$

$$\times \sqrt{2\ S(\omega,\theta)\ d\omega\,d\theta} \tag{1.12}$$

The integration in the above equation is not an integral in the Riemann sense. It represents simply a mathematical abstraction, and the equation should be interpreted in the following manner:

$$\eta(x,y,t) = \sum_{j=-s}^{s} \sum_{i=0}^{r} \left\{ \cos\frac{\omega_{2i+1}^2}{g}(x\cos\theta_{2j+1} + y\sin\theta_{2j+1}) \right.$$

$$\left. - \omega_{2i+1}t + \epsilon(\omega_{2i+1},\ \theta_{2j+1})\right\}$$

$$\times \sqrt{2\ S(\omega_{2i+1},\theta_{2j+1})(\omega_{2i+2}-\omega_{2i})(\theta_{2j+2}-\theta_{2j})} \tag{1.13}$$

with the limits $\omega_{2r}\to\infty$, $\omega_{2i+2}-\omega_{2i}=0$, $\theta_{2s+1}\to\pi$, $\theta_{-2s-1}\to-\pi$ and $\theta_{2j+2}-\theta_{2j}=0$.

It is often convenient to express the wave profile given in Eq. (1.12) in a vector form by writing $\omega^2/g=k$ and by using the following definition:

$$\mathbf{r} = x\mathbf{i} + y\mathbf{j}$$
$$\mathbf{k} = k\cos\theta\mathbf{i} + k\sin\theta\mathbf{j} \tag{1.14}$$
$$= k_x\mathbf{i} + k_y\mathbf{j}$$

Equation (1.12) then may be written as

$$\eta(\mathbf{r},t) = \int_{-\infty}^{\infty} \int_{-\pi}^{\pi} \cos(\mathbf{k}\cdot\mathbf{r} - \omega t + \epsilon)\mathrm{d}A(\omega,\theta)$$

$$= \mathrm{Re}\int_{-\infty}^{\infty} \int_{-\pi}^{\pi} \mathrm{e}^{\mathrm{i}(\mathbf{k}\cdot\mathbf{r} - \omega t + \epsilon)}\mathrm{d}A(\omega,\theta) \tag{1.15}$$

1.3 STOCHASTIC PREDICTION OF WAVE CHARACTERISTICS

The major portion of this text will be devoted to the presentation and discussion of stochastic analysis and probabilistic prediction of ocean waves. We therefore briefly summarize here the principle and approach supporting the currently available prediction methodologies.

For stochastic prediction of various properties (height, period, etc.) of random waves, the development of a probability function applicable for each of these properties is prerequisite. At the same time, evaluation of parameters involved in the probability function is mandatory; this may be called input information to the probability function.

There are two different approaches to acquire input information; one in the time domain, the other through the frequency domain. If, for example, one is interested in the statistical properties of wave height, the necessary input to the probability function can be evaluated by reading all individual wave heights from measured data, and estimating the parameters of the probability function based on statistical inference theory. This approach for evaluating the parameter(s) of the probability density function is called the random observation method in the time domain (see Section 3.9). The method is simple, and carries the merit of expediency; however, the reliability of estimated results is not high unless the number of observations is sufficiently large.

On the other hand, the spectral analysis approach in the frequency domain primarily discussed in this text is much more mathematically rigorous than the random observation method. This approach is based on stochastic theory developed by Rice in communication engineering (Rice 1944, 1945).

The stochastic analysis and prediction of random waves spans three domains; time, frequency and probability, as shown in Figure 1.7. Let us consider the time history of a wave measured at a certain location in the ocean. The measured record represents the time history of all waves passing that location irrespective of the direction the waves are coming from. This type of wave measurement, from which wave characteristics are evaluated without reference to wave directionality, is called *point measurement*.

It is assumed that waves are a weakly steady-state ergodic random process which will be defined in Section 2.1.1. It suffices here to state that this assumption is generally satisfied for waves in deep water.

In stochastic analysis, we define the auto-correlation function which yields the variance of random waves in the time domain, but more importantly it provides information necessary for transferring from the time domain to the frequency domain. That is, from information on the auto-correlation function, we can evaluate the wave spectral density function, which is often simply called the wave spectrum in the frequency domain. This can be done by applying the Wiener–Khintchine theorem which states that the correlation function and the spectral density function are Fourier transforms of each other.

Although the spectrum can be directly obtained from wave records without evaluating the auto-correlation function, the above-mentioned procedure provides the mathematical background on which the probabilistic prediction of wave characteristics is based.

Sometimes, several wave records are taken simultaneously. In this case, auto- as well as cross-correlation function can be evaluated from the records. The cross-correlation function in the time domain can also be transferred to the cross-spectral density function through the Wiener–Khintchine theorem; however, the significant difference between auto- and cross-spectra is that the latter is a complex-valued function in contrast to the former which is a real-valued function. The directional characteristics of waves can be obtained through analysis of cross-spectra evaluated for several wave height records, as will be discussed in detail in Chapter 7.

The wave spectrum is a source of information from which the probabilistic prediction of various wave properties can be achieved in the

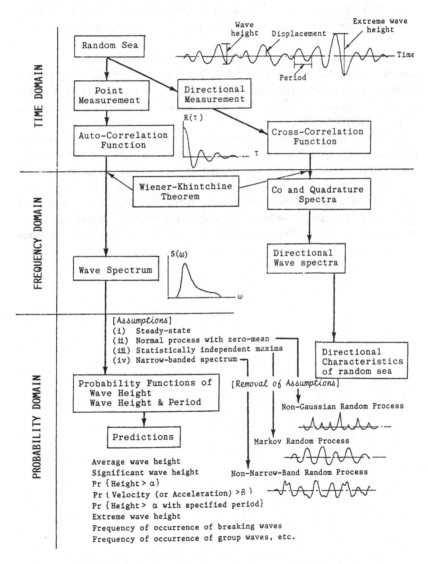

Fig. 1.7. Principle and procedure for predicting stochastic properties of waves in random seas.

probability domain. Assumptions most commonly introduced at this stage are:

    (i) waves are considered to be a steady-state ergodic random process;

    (ii) waves are a Gaussian random process; namely, wave profiles are distributed following the normal probability distribution with zero mean and a variance representing the sea severity;

    (iii) the wave spectral density function is narrow-banded; i.e. the spectrum is sharply concentrated at a particular frequency; and

    (iv) wave peaks and troughs are statistically independent.

Under these conditions, the probability functions applicable to wave height, wave period, combined wave height and period as well as the frequency of occurrence of breaking waves, group waves, etc., in a given sea can be analytically derived.

It should be noted that many studies on the statistical properties of random waves have been carried out by removing some of the assumptions cited above. The results of these studies will be discussed later in appropriate chapters. The assumption of a Gaussian random process (item (ii)) is almost always considered in the analytical development of probability functions associated with waves where water depth is sufficiently deep, but this assumption is not valid for waves in finite water depth. This subject will be discussed in Chapter 9.

2 SPECTRAL ANALYSIS

2.1 SPECTRAL ANALYSIS OF RANDOM WAVES

2.1.1 Fundamentals of stochastic processes

Throughout this chapter as well as others, we will evaluate various characteristics of wind-generated waves based on the stochastic process concept. The fundamentals of the stochastic process concept are outlined here.

First, the *stochastic process* (or *random process*), $x(t)$, is defined as a family of random variables. In the strict sense, $x(t)$ is a function of two arguments, time and sample space. To elaborate on this definition of a stochastic process, let us consider a set of n wave recorders $(^1x, {}^2x, {}^3x, ..., {}^nx)$ dispersed in a certain area in the ocean as illustrated in Figure 2.1(a). Let us consider a set of time histories of wave records $\{^1x(t), {}^2x(t), ..., {}^nx(t)\}$ as illustrated in Figure 2.1(b). It is recognized that at any time t_j, $x(t_j)$ is a random variable, and a set $\{^1x(t_j), {}^2x(t_j), ..., {}^nx(t_j)\}$ can be considered as a random sample of size n. This simultaneous collection of wave data observed at a specified time is called an *ensemble*. If we construct a histogram from a set of ensembles of wave records, it may be normally distributed with zero mean and a certain variance as shown in Figure 2.1(c).

As the example in the previous paragraph demonstrates, the statistical properties of random waves $x(t)$ must be obtained with respect to an ensemble, in principle. However, an ensemble of wave records has never been considered in practice; instead, the statistical properties are usually evaluated from analysis of a single wave record. This is permissible if the waves are assumed to satisfy the ergodic property. This subject will be discussed later in this section.

Referring to Figure 2.1, an ensemble of wave records obtained in deep water may obey the normal probability law with zero mean and certain variance. This situation may or may not remain constant as time progresses. In particular, there is no guarantee that the magnitude of variance is always constant. If all statistical properties, such as all moments of the ensemble wave profile distribution, are invariant under translation of time, then the waves can be called a *steady-state* (or *stationary*) *random process*. This condition is very severe, hence, a somewhat relaxed condition is used for a steady-state random process. That is, consider the covariance of ensembles at time t_j and $(t_j + \tau)$. If the covariance between two ensembles depends on the time difference τ for all time t, then the random process (waves in the present case) is called a *weakly* (or *covariance*) *steady-state random process*. We may write this condition as follows:

$$\mathrm{Cov}[x(t_j), x(t_j + \tau)] = \frac{1}{n} \sum_{k=1}^{n} \{^k x(t_j) - \bar{x}(t_j)\} \{^k x(t_j + \tau) - \bar{x}(t_j + \tau)\}$$

$$= R(\tau) \tag{2.1}$$

where

$$\bar{x}(t_j) = \frac{1}{n} \sum_{k=1}^{n} {}^k x(t_j) \qquad \bar{x}(t_j + \tau) = \frac{1}{n} \sum_{k=1}^{n} {}^k x(t_j + \tau)$$

Thus far stochastic processes have been discussed in terms of ensembles. We now introduce the ergodic theorem. In general, a random process $x(t)$ is *ergodic* if all statistics associated with the ensemble can also be determined from a single time history $x(t)$. It is taken for granted that the ergodic property holds for random waves. If this is the case, then the

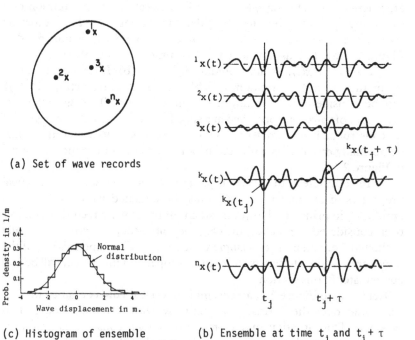

(a) Set of wave records

(c) Histogram of ensemble

(b) Ensemble at time t_j and $t_j + \tau$

Fig. 2.1. Definition of ensemble of random waves: (a) set of wave records; (b) ensemble at time t_j and $t_j + \tau$; (c) histogram of ensemble.

ensemble mean of the set shown in Figure 2.1(a) must be equal to zero, and the variance evaluated for any member of the ensemble, $x(t)$, should be equal to that of the ensemble. Furthermore, the probability density function applicable for the ensemble must also hold for the time history. With these conditions in mind, ocean waves are considered to be a weakly steady-state, ergodic random process, and hence the statistical properties can be evaluated by analysis of the time history of a single wave record.

2.1.2 Auto-correlation function

The *auto-correlation function*, denoted by $R_{xx}(\tau)$, is defined as follows:

$$R_{xx}(\tau) = \lim_{T \to \infty} \frac{1}{2T} \int_{-T}^{T} x(t)\, x(t+\tau)\mathrm{d}t \qquad (2.2)$$

As can be seen in Eq. (2.2), the auto-correlation function is evaluated as the product of two readings taken from the same record where the reading points are separated by a shift in time τ (see Figure 2.2). It is denoted by $R_{xx}(\tau)$ in order to distinguish it from the cross-correlation function between two wave records $x(t)$ and $y(t)$, denoted by $R_{xy}(\tau)$.

The auto-correlation function is essentially a covariance function; however, because of the weakly steady-state condition, the covariance function for time t and $(t+\tau)$ depends only on the time difference τ, and because of the zero mean, the covariance function becomes the auto-correlation function.

Properties of the auto-correlation function $R_{xx}(\tau)$ are listed below without proof. Readers who are interested in details of the auto-correlation function may refer to any text book on stochastic processes.

1. $R_{xx}(\tau)$ is an even function.
2. $R_{xx}(\tau)$ is maximum at $\tau=0$ and $R_{xx}(0)$ is equal to the second

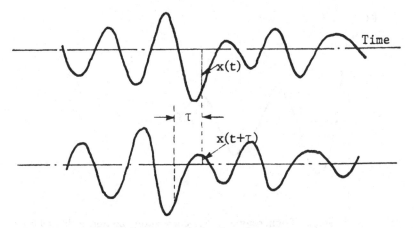

Fig. 2.2. Definition of auto-correlation function.

moment of the wave data $E[x^2(t)]$. Since the mean value of $x(t)$ is zero, $R_{xx}(0)$ represents the variance of the wave record.

3. $R_{xx}(0)$ represents the time average of wave energy $\overline{P}$. That is, from Eq. (2.2), we have

$$R_{xx}(0)=\lim_{T\to\infty} \frac{1}{2T}\int_{-T}^{T} \{x(t)\}^2 dt=\overline{P} \tag{2.3}$$

4. The auto-correlation of waves consisting of seas and swell, which are considered to be statistically independent, is equal to the sum of the individual auto-correlation functions.

An example of the auto-correlation function obtained for waves measured in the ocean is shown in Figure 2.3.

2.1.3 Spectral density function (spectrum)

It was shown in Eq. (2.3) that the magnitude of the auto-correlation function for $\tau=0$ represents the time average of the wave energy, $\overline{P}$. Let us express the average energy in terms of wave frequency ω in radians per second. This can be done by applying the following *Parseval theorem*:

$$\int_{-\infty}^{\infty} \{x(t)\}^2 dt=\frac{1}{2\pi}\int_{-\infty}^{\infty} |X(\omega)|^2 d\omega \tag{2.4}$$

where $X(\omega)$ is the Fourier transform of $x(t)$. That is,

$$X(\omega)=\int_{-\infty}^{\infty} x(t)e^{-i\omega t}dt \tag{2.5}$$

If the Fourier transform is carried out in terms of frequency, f, in cycles per second (hertz, Hz), then the Parseval theorem may be written as

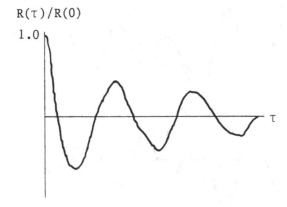

Fig. 2.3. Example of dimensionless auto-correlation function evaluated from measured waves.

$$\int_{-\infty}^{\infty} \{x(t)\}^2 dt = \int_{-\infty}^{\infty} |X(f)|^2 df \tag{2.6}$$

With the aid of Eq. (2.4) or Eq. (2.6), the average wave energy given in Eq. (2.3) can be written as

$$\overline{P} = \begin{cases} \lim_{T \to \infty} \dfrac{1}{4\pi T} \displaystyle\int_{-\infty}^{\infty} |X_T(\omega)|^2 d\omega & \text{for frequency } \omega \\[3ex] \lim_{T \to \infty} \dfrac{1}{2T} \displaystyle\int_{-\infty}^{\infty} |X_T(f)|^2 df & \text{for frequency } f \end{cases} \tag{2.7}$$

We now define the *spectral density function* of random waves $x(t)$ as follows:

$$S_{xx}(\omega) = \lim_{T \to \infty} \frac{1}{2\pi T} |X_T(\omega)|^2 \tag{2.8}$$

and in terms of the frequency f,

$$S_{xx}(f) = \lim_{T \to \infty} \frac{1}{T} |X_T(f)|^2 \tag{2.9}$$

Since the spectral density function is an even function, the time average of wave energy $\overline{P}$ can be expressed from Eqs. (2.7) and (2.8) or (2.9) as

$$\overline{P} = \begin{cases} \dfrac{1}{2} \displaystyle\int_{-\infty}^{\infty} S_{xx}(\omega) \, d\omega = \displaystyle\int_{0}^{\infty} S_{xx}(\omega) \, d\omega \\[3ex] \dfrac{1}{2} \displaystyle\int_{-\infty}^{\infty} S_{xx}(f) \, df = \displaystyle\int_{0}^{\infty} S_{xx}(f) \, df \end{cases} \tag{2.10}$$

The above equations show that the area under the spectral density function represents the average energy of random waves with respect to time. Furthermore, from the properties of the auto-correlation function, the area under the spectrum is also equal to the variance of waves $x(t)$.

It is noted that the spectral density function is often defined as

$$S_{xx}(\omega) = \lim_{T \to \infty} \frac{1}{2T} |X_T(\omega)|^2 \tag{2.11}$$

This definition of the spectral density function is certainly correct. Care has to be taken, however, that for the definition given in Eq. (2.11), the area under the spectral density function is equal to π times the average energy as well as π times the variance.

2.1.4 Wiener–Khintchine theorem

The *Wiener–Khintchine theorem* plays an extremely significant role in the stochastic analysis of random waves in that is presents the relationship between the auto-correlation function defined in the time domain and the spectral density function defined in the frequency domain. The theorem states that for a weakly steady-state random wave $x(t)$, the auto-correlation function $R_{xx}(\tau)$ and spectral density function $S_{xx}(\omega)$ are a Fourier transform pair. That is,

$$S_{xx}(\omega)=\frac{1}{\pi}\int_{-\infty}^{\infty} R_{xx}(\tau)\,e^{-i\omega\tau}d\tau$$

$$R_{xx}(\tau)=\frac{1}{2}\int_{-\infty}^{\infty} S_{xx}(\omega)\,e^{-i\omega\tau}d\omega \qquad (2.12)$$

The proof of this theorem is as follows:

$$\int_{-\infty}^{\infty} R_{xx}(\tau)\,e^{-i\omega\tau}d\tau=\lim_{T\to\infty}\frac{1}{2T}\int_{-\infty}^{\infty}\int_{-\infty}^{\infty} x(t)\,x(t+\tau)\,e^{-i\omega\tau}dt\,d\tau$$

$$=\lim_{T\to\infty}\frac{1}{2T}\int_{-\infty}^{\infty}\int_{-\infty}^{\infty} x(t)\,x(t+\tau)\,e^{-i\omega(t+\tau)}\,e^{i\omega t}dt\,d\tau$$

$$=\lim_{T\to\infty}\frac{1}{2T}X(\omega)\,X^{\star}(\omega)=\lim_{T\to\infty}\frac{1}{2T}|X(\omega)|^2=\pi S_{xx}(\omega)$$

where $X^{\star}(\omega)$ is the conjugate of $X(\omega)$. Then, by taking the inverse Fourier transform, we have

$$R_{xx}(\tau)=\frac{1}{2}\int_{-\infty}^{\infty} S_{xx}(\omega)\,e^{i\omega\tau}d\omega$$

Since the auto-correlation function and the spectral density function are both real and even functions, the Wiener–Khintchine theorem can also be written as follows:

$$S_{xx}(\omega)=\frac{1}{\pi}\int_{-\infty}^{\infty} R_{xx}(\tau)\,e^{-i\omega\tau}d\tau=\frac{2}{\pi}\int_{0}^{\infty} R_{xx}(\tau)\cos\omega\tau\,d\tau$$

$$R_{xx}(\tau)=\frac{1}{2}\int_{-\infty}^{\infty} S_{xx}(\omega)\,e^{i\omega\tau}d\omega=\int_{0}^{\infty} S_{xx}(\omega)\cos\omega\tau\,d\omega \qquad (2.13)$$

In terms of frequency f in hertz, the Wiener–Khintchine theorem may be presented as

$$S_{xx}(f) = 2 \int_{-\infty}^{\infty} R_{xx}(\tau)\, e^{-i2\pi f\tau} d\tau = 4 \int_{0}^{\infty} R_{xx}(\tau) \cos 2\pi f\tau \, d\tau$$

$$R_{xx}(\tau) = \frac{1}{2} \int_{-\infty}^{\infty} S_{xx}(f)\, e^{i2\pi f\tau} df = \int_{0}^{\infty} S_{xx}(f) \cos 2\pi f\tau \, df \qquad (2.14)$$

We summarize here the various functional relationships derived in the preceding sections as follows:

- Assume that wind-generated waves $x(t)$ are a steady-state, ergodic random process. Let the mean value $E[x(t)]$ be zero.
- Evaluate the auto-correlation function $R_{xx}(\tau)$ from the measured waves by Eq. (2.2). $R_{xx}(0)$ represents the average wave energy with respect to time, $\bar{P}$. $R_{xx}(0)$ is also equal to the variance of the waves.
- The Fourier transform of the auto-correlation function $R_{xx}(\tau)$ yields the wave spectral density function $s_{xx}(\omega)$. However, care must be taken in the definition of the spectral density function such that the area under $S_{xx}(\omega)$ is equal to the variance of $x(t)$.
- If $S_{xx}(\omega)$ is defined as given in Eq. (2.8), then we have the following functional relationships:

$$\left.\begin{array}{l} \displaystyle\int_{0}^{\infty} S_{xx}(\omega)\, d\omega \\[2em] \displaystyle\int_{0}^{\infty} S_{xx}(f)\, df \end{array}\right\} = \bar{P} = R_{xx}(0) = \mathrm{Var}[x(t)] \qquad (2.15)$$

2.1.5 Spectral analysis of two wave records

The spectral analysis technique developed for a single wave record can be equally applicable to two wave records $x(t)$ and $y(t)$. The analysis is called cross-spectral analysis.

Let us first consider the *cross-correlation function* denoted by $R_{xy}(\tau)$ which is defined as

$$R_{xy}(\tau) = E\left[x(t)y(t+\tau)\right] = \lim_{T\to\infty} \frac{1}{2T} \int_{-T}^{T} x(t)y(t+\tau)dt \qquad (2.16)$$

It is noted that in evaluating $R_{xy}(\tau)$ as defined in Eq. (2.16), the wave record $y(t)$ should be shifted backward by time τ. If $y(t)$ is shifted forward by time τ, then the cross-correlation function is denoted by $R_{yx}(\tau)$. An important property of the cross-correlation function is that $R_{xy}(\tau) \neq R_{yx}(\tau)$; instead,

$$R_{xy}(\tau) = R_{yx}(-\tau) \tag{2.17}$$

The cross-correlation function $R_{xy}(\tau)$ is not necessarily maximum at $\tau=0$, and unlike the auto-correlation function, $R_{xy}(0)$ does not have any significant meaning.

The *cross-spectral density function* of two wave records $x(t)$ and $y(t)$ is defined as follows:

$$S_{xy}(\omega) = \lim_{T \to \infty} \frac{1}{2\pi T} X^{\star}(\omega)\, Y(\omega) \qquad \text{for frequency } \omega$$

$$S_{xy}(f) = \lim_{T \to \infty} \frac{1}{T} X^{\star}(f)\, Y(f) \qquad \text{for frequency } f \tag{2.18}$$

where $X^{\star}(\omega)$ and $X^{\star}(f)$ are conjugate functions of $X(\omega)$ and $X(f)$, respectively.

The Wiener–Khintchine theorem defined for a single random variable is equally applicable for stochastic analysis of two random waves. That is, the cross-correlation function and cross-spectral density function are a Fourier transform pair. Here, the cross-spectral function is a complex function in contrast to a real-valued function for the auto-spectral density function.

Let us derive the cross-spectral density function in detail in the following through the Fourier transform of the cross-correlation function.

$$S_{xy}(\omega) = \frac{1}{\pi} \int_{-\infty}^{\infty} R_{xy}(\tau)\, e^{-i\omega\tau} d\tau$$

$$= \frac{1}{\pi} \left\{ \int_{-\infty}^{0} R_{yx}(\tau)\, e^{-i\omega\tau} d\tau + \int_{0}^{\infty} R_{xy}(\tau)\, e^{-i\omega\tau} d\tau \right\}$$

$$= \frac{1}{\pi} \left\{ \int_{0}^{\infty} R_{yx}(\tau)\, e^{i\omega\tau} d\tau + \int_{0}^{\infty} R_{xy}(\tau)\, e^{-i\omega\tau} d\tau \right\}$$

$$= \frac{1}{\pi} \int_{0}^{\infty} \left\{ R_{xy}(\tau) + R_{yx}(\tau) \right\} \cos \omega\tau \, d\tau$$

$$+ i\frac{1}{\pi} \int_{0}^{\infty} \left\{ -R_{xy}(\tau) + R_{yx}(\tau) \right\} \sin \omega\tau \, d\tau$$

$$= C_{xy}(\omega) + i\, Q_{xy}(\omega) \tag{2.19}$$

where

$$C_{xy}(\omega) = \frac{1}{\pi} \int_0^\infty \left\{ R_{xy}(\tau) + R_{yx}(\tau) \right\} \cos \omega \tau \, d\tau$$

$$Q_{xy}(\omega) = \frac{1}{\pi} \int_0^\infty \left\{ -R_{xy}(\tau) + R_{yx}(\tau) \right\} \sin \omega \tau \, d\tau \qquad (2.20)$$

The real part $C_{xy}(\omega)$ of the cross-spectral density function is referred to as the *co-spectrum*, while the imaginary part $Q_{xy}(\omega)$ is referred to as the *quadrature-spectrum*. The amplitude spectrum of $S_{xy}(\omega)$ can be evaluated from Eq. (2.19) as

$$S_{xy}(\omega) = \sqrt{ \{ C_{xy}(\omega) \}^2 + \{ Q_{xy}(\omega) \}^2 } \qquad (2.21)$$

and the phase spectrum becomes

$$\epsilon(\omega) = \tan^{-1} \left\{ \frac{Q_{xy}(\omega)}{C_{xy}(\omega)} \right\} \qquad (2.22)$$

The co-spectrum $C_{xy}(\omega)$ is an even function, while the quadrature spectrum $Q_{xy}(\omega)$ is an odd function. Properties of these spectra are summarized below:

$$\left. \begin{array}{l} C_{xy}(-\omega) \\[4pt] C_{yx}(\omega) \\[4pt] C_{xy}(-\omega) \end{array} \right\} = C_{xy}(\omega) \qquad (2.23)$$

and

$$\left. \begin{array}{l} Q_{xy}(-\omega) \\[4pt] Q_{yx}(\omega) \end{array} \right\} = -Q_{xy}(\omega)$$

$$Q_{yx}(-\omega) \quad = Q_{xy}(\omega) \qquad (2.24)$$

From the above properties, we have the following relationship:

$$\begin{aligned} S_{xy}(-\omega) &= C_{xy}(-\omega) + i Q_{xy}(-\omega) \\ &= C_{xy}(\omega) - i Q_{xy}(\omega) = S_{xy}^{\star}(\omega) \end{aligned} \qquad (2.25)$$

and

$$\begin{aligned} S_{xy}(-\omega) &= S_{yx}(\omega) \\ S_{yx}(\omega) &= S_{xy}^{\star}(\omega) \\ S_{yx}(-\omega) &= S_{xy}(\omega) \end{aligned} \qquad (2.26)$$

where $S_{xy}^{\star}(\omega)$ is the complex conjugate of $S_{xy}(\omega)$.

The cross-spectral density function presented in terms of the frequency f in hertz is as follows:

$$S_{xy}(f)=2\int_{-\infty}^{\infty} R_{xy}(\tau)\ ^{-\mathrm{i}2\pi f\tau}\mathrm{d}\tau=C_{xy}(f)+\mathrm{i}Q_{xy}(f) \tag{2.27}$$

where

$$C_{xy}(f)=2\int_{0}^{\infty} \left\{ R_{xy}(\tau)+R_{yx}(\tau) \right\} \cos 2\pi f\tau\,\mathrm{d}\tau$$

$$Q_{xy}(f)=2\int_{0}^{\infty} \left\{ -R_{xy}(\tau)+R_{yx}(\tau) \right\} \sin 2\pi f\tau\,\mathrm{d}\tau \tag{2.28}$$

A comprehensive application of cross-spectral analysis will be seen in Chapter 7 where the directional characteristics of ocean waves is evaluated through cross-spectral analysis of wave records.

2.1.6 Wave-number spectrum

The wave spectrum discussed thus far is for wave displacement measured at a certain location (or locations) in the sea. We now broaden the analysis to displacement of an irregular sea surface in a stationary, homogeneous wave field. In this case, the one-dimensional Fourier transform pair, $X(\omega)$ and $x(t)$, shown in Eq. (2.5) is extended to the three-dimensional Fourier transform. By introducing the coordinate vector $\mathbf{r}$ and wave-number vector $\mathbf{k}$, we may write the three-dimensional Fourier transform pair in the following vector form:

$$Z(\mathbf{k},\omega)=\int_{t}\int_{\mathbf{r}} z(\mathbf{r},t)\ \mathrm{e}^{\mathrm{i}(\mathbf{k}\cdot\mathbf{r}-\omega t)}\mathrm{d}\mathbf{r}\,\mathrm{d}t$$

$$z(\mathbf{r},t)=\frac{1}{(2\pi)^3}\int_{\omega}\int_{\mathbf{k}} Z(\mathbf{k},\omega)\ \mathrm{e}^{-\mathrm{i}(\mathbf{k}\cdot\mathbf{r}-\omega t)}\mathrm{d}\mathbf{k}\,\mathrm{d}\omega \tag{2.29}$$

where

$$\mathbf{r}=x\mathbf{i}+y\mathbf{j} \qquad \mathrm{d}\mathbf{r}=\mathrm{d}x\,\mathrm{d}y$$

$$\mathbf{k}=k_x\mathbf{i}+k_y\mathbf{j} \qquad \mathrm{d}\mathbf{k}=\mathrm{d}k_x\,\mathrm{d}k_y$$

Let the displacement of the sea surface be $\eta(\mathbf{r},t)$ and define the auto-correlation as

$$R(\boldsymbol{\rho},\tau)=E[\eta(\mathbf{r},t)\eta(\mathbf{r}+\boldsymbol{\rho},\,t+\tau)] \tag{2.30}$$

Then, following the Wiener–Khintchine theorem, the wave spectrum $S(\mathbf{k},\omega)$ and auto-correlation function $R(\boldsymbol{\rho},\tau)$ are written as

$$S(\mathbf{k},\omega)=\frac{1}{\pi}\int_{\tau}\int_{\boldsymbol{\rho}} R(\boldsymbol{\rho},\tau)\ \mathrm{e}^{\mathrm{i}(\mathbf{k}\cdot\boldsymbol{\rho}-\omega\tau)}\mathrm{d}\boldsymbol{\rho}\,\mathrm{d}\tau$$

$$R(\boldsymbol{\rho},\tau)=\frac{1}{2(2\pi)^2}\int_\omega\int_\mathbf{k} S(\mathbf{k},\omega)\,e^{-i(\mathbf{k}\cdot\boldsymbol{\rho}-\omega\tau)}d\mathbf{k}\,d\omega \qquad (2.31)$$

Note that $S(\mathbf{k},\omega)$ carries the constant $1/\pi$ because of the definition of the spectral density function given in Eq. (2.8). $S(\mathbf{k},\omega)$ is called the *wave-number frequency spectrum*, and integration with respect to wave number and frequency yields the variance of random waves.

By integrating $S(\mathbf{k},\omega)$ with respect to frequency, we have the *wave-number spectrum*,

$$S(\mathbf{k})=\int_{-\infty}^\infty S(\mathbf{k},\omega)\,d\omega \qquad (2.32)$$

On the other hand, by letting $\tau=0$ and by integrating the auto-correlation function given in Eq. (2.31) with respect to ω, we have

$$R(\boldsymbol{\rho})=\frac{1}{(2\pi)^2}\int_\mathbf{k} S(\mathbf{k})\,e^{-i\mathbf{k}\cdot\boldsymbol{\rho}}d\mathbf{k} \qquad (2.33)$$

and its inverse yields

$$S(\mathbf{k})=\int_\rho R(\boldsymbol{\rho})\,e^{i\mathbf{k}\cdot\boldsymbol{\rho}}d\boldsymbol{\rho} \qquad (2.34)$$

The wave-number spectrum is a density function per unit area of the **k**-space irrespective of the frequency associated with the wave number. Because of this, a stereoscopic photograph of the sea surface provides information on the wave characteristics at the moment when the picture is taken (Holthuijsen 1981).

If we integrate $S(\mathbf{k},\omega)$ with respect to wave-number **k**, we have the frequency spectrum

$$S(\omega)=\int_\mathbf{k} S(\mathbf{k},\omega)\,d\mathbf{k} \qquad (2.35)$$

and by letting $\rho=0$ and by integrating with respect to **k** in Eq. (2.31), we have

$$R(\tau)=\frac{1}{2}\int_\omega S(\omega)\,e^{i\omega\tau}d\omega \qquad (2.36)$$

and its inverse

$$S(\omega)=\frac{1}{\pi}\int_\tau R(\tau)\,e^{-i\omega\tau}d\tau \qquad (2.37)$$

which is the relationship derived in Eq. (2.12). As seen in the above, $S(\omega)$ is a spectral density function irrespective of wave number or wave direction, and this is the usual situation in obtaining the time history of waves recorded at a single point ($\mathbf{k}=0$).

The conversion of the frequency spectrum $S(\omega)$ to the wave-number spectrum $S(\mathbf{k})$ and vice versa can be done for a specified directional angle by using the dispersion relationship given by $\omega^2 = kg \tanh kh$, where h is water depth. That is,

(a) From $S(\omega)$ to $S(k)$

$$S(k) = \left[S(\omega) \right]_{\omega = \sqrt{kg \tanh kh}} \times \frac{d\omega}{dk} \qquad (2.38)$$

Here, $d\omega/dk$ is the wave group velocity given by

$$\frac{d\omega}{dk} = \frac{1}{2}\left(1 + \frac{2kh}{\sinh 2kh} \right) \sqrt{\frac{g}{k} \tanh kh}$$

$$= \frac{1}{2}\left(1 + \frac{2kh}{\sinh 2kh} \right) \times (\text{wave celerity}) \qquad (2.39)$$

For deep water waves, we have $\omega = \sqrt{kg}$ and $d\omega/dk = (1/2)\sqrt{g/k}$.

(b) From $S(k)$ to $S(\omega)$

$$S(\omega) = \left[s(k) \right]_{k = f(\omega)} \times \frac{dk}{d\omega} \qquad (2.40)$$

For deep water waves, $k = \omega^2/g$. For waves of finite water depth, however, k can be evaluated for a specified ω but cannot be presented in closed form. Examples of comparisons of wave spectra given in the frequency space, $S(f)$, and that given in the wave-number space, $S(k)$, are shown in Figures 2.21(a) and (b) in connection with the TMA spectrum.

Next, we may write the surface elevation of random waves in terms of the wave-number spectrum. The formula representing the surface wave elevation given in Eq. (1.12) may be written as a function of the wave-number spectrum as follows:

$$\eta(\mathbf{r},t) = \int_\omega \int_\mathbf{k} e^{i(\mathbf{k}\cdot\mathbf{r} - \omega t)} dS(\mathbf{k},\omega) \qquad (2.41)$$

where

$$E[dS(\mathbf{k},\omega)\, dS^\star(\mathbf{k}',\omega')] = \begin{cases} S(\mathbf{k},\omega)\, dk\, d\omega & \text{if } \mathbf{k}=\mathbf{k}',\ \omega=\omega' \\ 0 & \text{otherwise} \end{cases} \qquad (2.42)$$

Equation (2.41) is a Fourier–Stieltjes integral, where the integration is over all wave-number and frequency space. The derivation of Eq. (2.42) may be found in Moyal (1949). The wave profile described by Eq. (2.41) will be used in the analysis of the directional wave spectrum discussed in Section 7.4.

2.1.7 Wave velocity and acceleration spectra

It is of considerable importance to obtain the spectral density functions of wave velocity or acceleration (both in the vertical direction) from the displacement spectrum or vice versa. In fact, the wave spectrum is often evaluated from data obtained by an accelerometer installed in a buoy. In this case, the wave displacement record can be acquired through double integration of the acceleration record assuming that there is no base-line shift in the original record. It is convenient, however, to evaluate the displacement spectrum by applying the following general formulation applicable to time derivatives of random processes.

Let $R_{xx}(\tau)$ and $S_{xx}(\omega)$ be the auto-correlation and spectral density function, respectively, of a wave record $x(t)$. Then, the auto-correlations of velocity and acceleration can be evaluated by

$$R_{\dot{x}\dot{x}}(\tau) = -\frac{d^2}{d\tau^2}R_{xx}(\tau)$$

$$R_{\ddot{x}\ddot{x}}(\tau) = \frac{d^4}{d\tau^4}R_{xx}(\tau) \tag{2.43}$$

Hence, the variances of velocity and acceleration can be obtained as

$$\text{Var}[\dot{x}(t)] = -\frac{d^2}{d\tau^2}R_{xx}(0)$$

$$\text{Var}[\ddot{x}(t)] = \frac{d^4}{d\tau^4}R_{xx}(0) \tag{2.44}$$

We may apply the spectral density functions of wave velocity and acceleration in the same fashion as for the displacement spectrum. Then, the relationship between them can be written as

$$S_{\dot{x}\dot{x}}(\omega) = \lim_{T\to\infty}\frac{1}{2\pi T}|\dot{X}(\omega)|^2 = \omega^2 \lim_{T\to\infty}\frac{1}{2\pi T}|X(\omega)|^2 = \omega^2 S_{xx}(\omega)$$

$$S_{\ddot{x}\ddot{x}}(\omega) = \lim_{T\to\infty}\frac{1}{2\pi T}|\ddot{X}(\omega)|^2 = \omega^4 \lim_{T\to\infty}\frac{1}{2\pi T}|X(\omega)|^2 = \omega^4 S_{xx}(\omega) \tag{2.45}$$

In terms of the frequency f in hertz, we have

$$S_{\dot{x}\dot{x}}(f) = (2\pi f)^2 S_{xx}(f)$$

$$S_{\ddot{x}\ddot{x}}(f) = (2\pi f)^4 S_{xx}(f) \tag{2.46}$$

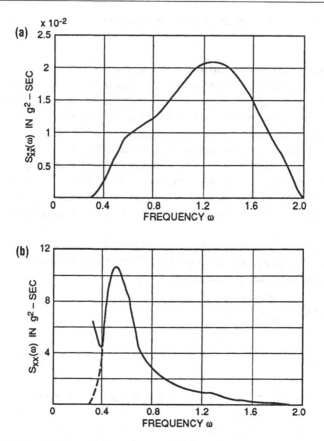

Fig. 2.4. Wave displacement spectrum $S_{xx}(\omega)$ computed from acceleration spectrum $S_{\ddot{x}\ddot{x}}(\omega)$.

In evaluating the wave displacement spectrum $S_{xx}(\omega)$ from the wave acceleration spectrum $S_{\ddot{x}\ddot{x}}(\omega)$ by applying Eq. (2.45), very large values of $S_{xx}(\omega)$ are commonly observed in the low frequency range as illustrated in Figure 2.4. This is because $S_{\ddot{x}\ddot{x}}(\omega)/\omega^4$ becomes very large for small ω. In this case, it is customary to bring $S_{xx}(\omega)$ to zero following the general trend of the low frequency portion of the spectrum.

2.2 CHARACTERISTICS OF WAVE SPECTRA

The shape of wave spectra, in general, varies considerably depending on the severity of wind velocity, time duration of wind blowing, fetch length, etc. This complicated situation was systematized by Phillips and Kitaigordoskii by applying similarity theory and dimensional analysis as presented below.

Let us consider the situation of winds blowing with constant velocity over the sea surface. First, waves of short length (high frequencies) are generated, and gradually waves of longer lengths will follow. However, the growth of waves under this situation cannot be continued indefinitely. At a certain time in the growing state, the wave profile reaches its limiting form beyond which it becomes locally unstable and breaking takes place. This results in an energy loss for restoring stability. Another phenomenon contributing to the limiting form of a wave profile is the translation of wave energy from one frequency to another, although its contribution is less than that of wave breaking.

Because the wave profile is of limited form, the wave spectrum reaches an upper limit under a given constant wind velocity; namely, a state of energy saturation in which a balance is set up between the rate at which energy is gained from the wind and the rate at which it is lost either by breaking or by nonlinear wave–wave interaction. This situation is called a *fully developed sea*, and the range of frequencies where wave energy exists for this state is defined as the *equilibrium range*, often called the *saturation range* (Phillips 1958, 1966, 1967). This is shown as the frequency $\omega > \omega_0$ in Figure 2.5. The magnitude of wave spectral density in the equilibrium range represents an upper limit of wave energy.

Based on the concept of an equilibrium range in a wave spectrum, Phillips carries out a dimensional analysis and derives the following formula applicable for higher frequencies of the spectrum:

$$S(\omega) = \alpha g^2 \omega^{-5} \tag{2.47}$$

where α is a constant.

Figure 2.6, taken from Phillips' 1958 paper, demonstrates the equilibrium range established in Burling's wind-generated wave spectra obtained from measurements in a reservoir. Burling analyzes a large number of records obtained over fetches from 400 m to 1350 m, with wind velocities ranging from 5 to 8 m/s at 10 m height (Burling 1959). Since Burling's wave data are taken in a reservoir, their frequencies ω are rather high, from 3 to 12 rps. In Figure 2.6 the broken lines indicate the upper and lower bounds of measured spectral density at each frequency,

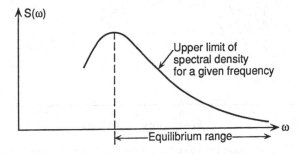

Fig. 2.5. Definition of equilibrium range of wave spectrum, $\omega > \omega_0$.

with the crosses representing the mean value. The heavy line is Phillips' formulation given in Eq. (2.47) with $\alpha = 7.6 \times 10^{-3}$.

From consideration of similarity theory, Kitaigorodskii (1961, 1973) shows that the spectral density function, in general, can be expressed as a function of two dimensionless quantities. That is,

$$S(\omega) = g^2 \omega^{-5} F\left(\frac{u_* \omega}{g}, \frac{gX}{u_*^2}\right) \tag{2.48}$$

where

$u_* \omega / g = \bar{\omega} =$ dimensionless frequency
$gX/u_*^2 = \bar{X} =$ dimensionless fetch
$u_* \quad =$ wind shear (friction) velocity $= \sqrt{\tau_0/\rho_a}$
$\tau_0 \quad =$ tangential wind friction
$\rho_a \quad =$ air density
$X \quad =$ fetch length.

In particular, for a fully developed sea, the spectrum is a function of dimensionless frequency only. That is,

$$S(\omega) = g^2 \omega^{-5} F\left(\frac{u_* \omega}{g}\right) \tag{2.49}$$

We now address the evaluation of wind shear velocity, u_*, from a knowledge of the mean wind velocity measured at an arbitrary height z, denoted by $\bar{U}_z$. Kitaigorodskii uses the following formula developed by Charnock (1958) and Ellison (1956) for u_*:

$$\bar{U}_z / u_* = 11.0 + 2.5 \ln(gz/u_*^2) \tag{2.50}$$

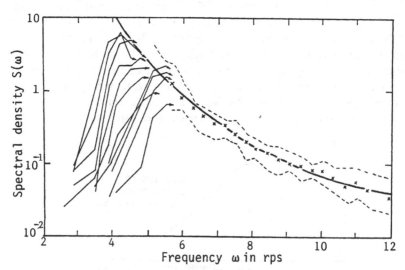

Fig. 2.6. Wind-generated wave spectrum obtained from data by Burling (Phillips 1958).

Wu (1969, 1980, 1982) carried out an extensive series of studies on wind characteristics over a sea surface, and obtained the shear velocity as a function of mean wind velocity at 10 m height as follows:

$$u_* = \sqrt{C_{10}}\, \overline{U}_{10} \tag{2.51}$$

where C_{10} = surface drag coefficient evaluated from wind velocity at 10 m height
$= (0.8 + 0.065\, \overline{U}_{10}) \times 10^{-3}$
The wind velocity at an arbitrary height, $\overline{U}_z$, can be evaluated as

$$\overline{U}_z = \overline{U}_{10} + u_* \ln(z/10) \tag{2.52}$$

From Eqs. (2.51) and (2.52), the shear velocity u_* can be evaluated from a knowledge of $\overline{U}_z$.

Following the analysis leading to Eq. (2.47), Kitaigorodskii presents Burling's data given in Figure 2.6 as a function of dimensionless frequency. The result is shown in Figure 2.7. As can be seen, the data follow very well Phillips' formula given in Eq. (2.47). The constant given in Eq. (2.47) is found to be 6.5×10^{-3} for this data. The vertical lines in the figure give the lower and upper bounds of the range of modal frequencies obtained for this data.

Another comparison of the high frequency range of wave spectra obtained in various environmental conditions made by Hess et al. (1969) is shown in Figure 2.8. The figure includes a variety of data obtained in laboratory experiments, in very large as well as limited fetches, near the center of a hurricane, etc. The solid line given in the figure is for the value $\alpha = 5.85 \times 10^{-3}$ in Eq. (2.47).

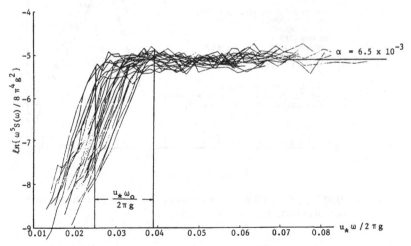

Fig. 2.7. High-frequency range of dimensionless spectra obtained from data by Burling (Kitaigorodskii 1973).

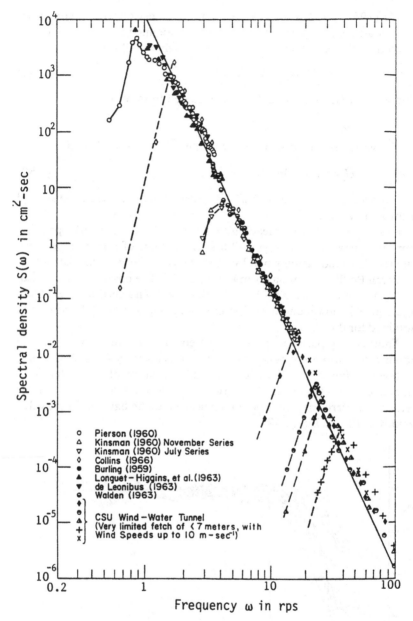

Fig. 2.8. High-frequency range of wave spectra obtained in various environmental conditions (Hess *et al.* 1969).

Although Eq. (2.47) contributed significantly to the analysis of wave spectra from data obtained in the ocean, results of field and laboratory measurements have indicated a need for modification of the formula. Garrett (1969) first pointed out that the shape of the high frequency regions of wave spectra is significantly different from that given by Eq. (2.47). Toba claims, based on a complete analysis of measured waves in the growing stage of wind-generated seas, that the shape of wave spectra in the equilibrium range is not proportional to ω^{-5}; instead, he proposes the following formula (Toba 1973, 1978);

$$S(\omega) = \alpha g u_* / \omega^4 \qquad\qquad (2.53)$$

The concept of Toba's formula is supported by Mitsuyasu et al. (1980) through the analysis of extensive field measurements. However, they consider the constant α in Eq. (2.53) to be fetch dependent although the dependency is very weak. Kahma (1981) proposes, through analysis of his measured data, a formula similar to Eq. (2.53) given by

$$S(\omega) = 4.5 \times 10^{-3} g \overline{U}_{10} / \omega^4 \qquad\qquad (2.54)$$

Figure 2.9 shows, on the other hand, an example of the relationship between dimensionless spectrum and frequency presented by Forristall

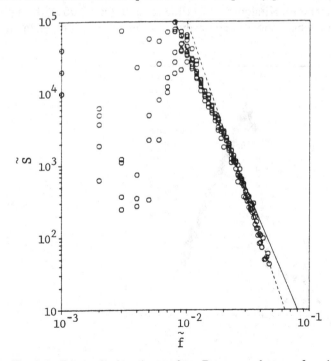

Fig. 2.9. Dimensionless spectra from Pacesetter data as a function of dimensionless frequency (Forristall 1981).

(1981) from analysis of data obtained 100 miles off the coast of New Jersey. The dimensionless spectral density function and frequency in the figure are defined as $\tilde{S}=S\,g^3/u_*^5$ and $\tilde{f}=f\,u_*/g$, respectively. The solid line in the figure has a slope of f^{-4} and the dashed line has a slope of f^{-5}. As can be seen, the dimensionless spectrum is proportional to f^{-4} for $0.01<\tilde{f}<0.025$, and is proportional to f^{-5} for $\tilde{f}>0.03$.

Another example of the relationship between the dimensionless spectrum and frequency from analysis of data obtained in the Gulf of Mexico is shown in Figure 2.10. The dimensionless spectrum and frequency in the figure are defined as $\hat{S}=S\,f_0/H_s^2$ and $\hat{f}=f/f_0$, where H_s is the significant wave height and f_0 is the mean frequency. In this data set, presumably the maximum dimensionless frequencies are all less than 2.5 $\hat{f}$, and thereby the data shows proportionality to the inverse of the 4th power of the frequency. Forristall gives the following formula in this case:

$$\hat{S}=0.051\,\hat{f}^{-4} \tag{2.55}$$

Donelan et al. (1985) also show, based on their experiments in Lake Ontario and in the laboratory, that the spectrum is proportional to ω^{-4} in the frequency region from $1.5\omega_p$ to $3.0\omega_p$, where ω_p is the peak frequency of the spectrum.

On the other hand, Kitaigorodskii (1983) and Phillips (1985), through different approaches, show analytically that the wave spectrum in the

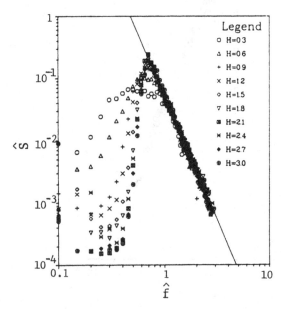

Fig. 2.10. Dimensionless spectra from ODGP in Gulf of Mexico data as a function of dimensionless frequency (Forristall 1981).

equilibrium range may be inversely proportional to the 4th power of the frequency.

Battjes *et al.* (1987) reanalyze wave spectra obtained during the JONSWAP Project (Hasselmann *et al.* 1973) and find that Toba's formula given in Eq. (2.53) provides a statistically better fit to the observed data. They also show that the proportionality factor α in Eq. (2.53) is not a constant; instead, there is a relatively large scatter in the observed data which may be partly attributed to the influence of tidal currents.

2.3 WAVE SPECTRAL FORMULATIONS

One of the most important applications of information on random seas in the naval, ocean and civil engineering fields is the prediction of responses (motions and wave-induced forces and bending moments) of ships and offshore structures in a seaway and the evaluation of hydro-dynamic forces acting on nearshore marine structures. In order to carry out stochastic prediction of linear and nonlinear responses for design of marine systems, wave spectra representing desired sea states are mandatory. In particular, it is highly desirable to prepare wave spectra expected in the sea(s) where the system will be operated; however, this is not possible in all cases. It is extremely important, therefore, to prepare wave spectral formulations representing various conditions at sea.

Many spectral formulations have been proposed since early 1950. Among others, Neumann's wave spectral formulation developed in 1953 was a noteworthy contribution. He developed the formulation based on a theoretical analysis together with wave data primarily visually observed. Although his formulation is not used now, it contributed significantly during the 1950s to the forecasting of sea state and the design of ships and marine structures.

In the following, several spectral formulations are summarized which satisfy the general characteristics of wave spectra presented in Section 2.2. Only those obtained from measured data in the ocean are included.

2.3.1 Pierson–Moskowitz spectrum

The Pierson–Moskowitz spectral formulation (Pierson and Moskowitz 1964) was developed from analysis of measured data obtained in the North Atlantic by Tucker wave recorders installed on weather ships. Analysis was carried out only on selected wave records considered to have been acquired in fully developed seas. Dimensionless spectra, $S(f)\,g^3/U^5$, are classified into five different groups for wind speeds ranging from 20 to 40 knots as shown in Figure 2.11(a). As seen in the figure, the magnitude of spectra around the modal frequency shows some scatter. Pierson and Moskowitz attribute the discrepancy to difficulty in determining the precise wind speed measured on the ship's deck at 19.5 m above sea level.

To overcome this difficulty, the wind speed of each group is modified by taking the 5th-root of the ratio of Sg^3/U^5 (where S is spectral density at the modal frequency) to the average value. That is, by evaluating the following factor k_j

$$k_j = \left(S_j \frac{g^3}{U_j^5} \middle/ \frac{1}{n} \sum_j S_j \frac{g^3}{U_j^5} \right)^{1/5} \tag{2.56}$$

and modifying the wind speed as $k_j U$ for each group. This results in good agreement for the five grouped dimensionless wave spectra as shown in Figure 2.11(b). The value of the modal frequency, Uf/g, is obtained as 6.14.

By drawing the average line in Figure 2.11(b), the following spectral formulation in terms of frequency ω is derived:

$$S(\omega) = \frac{Ag^2}{\omega^5} \exp\left\{ -B\left(\frac{g/U}{\omega} \right)^4 \right\} \tag{2.57}$$

where $A = 8.10 \times 10^{-3}$, $B = 0.74$, and U is the wind speed at 19.5 m above the sea surface.

As shown in Eq. (2.57), the Pierson–Moskowitz spectral formulation depends on a single parameter, the wind speed U. The modal frequency, ω_m, is a fixed value and is also given as a function of wind speed. That is,

$$\omega_m = 0.87 \ (g/U) \tag{2.58}$$

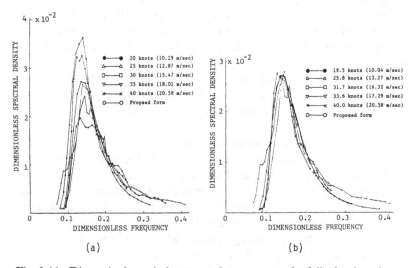

Fig. 2.11. Dimensionless wind-generated wave spectra for fully developed seas: (a) unmodified wind speed; (b) modified wind speed (Pierson and Moskowitz 1964).

It is very convenient, in practice, that the wave spectrum be given as a function of significant wave height, H_s, instead of wind speed. For this, the spectral density function given in Eq. (2.57) is integrated to yield,

$$\int_0^\infty S(\omega)\, d\omega = \frac{A}{4B} \frac{U^4}{g^2} \qquad (2.59)$$

On the other hand, by assuming that the spectrum is narrow-banded, the area under the spectral density function is equal to $(H_s/4)^2$ (see Section 3.6):

$$\int_0^\infty S(\omega)\, d\omega = (H_s/4)^2 \qquad (2.60)$$

Hence, from Eqs. (2.59) and (2.60), we can derive the following relationship between the wind speed and significant wave height for fully developed seas:

$$H_s = 2\sqrt{A/B}\,(U^2/g) = 0.21\,(U^2/g) \qquad (2.61)$$

where $A = 8.10 \times 10^{-3}$, $B = 0.74$.

Then, from Eqs. (2.57) and (2.61), the Pierson–Moskowitz spectrum can be presented in terms of significant wave height. That is,

$$S(\omega) = \frac{8.10}{10^3} \frac{g^2}{\omega^5} e^{-0.032\,(g/H_s)^2/\omega^4} \qquad (2.62)$$

Furthermore, by differentiating Eq. (2.62) with respect to ω, and by letting the derivative be zero, we have the relationship between the modal frequency, ω_m, and significant wave height as follows:

$$\omega_m = 0.4\,\sqrt{g/H_s} \qquad (2.63)$$

By using the relationship given in Eq. (2.63), the spectrum can also be written as

$$S(\omega) = \frac{8.10}{10^3} \frac{g^2}{\omega^5} e^{-(5/4)(\omega_m/\omega)^4} \qquad (2.64)$$

2.3.2 Two-parameter spectrum
In order to represent fully as well as partially developed sea states, Bretschneider (1959) developed the following spectral formulation:

$$S(\omega) = 3.437 \frac{F_1^2\, g^2}{F_2^4\, \omega^5} \exp\left\{ -0.675 \left(\frac{g}{F_2 U\omega} \right)^4 \right\} \qquad (2.65)$$

where $F_1 = g\overline{H}/U^2$
$\qquad F_2 = g\overline{T}/(2\pi U)$
$\qquad U = $ wind speed
$\qquad \overline{H} = $ average wave height

$$\overline{T}=\text{average wave period}=\int_0^\infty T\,S(T)\mathrm{d}T\int_0^\infty S(T)\mathrm{d}T$$

$S(T)=$ period spectrum

The spectral formulation given in Eq. (2.65) has some inexpediences in practical application as listed below. However, Eq. (2.65) can be reduced to a spectral formulation with two parameters; significant wave height and modal frequency. This certainly yields a variety of spectral shapes, although with some restrictions, for a specified sea severity. The inexpediences involved in Eq. (2.65) are as below:

(i) The area under the spectrum is eight times the variance of that commonly defined for random waves.

(ii) The average wave period, $\overline{T}$, is given based on a period spectrum; hence, it is not equal to $2\pi/\overline{\omega}$ where $\overline{\omega}$ is the mean frequency commonly defined based on a frequency spectrum $S(\omega)$.

If the area under the spectrum and the average wave period are modified as stated above, then Eq. (2.65) may be expressed in terms of the average wave height, $\overline{H}$, and the mean frequency, $\overline{\omega}$, as

$$S(\omega)=0.278\,\frac{\overline{\omega}^4}{\omega^5}\,\overline{H}^2\,e^{-0.437(\overline{\omega}/\omega)^4}\qquad(2.66)$$

Furthermore, we may express $S(\omega)$ in terms of significant wave height, H_s, and modal frequency ω_m, by using the conversion formulation applicable for narrow-band wave spectra; namely, $\omega_m/\overline{\omega}=0.77$ and $H_s/\overline{H}=1.60$ (see Section 2.3.3). Then, Bretschneider's spectrum can be presented as follows:

$$S(\omega)=\frac{1.25}{4}\,\frac{\omega_m^4}{\omega^5}\,H_s^2\,e^{-1.25(\omega_m/\omega)^4}\qquad(2.67)$$

This is called the *two-parameter wave spectrum* and is widely used for the design of marine systems. It is noted that the above spectral formulation is reduced to the Pierson–Moskowitz spectrum (Eq. 2.62) by letting $\omega_m=0.4\sqrt{g/H_s}$ given in Eq. (2.63).

By using the two-parameter spectral formulation, we can generate a family of wave spectra consisting of several members for a specified sea severity. As an example, a family is developed from statistical analysis of wave data obtained in the North Atlantic in which several modal frequencies for a specified significant wave height are evaluated with a 95 percent confidence coefficient. Examples of families of spectra obtained for significant wave heights of 3.0 m and 9.0 m are shown in Figure 2.12.

2.3.3 Spectral formulation as a function of ω^{-5}

It is stated at the end of Section 2.2 that wave spectra obtained from measured data in the ocean indicate the equilibrium range of spectra to be proportional to ω^{-4} in contrast to ω^{-5} as obtained by applying similarity theory. Nevertheless, the spectral formulations commonly considered for the design of marine systems are categorized in the following form:

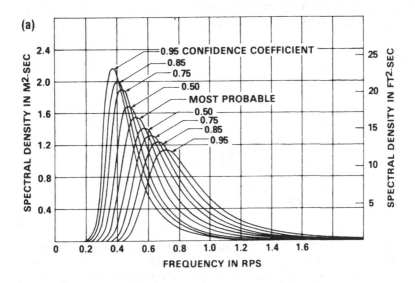

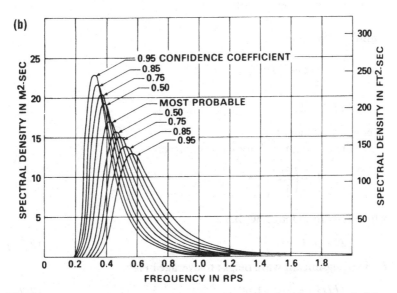

Fig. 2.12. Family of two-parameter wave spectra for significant wave heights (a) 3.0 m (9.8 ft) and (b) 9.0 m (29.5 ft).

$$S(\omega) = \frac{A}{\omega^5} e^{-B/\omega^4}$$ (2.68)

We now summarize the various statistical properties associated with a spectrum having the form of Eq. (2.68). It is assumed that the spectrum is narrow-banded, and therefore the wave height obeys the Rayleigh probability distribution (see Section 3.2).

(a) Moments

$$m_j = \int_0^\infty \omega^j S(\omega) d\omega$$

$$m_0 = A/4B$$

$$m_1 = \frac{\Gamma(3/4)}{4} \frac{A}{B^{3/4}} = 0.306 \frac{A}{B^{3/4}}$$

$$m_2 = \frac{\sqrt{\pi}}{4} \frac{A}{\sqrt{B}}$$

$$m_4 = \infty$$ (2.69)

As shown above, the 4th moment is not finite. Therefore, the band-width parameter ϵ of the spectrum is unity which implies that the spectrum is wide-banded. This contradicts the assumption of the spectrum is being narrow-banded. Nevertheless, the formulae listed below are often conveniently used in practice.

(b) Significant wave height, H_s

From $\int_0^\infty S(\omega) \, d\omega = (H_s/4)^2$, we have

$$H_s = (4A/B)^{1/2}$$ (2.70)

(c) Average wave height, $\overline{H}$

From the Rayleigh probability distribution,

$$\overline{H} = \int_0^\infty x \left(\frac{2x}{R} e^{-x^2/R} \right) dx = \frac{\sqrt{\pi}}{2} \sqrt{R}$$

where $R = 8m_0$. Hence, we have

$$\overline{H} = (\pi A/2B)^{1/2}$$ (2.71)

(d) Ratio of significant wave height to average height

$$H_s/\overline{H} = \sqrt{8/\pi} = 1.60$$ (2.72)

(e) Modal frequency and period

From

$$\frac{d}{d\omega}S(\omega)=4B\omega^{-4}-5=0$$

we have

$$\omega_m=(4B/5)^{1/4}=0.95B^{1/4}$$
$$T_m=2\pi(4B/5)^{-1/4}=6.64B^{-1/4} \qquad (2.73)$$

(f) Mean (average) frequency and period

$$\overline{\omega}=m_1/m_0=\Gamma(3/4)B^{1/4}=1.23B^{1/4}$$
$$\overline{T}=2\pi/\overline{\omega}=5.13B^{-1/4} \qquad (2.74)$$

(g) Average zero-crossing frequency and period

$$\overline{\omega}_0=\sqrt{m_2/m_0}=(\pi B)^{1/4}=1.33B^{1/4}$$
$$\overline{T}_0=2\pi/\overline{\omega}_0=4.72B^{-1/4} \qquad (2.75)$$

(h) Ratio of frequencies and periods

$$\omega_m/\overline{\omega}=0.71 \qquad T_m/\overline{T}_0=1.41$$
$$\omega_m/\overline{\omega}=0.77 \qquad T_m/\overline{T}=1.29$$
$$\overline{\omega}_0/\overline{\omega}=1.08 \qquad \overline{T}_0/\overline{T}=0.92 \qquad (2.76)$$

2.3.4 Six-parameter spectrum

The shape of wave spectra obtained from measured data in the ocean varies considerably (even though the significant wave heights are the same) depending on the duration of wind blowing, fetch length, stage of storm growth and decay and existence of swell. In addition, wave spectra often have double peaks (bimodal frequencies). For example, Figure 2.13 shows a variety of shapes of wave spectra obtained in the North Atlantic for the same significant wave height of 3.5 m and wind speeds from 20 to 25 knots. Of the seven spectra shown in the figure, three (JHC 113, NW 23, NW 38) have double peaks, and three (JHC 113, NW 228, JHC 128) have the same modal frequency of 0.58 but have quite different shapes.

In order to cover a variety of spectral shapes including bimodal spectra observed in the ocean, the entire spectral shape is decomposed into two parts, the low frequency components and the high frequency components as shown in Figure 2.14. Each part is expressed by three parameters; significant wave height H_s, modal frequency ω_m and shape parameter λ. Here, the shape parameter is introduced through the following procedure.

First, the fundamental spectral formulation given in Eq. (2.68) is divided by its area $m_0=A/4B$ so that the spectrum has unit area. That is,

$$S'(\omega)=\frac{4B}{\omega^5}e^{-B/\omega^4} \tag{2.77}$$

The above unit-area spectrum can be considered as if it were a probability density function since it satisfies all the conditions required for the probability density function. In fact, $S'(\omega)$ yields the exponential probability density function $f(x)$ by letting $\omega^4=1/x$. Hence, it can be generalized into the form of a gamma probability density function with an additional parameter λ which controls the shape of the density function. That is,

$$S'(\omega)=\frac{4}{\Gamma(\lambda)}\frac{B^\lambda}{\omega^{4\lambda+1}}e^{-B/\omega^4} \tag{2.78}$$

The spectrum $S'(\omega)$ is now converted to a dimensional wave spectrum $S(\omega)$ having an area equal to $(H_s/4)^2$ under the assumption that the spectrum is narrow-banded. This results in the spectrum

$$S(\omega)=\frac{1}{4}\frac{B^\lambda}{\Gamma(\lambda)}\frac{H_s^2}{\omega^{4\lambda+1}}e^{-B/\omega^4} \tag{2.79}$$

The constant B in Eq. (2.79) can be determined in terms of the modal frequency, ω_m, by setting the derivative of $S(\omega)$ with respect to ω to be zero. That is,

$$B=\left(\frac{4\lambda+1}{4}\right)\omega_m^4 \tag{2.80}$$

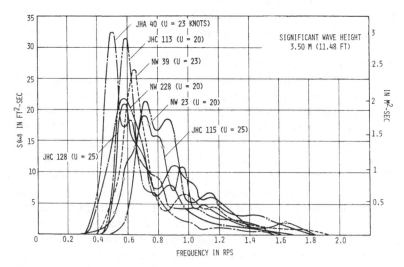

Fig. 2.13. Variety of wave spectra obtained from measured data in North Atlantic for significant wave height of 3.5 m (11.48 ft) (Ochi and Hubble 1976).

Thus, from Eqs. (2.79) and (2.80) we have the following spectral formulation carrying three parameters:

$$S(\omega) = \frac{1}{4} \frac{\left(\dfrac{4\lambda+1}{4}\,\omega_m^4\right)^{\lambda}}{\Gamma(\lambda)} \frac{H_s^2}{\omega^{4\lambda+1}} \, e^{-\left(\frac{4\lambda+1}{4}\right)\left(\frac{\omega_m}{\omega}\right)^4} \qquad (2.81)$$

Note that by letting $\lambda = 1$, Eq. (2.81) is reduced to the two-parameter wave spectrum given in Eq.(2.67). By connecting the two sets of spectra, one representing the low frequency components and the other the high frequency components of the wave energy, the following six-parameter spectral representation can be derived (Ochi and Hubble 1976):

$$S(\omega) = \frac{1}{4} \sum_{j=1,2} \frac{\left(\dfrac{4\lambda_j+1}{4}\,\omega_{mj}^4\right)^{\lambda_j}}{\Gamma(\lambda_j)} \frac{H_{sj}^2}{\omega^{4\lambda_j+1}} \, e^{-\left(\frac{4\lambda_j+1}{4}\right)\left(\frac{\omega_{mj}}{\omega}\right)^4} \qquad (2.82)$$

As seen in the above equation, the spectral formulation carries a total of six parameters. In reality, however, it is a one-parameter spectrum since all six parameters are given as a function of significant wave height, as will be shown later. The values of these parameters are determined from analysis of data obtained in the North Atlantic. The source of data is the same as that for the development of the Pierson–Moskowitz spectrum, but analysis is carried out on over 800 spectra including those in

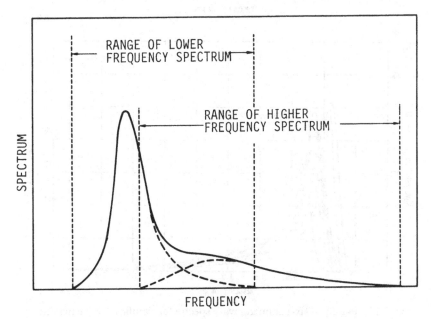

Fig. 2.14. Decomposition of wave spectrum into two parts.

partially developed seas and those having a bimodal shape. From a statistical analysis of the data, a family of wave spectra consisting of eleven members is generated for a desired sea severity (significant wave height) with the confidence coefficient of 0.95. The parametric values of these spectra are shown in Table 2.1. Examples of families of spectra for significant wave heights of 3 m and 9 m are shown in Figure 2.15.

A significant advantage of using a family of spectra for design of

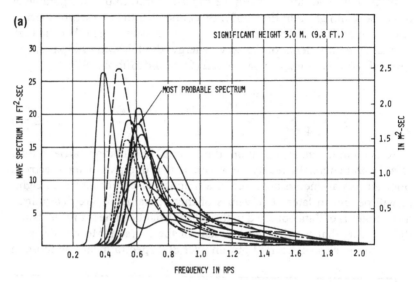

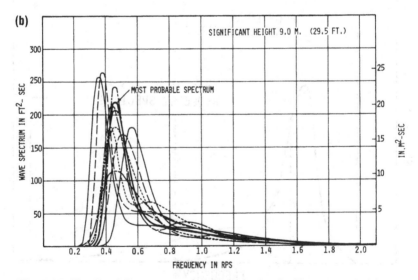

Fig. 2.15. Family of six-parameter wave spectra for significant wave heights
(a) 3.0 m (9.8 ft) and (b) 9.0 m (29.5 ft) (Ochi 1978a).

Table 2.1. *Values of the six parameters as a function of significant wave height in meters (Ochi and Hubble 1976).*

	H_{s1}	H_{s2}	ω_{m1}	ω_{m2}	λ_1	λ_2
Most probable spectrum	$0.84\,H_s$	$0.54\,H_s$	$0.70\,e^{-0.046\,H_s}$	$1.15\,e^{-0.039\,H_s}$	3.00	$1.54\,e^{-0.062\,H_s}$
0.95 confidence spectra	$0.95\,H_s$	$0.31\,H_s$	$0.70\,e^{-0.046\,H_s}$	$1.50\,e^{-0.046\,H_s}$	1.35	$2.48\,e^{-0.102\,H_s}$
	$0.65\,H_s$	$0.76\,H_s$	$0.61\,e^{-0.039\,H_s}$	$0.94\,e^{-0.036\,H_s}$	4.95	$2.48\,e^{-0.102\,H_s}$
	$0.84\,H_s$	$0.54\,H_s$	$0.93\,e^{-0.056\,H_s}$	$1.50\,e^{-0.046\,H_s}$	3.00	$2.77\,e^{-0.112\,H_s}$
	$0.84\,H_s$	$0.54\,H_s$	$0.41\,e^{-0.016\,H_s}$	$0.88\,e^{-0.026\,H_s}$	2.55	$1.82\,e^{-0.089\,H_s}$
	$0.90\,H_s$	$0.44\,H_s$	$0.81\,e^{-0.052\,H_s}$	$1.60\,e^{-0.033\,H_s}$	1.80	$2.95\,e^{-0.105\,H_s}$
	$0.77\,H_s$	$0.64\,H_s$	$0.54\,e^{-0.039\,H_s}$	0.61	4.50	$1.95\,e^{-0.082\,H_s}$
	$0.73\,H_s$	$0.68\,H_s$	$0.70\,e^{-0.046\,H_s}$	$0.99\,e^{-0.039\,H_s}$	6.40	$1.78\,e^{-0.069\,H_s}$
	$0.92\,H_s$	$0.39\,H_s$	$0.70\,e^{-0.046\,H_s}$	$1.37\,e^{-0.039\,H_s}$	0.70	$1.78\,e^{-0.069\,H_s}$
	$0.84\,H_s$	$0.54\,H_s$	$0.74\,e^{-0.052\,H_s}$	$1.30\,e^{-0.039\,H_s}$	2.65	$3.90\,e^{-0.085\,H_s}$
	$0.84\,H_s$	$0.54\,H_s$	$0.62\,e^{-0.039\,H_s}$	$1.03\,e^{-0.030\,H_s}$	2.60	$0.53\,e^{-0.069\,H_s}$

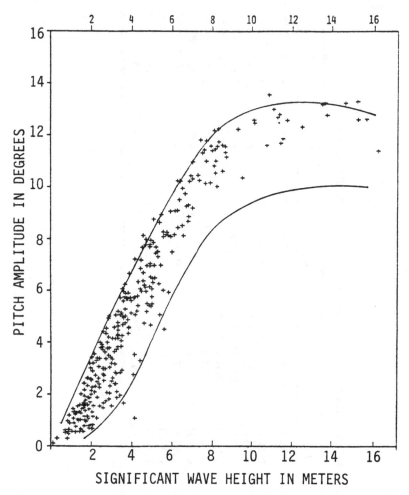

Fig. 2.16. Upper and lower bounds of extreme pitching motion of the MARINER computed by using a family of six-parameter wave spectra. (Crosses are computed using spectra obtained from measured data in the North Atlantic.)

marine systems is that one of the family members yields the largest response such as motions or wave-induced forces for a specified sea severity, while another yields the smallest response with confidence coefficient of 0.95. Hence, by connecting the largest and smallest values, respectively, obtained in each sea severity, the upper and lower-bounded response can be established. As an example, the lines in Figure 2.16 indicate the upper and lower-bounded values of the probable extreme pitching motion of the MARINER computed by using a family of six-parameter spectra. The crosses in the figure show values computed using spectra obtained from measured data in the North Atlantic (Ochi and Bales 1977). As can be seen, the upper and lower bound lines cover reasonably well the variation in responses computed using the measured spectra. It is also found that the bounds cover the variation of responses computed using measured spectra in various locations in the world, although the formulation is developed based on data measured in the North Atlantic. Thus, it may safely be concluded that the upper bound of the response evaluated by applying Eq. (2.82) can be used when designing marine systems.

2.3.5 JONSWAP spectrum

The JONSWAP formulation is based on an extensive wave measurement program known as the Joint North Sea Wave Project carried out in 1968 and 1969 along a line extending over 160 km into the North Sea from Sylt Island (Hasselmann *et al.* 1973). The spectrum represents wind-generated seas with fetch limitation, and wind speed and fetch length are inputs to this formulation. The original formulation is given as follows:

$$S(f) = \alpha \frac{g^2}{(2\pi)^4} \frac{1}{f^5} \exp\left\{-1.25(f_m/f)^4\right\} \gamma^{\exp\{-(f-f_m)^2/2(\sigma f_m)^2\}} \quad (2.83)$$

where γ = parameter, 3.30 as an average

$\alpha = 0.076\bar{x}^{-0.22}$

$\sigma = 0.07$ for $f \leqslant f_m$ and 0.09 for $f > f_m$

$f_m = 3.5 \, (g/\bar{U})\bar{x}^{-0.33}$

$\bar{x}$ = dimensionless fetch = $g \, x/\bar{U}^2$

x = fetch length

$\bar{U}$ = mean wind speed

g = gravity constant.

The formula may be expressed in terms of the frequency ω in rps as

$$S(\omega) = \alpha \frac{g^2}{\omega^5} \exp\left\{-1.25(\omega_m/\omega)^4\right\} \gamma^{\exp\{-(\omega-\omega_m)^2/2(\sigma\omega_m)^2\}} \quad (2.84)$$

where $\omega_m = 2\,\pi f_m$.

The parameter γ is called the *peak-shape parameter*, and it represents the ratio of the maximum spectral energy density to the maximum of the corresponding Pierson–Moskowitz spectrum. The term associated with the exponential power of γ is called the *peak enhancement factor*, and the JONSWAP spectrum is the product of the Pierson–Moskowitz spectrum and the peak enhancement factor.

The value of the peak-shape parameter γ is usually chosen as 3.30, and the spectrum is called the *JONSWAP spectral formulation*. The values of γ obtained from analysis of the original data, however, vary approximately from 1 to 6 even for a constant wind speed. γ is actually a random variable which is approximately normally distributed with mean 3.30 and variance 0.62 as shown in Figure 2.17 (Ochi 1979a).

Several studies have been carried out in detail on the scale parameter, α, and the peak-shape parameter, γ. Among others, Mitsuyasu *et al.* (1980) show that both parameters can be presented as a function of the dimensionless modal frequency, defined as $\bar{f}_m = f_m \overline{U}/g$. That is, from analysis of wave data obtained by Hasselmann as well as their own experimental data, Mitsuyasu *et al.* derive the following relationship:

$$\alpha = \frac{3.26}{10^2} \bar{f}_m^{0.857} \qquad (2.85)$$

This is shown in Figure 2.18. They also show that the peak-shape parameter γ can be presented as

$$\gamma = \frac{1.53}{10^4} \bar{f}_m^3 \alpha^{-3} \qquad (2.86)$$

By inserting Eq. (2.85) into Eq. (2.86), γ can be expressed as

$$\gamma = 4.42 \bar{f}_m^{0.429} \qquad (2.87)$$

This relationship reasonably agrees with measured data as shown in Figure 2.19.

Furthermore, from analysis of wave data obtained from measurements in a bay and from experiments in a wind-wave tank, Mitsuyasu *et al.* present the dimensionless frequency as a function of dimensionless fetch length as follows:

$$\bar{f}_m = 2.92 \, \bar{x}^{-0.33} \qquad (2.88)$$

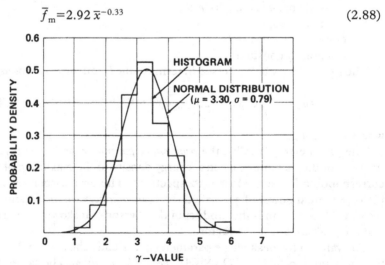

Fig. 2.17. Histogram of peak-shape parameter γ and associated normal probability distribution (Ochi, 1979a).

Thus, from Eqs. (2.85), (2.87) and (2.88), they show that the parameters α and γ can also be presented as a function of dimensionless fetch. That is,

$$\alpha = \frac{8.17}{10^2} \bar{x}^{-0.283}$$

$$\gamma = 7.0\, \bar{x}^{-0.142} \tag{2.89}$$

The JONSWAP spectral formulation is given as a function of wind speed. It is very convenient, in practice, for the spectrum to be presented in terms of sea severity. For this, a series of computations is carried out

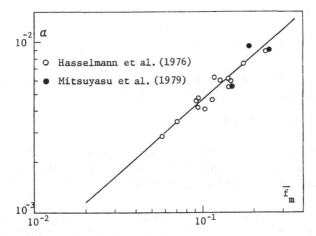

Fig. 2.18. Scale parameter α versus dimensionless modal frequency $\bar{f}_m$
(Mitsuyasu *et al.* 1980).

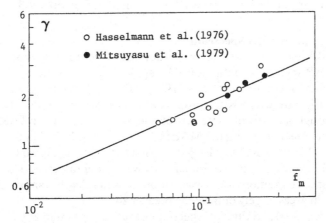

Fig. 2.19. Peak shape parameter γ versus dimensionless parameter $\bar{f}_m$
(Mitsuyasu *et al.* 1980).

Table 2.2. *k-value for
evaluating equivalent wind
speed for specified γ-value of
the JONSWAP spectrum
(Ochi 1979a).*

γ	k
1.75	96.2
2.64	88.3
3.30 (mean value)	83.7
3.96	80.1
4.85	76.4

using Eq. (2.83) for various combinations of fetch length and wind speed, and the following relationship derived (Ochi 1979a):

$$\overline{U} = k\, x^{-0.615}\, H_s^{1.08} \tag{2.90}$$

where k is a constant depending on γ-value as given in Table 2.2. $\overline{U}$ is in m/s, x in km and H_s in m. From Eqs. (2.83) and (2.90), the JONSWAP spectrum can be presented for a specified sea severity, H_s, and fetch length, x.

It is of interest to note that the shape of wave spectra during the growing stage of hurricane-generated seas is well represented by the JONSWAP spectral formulation; however, the values of parameters involved in the formulation are substantially different from those originally given in Eq. (2.83). This will be presented in Section 5.2.3.

2.3.6 TMA Spectrum

The TMA spectral formulation is developed as an extension of the JONSWAP spectrum so that it can be applied to wind-generated seas in finite water depth (Bouws *et al.* 1985). The concept is based on the similarity law shown by Kitaigorodskii *et al.* (1975), and its validity is verified through analysis of three sets of data obtained (a) near TEXEL in the North Sea, (b) during the MARSEN Project conducted in the North Sea and (c) in the ARSLOE Project carried out at Duck, North Carolina, USA. The data are acquired primarily by wave-rider buoys, and include a variety of sea conditions with observations at water depths ranging from 6 to 42 m, in areas largely outside the breaker zone.

In order to extend Phillips' spectral formulation given in Eq. (2.47) to be inclusive of waves in finite water depth, Kitaigorodskii *et al.* (1975) developed the following spectrum:

$$S(\omega) = \alpha \, g^2 \, \omega^{-5} \, \phi(\omega_h) \tag{2.91}$$

where $\phi(\omega_h)$ is a transformation factor given by

$$\phi(\omega_h) = \left[\frac{\left(k(\omega,h)\right)^{-3} \dfrac{\partial}{\partial \omega} k(\omega,h)}{\left(k(\omega,\infty)\right)^{-3} \dfrac{\partial}{\partial \omega} k(\omega,\infty)} \right]_{\omega_h = \omega\sqrt{h/g}} \tag{2.92}$$

In the above formula ω_h is a dimensionless frequency defined by $\omega\sqrt{h/g}$ (h=water depth), and $k(\omega,h)$ is the wave number associated with the dispersion relationship for waves in finite water depth. Figure 2.20 shows the transformation factor $\phi(\omega_h)$ taken from Kitaigorodskii *et al.* (1975).

Bouws *et al.* (1985) apply the transformation factor to the JONSWAP spectrum and present the TMA spectrum as follows:

$$S(\omega) = [\text{JONSWAP spectrum } (\omega)] \cdot \phi(\omega_h)$$

$$S(f) = [\text{JONSWAP spectrum } (f)] \cdot \phi(\omega_h)$$
$$\text{with } \omega_h = 2\pi f \sqrt{h/g} \tag{2.93}$$

where JONSWAP spectra $S(\omega)$ and $S(f)$ are given in Eqs. (2.84) and (2.83), respectively. Figure 2.21(a) and (b) taken from Bouws *et al.* (1985) are examples of a TMA spectrum which indicates clearly the effect of water depth on the magnitude as well as the shape of the JONSWAP spectrum. Note that Figure 2.21(a) is given in frequency space, while that in (b) is given in wave-number space.

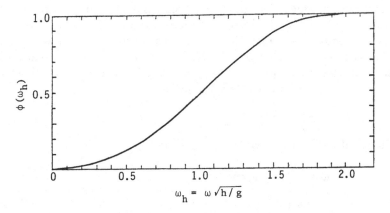

Fig. 2.20. Transformation factor as a function of dimensionless frequency ω_h (Kitaigorodskii *et al.* 1975).

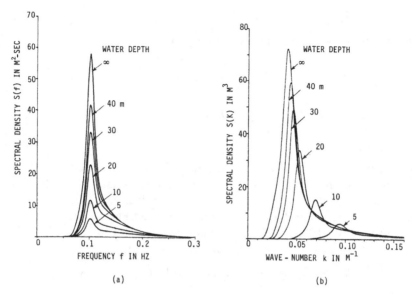

Fig. 2.21. TMA spectra for different water depths: (a) in frequency space; (b) in wave-number space. (JONSWAP parameters; $f_m=0.1$ Hz, $\alpha=0.01$, $\gamma=3.3$, $\sigma_a=0.07$, $\sigma_b=0.09$) (Bouws *et al.* 1985).

2.4 MODIFICATION OF WAVE SPECTRUM FOR MOVING SYSTEMS

The wave spectrum, in general, is obtained from measurement at a fixed location in the ocean. Let us consider the situation when a marine system (a ship for example) is moving in random seas. From the ship's viewpoint under this situation, the frequencies of encounters with waves are quite different from wave frequencies. They can be evaluated, however, by accounting for the ship's speed and direction relative to the waves by the following procedure (St. Denis and Pierson 1953).

Let a ship be moving with speed V_s in a certain direction, with waves approaching from an angle χ, as shown in Figure 2.22. For regular waves with wavelength λ and celerity V_w, the period of encounters with a ship moving with a speed V_s can be obtained as

$$T_e=\frac{\lambda}{V_w-V_s\cos\chi} \tag{2.94}$$

Hence, the encounter frequency is given by

$$\omega_e=\frac{2\pi}{\lambda}(V_w-V_s\cos\chi)=\omega-\frac{\omega^2}{g}V_s\cos\chi$$

$$=\omega(1-\alpha) \tag{2.95}$$

where

$$\alpha=\frac{\omega}{g}V_s\cos\chi$$

Consider the following four cases:

(i) $\alpha < 0$: in this case, waves approach the ship from the bow, and $\omega_e > \omega$.

(ii) $0 < \alpha < 1/2$: waves approach the ship from the stern, and $\omega_e < \omega$. Waves travel and pass the ship quickly. ω_e is maximum when $\alpha = 1/2$.

(iii) $1/2 < \alpha < 1$: waves approach the ship from the stern, and $\omega_e < \omega$. Waves pass the ship slowly.

(iv) $\alpha = 1$: wave velocity is the same as that of the ship and they maintain a constant position relative to one another.

(v) $\alpha > 1$: waves approach the ship from the stern, but the ship outstrips the waves. ω_e is negative referred to the moving ship; i.e. waves appear to come from a direction opposite to that of their origin.

In order to cover all these five cases, the wave frequency should be modified by taking the absolute value of Eq. (2.95). Further, the spectrum $S(\omega)$ should be modified by multiplying by the Jacobian, $|d\omega/d\omega_e|$, so that the area under the encounter wave spectrum is the same as that of the wave spectrum. Thus, in summary, the frequency and wave spectrum should be modified by applying the following formulae:

$$\omega_e = \left| \omega \left(1 - \frac{\omega}{g} V_s \cos \chi \right) \right| \tag{2.96}$$

$$\mathcal{J} = \left| \frac{d\omega}{d\omega_e} \right| = \left| \frac{1}{\sqrt{1 - \dfrac{4\omega_e V_s \cos\chi}{g}}} \right| \tag{2.97}$$

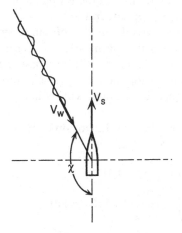

Fig. 2.22. Relative location of waves and a ship moving with speed V_s.

Fig. 2.23. Modification of the wave spectrum for a moving system.

Figure 2.23 shows a sketch indicating the modification of the wave spectrum for a moving system.

2.5 HIGHER-ORDER SPECTRAL ANALYSIS

Statistical analysis of the time history of wave records taken in the ocean at deep water depths indicates the waves to be a Gaussian random process; namely, the wave profiles are normally distributed. On rare occasions, very small deviations from normality exist; however, the deviations are not of such magnitude that nonlinearity need be considered in predicting statistical properties of the waves or for predicting the responses of marine systems in a seaway. Therefore, the spectral analysis methodologies presented in Section 2.1 are valid as far as random waves in deep water are concerned. However, this is not the case for random waves in finite water depths where the frequency components of the wave energy are not necessarily independent; instead, energy transfer between wave components takes place. This implies that the waves are no longer Gaussian and therefore the stochastic properties cannot be evaluated by the first two moments which yield the mean and variance. The third moment, or even higher moments, may be required to describe these waves adequately.

It will be shown that two-dimensional spectral analysis, called *bispectral analysis*, yields the third moment, and three-dimensional analysis, called *trispectral analysis*, yields the fourth moment, and so on.

One of the most important pieces of information obtained by carrying out higher-order spectral analysis is the clarification of the interaction between wave components. This is demonstrated by Hasselmann *et al.* (1962) who show that the nonlinear interaction of wave components is clarified through bispectral analysis. This subject will be discussed later in detail.

For bispectral analysis, consider the two-dimensional auto-correlation function, denoted by $M_{xx}(\tau_1,\tau_2)$, and defined as

$$M_{xx}(\tau_1,\tau_2) = \mathrm{E}[x(t)x(t+\tau_1)\,x(t+\tau_2)]$$

$$= \lim_{T\to\infty} \frac{1}{2T} \int\limits_{-T}^{T} x(t)x(t+\tau_1)x(t+\tau_2)\,\mathrm{d}t \qquad (2.98)$$

Analogous to the definition of the auto-correlation function $R_{xx}(\tau)$ shown in Figure 2.2, $M_{xx}(\tau_1,\tau_2)$ can be evaluated by shifting the time history of a wave record $x(t)$ by time τ_1 and τ_2, and then by integrating the product $x(t)x(t+\tau_1)x(t+\tau_2)$. It is important to note that there are six auto-correlation functions having the same value for a given τ_1 and τ_2. These are

$$M_{xx}(\tau_1,\tau_2)=M_{xx}(\tau_2,\tau_1)=M_{xx}(-\tau_2,\tau_1-\tau_2)=M_{xx}(\tau_1-\tau_2,-\tau_2)$$

$$=M_{xx}(-\tau_1,\tau_2-\tau_1)=M_{xx}(\tau_2-\tau_1,-\tau_1) \tag{2.99}$$

The two-dimensional spectral density function is called the *bispectrum* and is defined as follows:

$$B_{xx}(\omega_1,\omega_2)=\lim_{T\to\infty}\frac{1}{2\pi T}X(\omega_1)\,X(\omega_2)\,X^\star(\omega_1+\omega_2) \tag{2.100}$$

where $X(\omega)$ is the Fourier transform of $x(t)$ and $X^\star(\omega_1+\omega_2)$ is the conjugate of $X(\omega_1+\omega_2)$.

The definition given in Eq. (2.100) is often written as

$$B_{xx}(\omega_1,\omega_2)=\lim_{T\to\infty}\frac{1}{2\pi T}X(\omega_1)\,X(\omega_2)\,X(\omega_3) \tag{2.101}$$

where $\omega_1+\omega_2+\omega_3=0$. This definition is of different form from that given for the spectral density function $S_{xx}(\omega)$. However, Eq. (2.101) is equivalent to writing $S_{xx}(\omega)$ in Eq. (2.8) as follows:

$$S_{xx}(\omega)=\lim_{T\to\infty}\frac{1}{2\pi T}|X(\omega)|^2=\lim_{T\to\infty}\frac{1}{2\pi T}X(\omega)\,X^\star(\omega)$$

$$=\lim_{T\to\infty}\frac{1}{2\pi T}X(\omega_1)\,X(\omega_2) \tag{2.102}$$

where $\omega_1+\omega_2=0$.

It can easily be proved that the Wiener–Khintchine theorem can also be applied to the two-dimensional case. That is, the bispectrum $B_{xx}(\omega_1,\omega_2)$ is the two-dimensional Fourier transform of the auto-correlation function $M_{xx}(\tau_1,\tau_2)$, and can be written as

$$B_{xx}(\omega_1,\omega_2)=\frac{1}{\pi^2}\int_{-\infty}^{\infty}\int_{-\infty}^{\infty}M_{xx}(\tau_1,\tau_2)\,e^{-i(\omega_1\tau_1+\omega_2\tau_2)}d\tau_1\,d\tau_2 \tag{2.103}$$

In reverse, we have

$$M_{xx}(\tau_1,\tau_2)=E[x(t)x(t+\tau_1)x(t+\tau_2)]$$

$$=\frac{1}{2^2}\int_{-\infty}^{\infty}\int_{-\infty}^{\infty}B_{xx}(\omega_1,\omega_2)\,e^{i(\omega_1\tau_1+\omega_2\tau_2)}d\omega_1\,d\omega_2 \tag{2.104}$$

By letting $\tau_1=\tau_2=0$ and integrating in the positive domain, we have

$$M_{xx}(0,0) = E[x^3(t)] = \lim_{T \to \infty} \frac{1}{2T} \int_{-\infty}^{\infty} \{x(t)\}^3 \, dt$$

$$= \int_0^\infty \int_0^\infty B(\omega_1, \omega_2) \, d\omega_1 \, d\omega_2 \qquad (2.105)$$

Thus, we have proved that the integrated volume of the bispectrum represents the third moment of random waves $x(t)$.

It is stated in connection with Eq. (2.99) that there are six auto-correlation functions for a given τ_1 and τ_2. Since the Wiener–Khinchine theorem holds for each of these auto-correlation functions, there are six bispectra for a given ω_1 and ω_2 having the same value. These are

$$B_{xx}(\omega_1, \omega_2) = B_{xx}(\omega_2, \omega_1) = B(\omega_1, -\omega_1 - \omega_2)$$
$$= B_{xx}(-\omega_1 - \omega_2, \omega_1) = B_{xx}(\omega_2, -\omega_1 - \omega_2)$$
$$= B_{xx}(-\omega_1 - \omega_2, \omega_2) \qquad (2.106)$$

In addition, as shown in Eq. (2.100), the bispectrum is a complex function; therefore, a conjugate function exists for each bispectrum given in Eq. (2.106). These conjugate functions are

$$B_{xx}^*(-\omega_1, -\omega_2) = B_{xx}^*(-\omega_2, -\omega_1) = B_{xx}^*(-\omega_1, \omega_1 + \omega_2)$$
$$= B_{xx}^*(\omega_1 + \omega_2, -\omega_1) = B_{xx}^*(-\omega_2, \omega_1 + \omega_2)$$
$$= B_{xx}^*(\omega_1 + \omega_2, -\omega_2) \qquad (2.107)$$

In order to provide a better understanding of the relationship between the twelve bispectra given in Eqs. (2.106) and (2.107), Figure 2.24 shows the symmetric characteristics of the bispectrum. As can be seen, the six bispectra given in Eq. (2.106) are pairwise symmetric with respect to the line $\omega_1 = \omega_2$, while six conjugate spectra given in Eq. (2.107) are pairwise symmetric with respect to the origin. Because of these symmetries, the bispectrum $B_{xx}(\omega_1, \omega_2)$ in the fundamental domain which is the octant defined by $0 \leqslant \omega_2 < \omega_1$ and $0 \leqslant \omega_1 < \infty$ (shaded area in Figure 2.24) represents all other spectra.

The third moment, $E[\{x(t)\}^3]$, is equal to the sum of the integrated volumes under the bispectra at six locations ($A, B, ..., F$ in Figure 2.24). Hence, six times the integrated volume under $B_{xx}(\omega_1, \omega_2)$ in the fundamental domain yields the third moment.

Formulae presented so far deal with the two-dimensional spectrum, but the principle can be generalized to the n-dimensional spectrum. That is, following the Wiener–Khintchine theorem we may write the n-dimensional spectrum as

$$B(\omega_1, \omega_2, ..., \omega_n) = \frac{1}{\pi^n} \int_0^\infty \int_0^\infty \cdots \int_0^\infty M(\tau_1, \tau_2, ..., \tau_n)$$
$$\times e^{-i(\omega_1 \tau_1 + \omega_2 \tau_2 + ... + \omega_n \tau_n)} d\tau_1 \, d\tau_2 ... d\tau_n \quad (2.108)$$

where

$$M(\tau_1, \tau_2, ..., \tau_n) = E[x(t)x(t+\tau_1)x(t+\tau_2)...(t+\tau_n)]$$

$$= \lim_{T \to \infty} \frac{1}{2T} \int_{-T}^{T} x(t)x(t+\tau_1)...x(t+\tau_n)dt$$

$$= \frac{1}{2^n} \int_{-\infty}^{\infty} \int_{-\infty}^{\infty} ... \int_{-\infty}^{\infty} B(\omega_1, \omega_2, ..., \omega_n)$$

$$\times e^{i(\omega_1\tau_1 + \omega_2\tau_2 + ... + \omega_n\tau_n)} d\omega_1 d\omega_2 ... d\omega_n \quad (2.109)$$

which yield

$$M(0, 0, ..., 0) = E[x^{n+1}(t)]$$

$$= \int_{0}^{\infty} \int_{0}^{\infty} ... \int_{0}^{\infty} B(\omega_1, \omega_2, ..., \omega_n) d\omega_1 d\omega_2 ... d\omega_n \quad (2.110)$$

Thus, it is clear that integration of the n-dimensional spectrum with respect to $\omega_1, \omega_2, ..., \omega_n$ yields the $(n+1)$th moment of random waves.

Bispectral analysis of ocean wave records was first carried out by Hasselman et al. (1962) in which the wave–wave interaction obtained through bispectral analysis was compared with theory developed by Hasselmann (1962). An example of their analysis is shown in Figure 2.25. In the figure, the contour lines at 10^3, 10^4, 10^5 and 10^6 cm^3 s^2 represent the absolute value of the bispectrum per unit frequency band

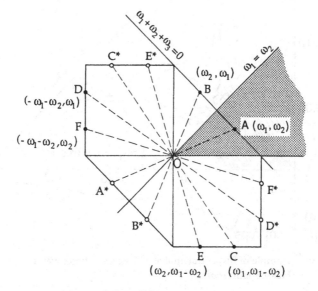

Fig. 2.24. System characteristics of bispectrum.

squared. The bispectrum obtained from the wave record is shown in the octant below the 45 degree line, while the theoretical results are shown in the upper octant. In the case of perfect agreement between theory and the evaluated bispectrum, the pattern would be completely symmetrical about the 45 degree line. For the example shown in the figure, good agreement is obtained between theory and bispectrum. The unispectrum of waves is also shown along the two axes in the figure.

The frequency in Figure 2.25 is given in Nyquist units, defined as the sampling frequency of the record, which is 0.25 Hz in this example. The same two plots along the two frequency axes are the wave spectral density function in cm/Hz units. As can be seen, the modal frequency of the spectrum is 0.22 Nyquist frequency (0.055 Hz, 0.35 rps), and the bispectrum indicates that wave–wave interactions take place between the modal frequency and many other frequencies. The largest value is on the 45 degree

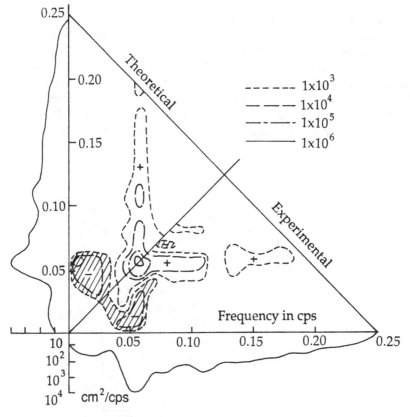

Fig. 2.25. Example of bispectral analysis of waves in finite water depth (Hasselmann *et al.* 1962).

line at the modal frequency, which indicates a strong interaction of the peak energy density with itself.

One interesting application of bispectral analysis is the assessment of the amount of nonlinearity involved in the wave spectrum in order to evaluate the statistical properties of waves in finite water depth. This subject will be discussed in Chapter 9.

3   WAVE AMPLITUDE AND HEIGHT

3.1 INTRODUCTION

In this chapter, the stochastic properties and probability distributions applicable to wave amplitude and height will be presented under various conditions.

First, the underlying assumptions considered throughout this chapter are that random waves are considered to be weakly steady-state, ergodic, and a normal (Gaussian) random process. The normal process assumption is valid only for waves in deep water. Waves in finite water depths are commonly treated as nonlinear, and considered to be a non-Gaussian random process, as will be discussed in detail in Chapter 9. With these basic assumptions, probability distribution functions which represent the statistical characteristics of random waves are analytically derived.

The most commonly considered probability distribution for wave amplitude is developed assuming that the wave spectrum is narrow-banded. The wave profile under this condition is slowly changing with constant period, and there exists a single peak or trough during each half-cycle. Waves generated by moderate wind speeds, an example of which is shown in Figure 3.1(a), demonstrate that the narrow-banded spectrum assumption is generally acceptable. In this case, wave amplitude follows the Rayleigh probability law as will be presented in Section 3.2. The Rayleigh probability distribution is most commonly considered for the design of marine system.

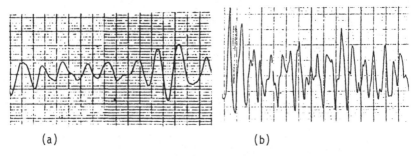

(a)   (b)

Fig. 3.1. Wind-generated wave profiles: (a) narrow-band random waves; (b) non-narrow-band random waves.

Waves generated by substantial wind speeds, on the other hand, demonstrate that the wave profile is more irregular and often two or three peaks or troughs exist during each cycle. An example of a wave record taken in relatively severe seas is shown in Figure 3.1(b). Note that because of double peaks or troughs, the wave amplitude cannot be clearly defined; hence, wave peaks and troughs are called the positive maxima and negative minima, respectively. The statistical characteristics of these quantities will be presented in Section 3.3.

The joint probability distribution of two wave amplitudes, in particular, amplitudes separated by a specified time interval or two successive wave amplitudes, is extremely important for predicting the possible occurrence of group waves, an event which causes a serious problem for the safety of marine systems. The derivation of the joint distribution of two wave amplitudes is discussed in Section 3.4.

The magnitude of wave heights (peak-to-trough excursions), in general, is not necessarily twice that of the amplitude in random seas; nevertheless, it is commonly assumed that heights are twice the amplitude. Strictly speaking, however, the probabilistic estimation of wave height should be made through the probability density function for the sum of positive and negative amplitudes assuming that these two are statistically independent random variables. This subject will be discussed in Section 3.5.

The severity of sea condition is most commonly presented by significant wave height which is defined as the average of the highest one-third observed or measured wave heights. The functional relationship between significant wave height and the wave spectrum is presented under the narrow-band assumption in Section 3.6.

One way to present a more comprehensive picture of the stochastic properties of random waves is to consider the probability distribution of every excursion, $x_1, x_2,...$ illustrated in Figure 3.2. Statistical analysis of the vertical distance (displacement) between successive peak-to-trough excursions or vice versa is called half-cycle excursion analysis. As can be seen in the figure, half-cycle excursions may be categorized as those crossing the zero-line and those located above or below the zero-line.

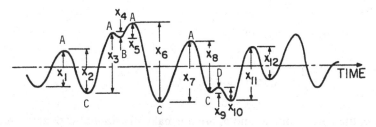

Fig. 3.2. Definition of half-cycle excursion.

Derivation of the probability density function applicable for half-cycle excursions is presented in Section 3.7.

The probability distribution of wave height over a long time period (the lifetime of a marine system, for example) is discussed in Section 3.8. Long-term statistics is an accumulation of short-term wave statistics, and plays a significant role in evaluating the fatigue loads of marine systems.

Probability distributions applicable to wave amplitude under the various conditions discussed in Sections 3.2 through 3.7 are all derived from the wave spectrum. It is sometimes convenient to obtain statistical properties (including extreme values) of wave height from information contained in the wave record without carrying out spectral analysis. This can be done by applying statistical inference theory, though most commonly the estimation is based on the assumption that waves are considered to be a narrow-band random process. This subject is discussed in Section 3.9.

3.2 PROBABILITY DISTRIBUTION OF AMPLITUDES WITH NARROW-BAND SPECTRUM

3.2.1 Derivation of probability density function

A *Narrow-band random process* is defined as one whose spectral density function is sharply concentrated in the neighborhood of a certain frequency ω_0. This implies that the random process $x(t)$ has a constant frequency, and may be written as

$$\begin{aligned} x(t) &= A(t) \cos\{\omega_0 t + \epsilon(t)\} \\ &= A(t)\{\cos \epsilon(t) \cos \omega_0 t - \sin \epsilon(t) \sin \omega_0 t\} \end{aligned} \quad (3.1)$$

where $A(t)$ is the amplitude and $\epsilon(t)$ is the phase; both are random variables.

On the other hand, assuming that the random process $x(t)$ is a normal random process with zero mean and variance σ^2, $x(t)$ may be written as

$$x(t) = \sum_{n=1}^{\infty} (a_n \cos n\omega t + b_n \sin n\omega t) \quad (3.2)$$

where

$$a_n = \frac{2}{T} \int_0^T x(t) \cos n\omega t \, \mathrm{d}t$$

$$b_n = \frac{2}{T} \int_0^T x(t) \sin n\omega t \, \mathrm{d}t$$

Here, the coefficients a_n and b_n are normally distributed with zero mean and variance σ^2.

By writing $n\omega t$ in Eq. (3.2) as $(n\omega - \omega_0)t + \omega_0 t$, Eq. (3.2) can be written as follows:

$$x(t) = x_c(t) \cos \omega_0 t - x_s(t) \sin \omega_0 t \qquad (3.3)$$

where

$$x_c(t) = \sum_{n=1}^{\infty} \{a_n \cos(n\omega - \omega_0)t + b_n \sin(n\omega - \omega_0)t\}$$

$$x_s(t) = \sum_{n=1}^{\infty} \{a_n \sin(n\omega - \omega_0)t - b_n \cos(n\omega - \omega_0)t\} \qquad (3.4)$$

From a comparison between Eq. (3.1) and (3.3), we have

$$\begin{aligned} x_c(t) &= A(t) \cos \epsilon(t) \\ x_s(t) &= A(t) \sin \epsilon(t) \end{aligned} \qquad (3.5)$$

We may write $x_c(t)$, $x_s(t)$, $A(t)$ and $\epsilon(t)$ as the random variables x_c, x_s, A and ϵ for a given time t. Since x_c and x_s are the summations of normal random variables, they are also normally distributed. It can be proved that x_c and x_s are statistically independent normal random variables with zero mean and variance which is equal to the area under the spectral density function of $x(t)$. That is,

$$\begin{aligned} E[x_c] &= E[x_s] = 0 \\ E[x_c x_s] &= 0 \\ E[x_c^2] &= E[x_s^2] = \sigma^2 = \int_0^{\infty} S(\omega) \, d\omega \end{aligned} \qquad (3.6)$$

where $S(\omega)$ = spectral density function of $x(t)$. For proofs, refer to Rice (1944) or Davenport and Root (1958).

We may write the joint probability density function of x_c and x_s as

$$f(x_c, x_s) = \frac{1}{2\pi\sigma^2} e^{-\frac{1}{2\sigma^2}(x_c^2 + x_s^2)} \qquad -\infty < x_c < \infty \qquad -\infty < x_s < \infty \quad (3.7)$$

Next, let us transform the joint probability density function $f(x_c, x_s)$ to the joint probability density function $f(A, \omega)$ by using the relationship given in Eq. (3.5). The transformation yields

$$f(A, \epsilon) = \frac{1}{2\pi\sigma^2} e^{-A^2/2\sigma^2} \qquad 0 \le A < \infty \qquad 0 \le \epsilon \le 2\pi \quad (3.8)$$

Then, the marginal probability density function of A can be obtained by

$$f(A) = \int_0^{2\pi} f(A, \epsilon) d\epsilon = \frac{A}{\sigma^2} e^{-A^2/2\sigma^2} \qquad 0 \le A < \infty \qquad (3.9)$$

The above probability density function is the Rayleigh probability distribution which is usually written in the form

$$f(A) = \frac{2A}{R} e^{A^2/R} \qquad 0 \leqslant A < \infty \qquad (3.10)$$

where R is a parameter.

It can be concluded therefore that wave amplitude obeys the Rayleigh probability law under the narrow-band random process assumption, and that the parameter of the distribution is equal to twice the area under the spectral density function. An example of a comparison between the probability density function and the histogram of amplitude constructed from measured data is shown in Figure 3.3.

The Rayleigh probability distribution was first introduced for predicting wave amplitude by Longuet-Higgins (1952). However, the parameter R of the distribution shown in Eq. (3.10) was not expressed in terms of the wave spectrum; instead, it was given in the following form of an estimator considered in statistical inference theory:

$$R = \frac{1}{n} \sum_{i=1}^{n} x_i^2 \qquad (3.11)$$

where x_i is the observed wave amplitude. This will be discussed in detail in Section 3.9.

The marginal probability density function of the phase ϵ can be derived from Eq. (3.8) as

$$f(\epsilon) = \int_0^\infty f(A, \epsilon) \, dA = \frac{1}{2\pi} \qquad 0 \leqslant \epsilon \leqslant 2\pi \qquad (3.12)$$

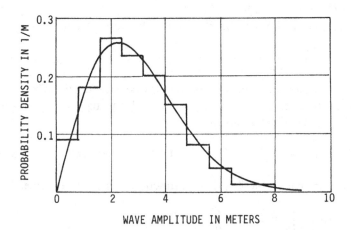

WAVE AMPLITUDE IN METERS

Fig. 3.3. Comparison between Rayleigh probability density function and histogram of wave amplitude constructed from measured data (significant wave height 9.20 m).

The probability density function $f(\epsilon)$ is a uniform distribution. This implies that the phase of a narrow-band Gaussian random process can take on any value between 0 and 2π.

The probability density function of the amplitude of a narrow-band Gaussian random process can also be developed approximately by the following procedure.

Let $x(t)$ and $\dot{x}(t)$ be the wave displacement and velocity, respectively, and consider the displacement $x(t)$ to be crossing a specified level x in the upward direction with velocity $\dot{x}$ as shown in Figure 3.4. Note that the number of upward-crossings of the level is equal to the number of peaks for a narrow-band random process. The time required for $x(t)$ to pass through the distance dx with velocity $\dot{x}$ is given by $dx/\dot{x}$. Hence, the average (expected) number of the up-crossings of a specified level ζ per unit time can be evaluated based on the joint probability density function $f(x,\dot{x})$ with $x=\zeta$ as follows:

$$E[N_\zeta] = \int_0^\infty \dot{x} f(\zeta,\dot{x})\, d\dot{x} \qquad (3.13)$$

The random variables x and $\dot{x}$ are statistically independent for a Gaussian random process. By letting the variances of x and $\dot{x}$ be σ_x^2 and $\sigma_{\dot{x}}^2$, respectively, Eq. (3.13) can be evaluated as

$$E[N_\zeta] = \frac{1}{2\pi} \frac{\sigma_{\dot{x}}}{\sigma_x} e^{-\zeta^2/2\sigma_x^2} \qquad (3.14)$$

The average (expected) number of up-crossing of the zero-line can be evaluated from Eq. (3.13) by letting $\zeta=0$ as

$$E[N_+] = \int_0^\infty \dot{x} f(0,\dot{x})\, d\dot{x} = \frac{1}{2\pi} \frac{\sigma_{\dot{x}}}{\sigma_x} \qquad (3.15)$$

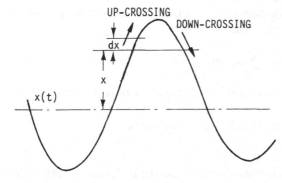

Fig. 3.4. Explanatory sketch of level-crossing of narrow-band random wave.

Assuming that the ratio of N_ζ/N_+ and N_+ are statistically independent, we can evaluate the probability that the peak exceeds a level ζ from Eqs. (3.14) and (3.15) as

$$\Pr\{\text{Peak}>\zeta\} = E[N_\zeta/N_+] = E[N_\zeta]/E[N_+] = e^{-\zeta^2/2\sigma_x^2} \qquad (3.16)$$

This probability is equal to $1-F(\zeta)$, where $F(\zeta)$ is the cumulative distribution function of the peak (amplitude). Hence, by differentiating Eq. (3.16) with respect to ζ, the probability density function of peak (amplitude) can be derived. That is,

$$f(\zeta) = \frac{\zeta}{\sigma_x^2} e^{-\zeta^2/2\sigma_x^2} \qquad 0 \leqslant \zeta < \infty \qquad (3.17)$$

This is the same probability density function given in Eq. (3.10) with $R = 2\sigma_x^2$.

3.2.2 Wave envelope process

It is assumed in the previous section that the wave spectrum is narrow-banded with a constant frequency ω_0, and the magnitude of wave amplitude varies slowly with time; its frequency is much smaller than ω_0. This implies that upper and lower envelopes can be drawn by connecting peaks and troughs, respectively, and thereby the statistical properties of peaks and troughs may be evaluated through the envelope process instead of dealing directly with the amplitude. That is, the *envelope process* is defined as a pair of symmetric curves that pass through the wave crests and troughs. It represents a measure of change of wave amplitude in the time domain.

The probabilistic analysis of random phenomena based on the envelope process was first introduced by Rice (1944, 1945), and numerous studies on waves and associated problems have been carried out by applying the concept of the envelope process. These include Longuet-Higgins (1952), Ewing (1973), Tung (1974), Yuen and Lake (1975) and Tayfun (1983), among others.

For a mathematical presentation of the envelope process, let us write the wave profile $x(t)$ as

$$x(t) = \sum_{n=0}^{\infty} a_n \cos(\omega_n t + \epsilon_n) \qquad (3.18)$$

where a_n is a normal random variable with zero mean and a certain variance.

Let $\bar{\omega}$ be the mean frequency defined as

$$\bar{\omega} = m_1/m_0 \qquad (3.19)$$

where $m_j = j$th moment of wave spectrum. Then, we may write Eq. (3.18) as

$$x(t) = \mathrm{Re} \sum_{n=0}^{\infty} a_n \, e^{i\{(\omega_n - \bar{\omega})t + \epsilon_n\} + i\bar{\omega}t}$$

$$= \mathrm{Re} \, \rho(t) \, e^{i\phi(t)} \, e^{i\bar{\omega}t} = \mathrm{Re} \, \rho(t) \, e^{i\chi(t)} \tag{3.20}$$

where

$$\chi(t) = \bar{\omega}t + \phi(t)$$

$$\rho \cos \phi = \sum_n a_n \cos \{(\omega_n - \bar{\omega})t + \epsilon_n\}$$

$$\rho \sin \phi = \sum_n a_n \sin \{(\omega_n - \bar{\omega})t + \epsilon_n\}$$

In the above equation, $\rho(t) \exp\{i\phi(t)\}$ is a slow amplitude modulation of the random process and $\exp\{i\bar{\omega}t\}$ is a carrier wave, as shown in Figure 3.5. Here, $\rho(t)$ is the *envelope process*, and $\chi(t)$ is called the *total phase*. It is noted that even when the spectrum is not narrow, it is possible to define the complex envelope function $\rho(t)$ by writing Eq. (3.18) in the form given in Eq. (3.20).

Let us write $x(t)$ in terms of the envelope process $\rho(t)$ and the total phase $\chi(t)$ as

$$x(t) = x_c(t) + i\, x_s(t) \tag{3.21}$$

where

$$x_c(t) = \rho(t) \cos \chi(t)$$
$$x_s(t) = \rho(t) \sin \chi(t)$$

Now, the envelope process $\rho(t)$ for a given wave record $x(t)$ can be evaluated by applying the concept of the Hilbert transform. The Hilbert transform of a function $x(t)$ is defined as follows (see Appendix C):

$$\tilde{x}(t) = \frac{1}{\pi} \int_{-\infty}^{\infty} \frac{x(\tau)}{t - \tau} d\tau \tag{3.22}$$

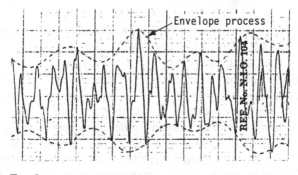

Fig. 3.5. Envelope process.

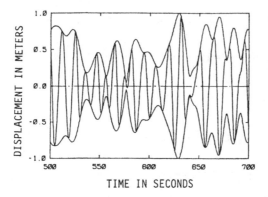

TIME IN SECONDS

Fig. 3.6. Example of envelope process of random waves obtained through application of Hilbert transformation technique (Shum and Melville 1984).

As stated in Appendix C, for a real-valued function $x(t)$, we may define a complex process given by

$$z(t) = x(t) + i\tilde{x}(t) \tag{3.23}$$

Then the envelope of $z(t)$ can be obtained as

$$\rho(t) = \sqrt{\{x(t)\}^2 + \{\tilde{x}(t)\}^2} \tag{3.24}$$

We may apply this method to Eq. (3.21). Note that $x_c(t)$ in Eq. (3.21) is known from the wave record, and furthermore it can be proved that the Hilbert transform of $\cos x(t)$ is $\sin x(t)$. Hence, by taking the Hilbert transform of the wave record, $x_s(t)$ can be evaluated and then the envelope process of a given wave record drawn using Eq. (3.24). An example of the envelope process of random waves obtained through application of the Hilbert transform is shown in Figure 3.6. In practice, the envelopes are generally obtained from a filtered record. The effect of lower and upper frequencies that are cut off when filtering a wave record is discussed in detail by Longuet-Higgins (1984).

3.3 PROBABILITY DISTRIBUTION OF WAVE MAXIMA WITH NON-NARROW-BAND SPECTRUM

The profile of waves, in particular those observed in relatively severe seas, is much more irregular than that shown under the assumption of the narrow-band random process. The profile is irregular in the sense that often two or more peaks (or troughs) exist during each half-cycle; and hence wave amplitude cannot be clearly defined. In this situation, Cartwright and Longuet-Higgins (1956) define the local maxima and minima of the wave profile and develop probability density functions.

Let us first define the local maxima and minima on both the positive (above the mean value) and negative (below the mean value) sides of a random process. As illustrated in Figure 3.7, the local peaks and troughs located on the positive side satisfy the condition $x(t)>0$ and $\dot{x}(t)=0$, and they are defined as the *positive maxima* for $\ddot{x}(t)<0$ and *positive minima* for $\ddot{x}(t)>0$, respectively. Similarly, the peaks and troughs on the negative side of a random process satisfy the condition $x(t)<0$ and $\dot{x}(t)=0$, and they are defined as *negative maxima* for $\ddot{x}(t)<0$ and *negative minima* for $\ddot{x}(t)>0$, respectively. For a Gaussian random process, the positive maxima and the negative minima follow the same probability law. The same is true for the positive minima and the negative maxima.

With regard to the statistical prediction of local maxima and minima, distributions of two different formulations are available; one considers the probability distribution of positive maxima (or negative minima) only, the other deals with the probability distribution of maxima (or minima) including positive as well as negative maxima (or minima). In the following, the derivation of the former probability distribution is discussed in detail, and then the latter probability distribution is outlined.

In order to develop the probability density function of the positive maxima, denoted as Z, consider the probability that Z will exceed a certain level ζ. Following the same concept as shown in Eq. (3.16), the probability is equivalent to the average value of the ratio of the number of Z above ζ per unit time, denoted by N_ζ, to the total number of Z per unit time, denoted by N_+. That is

$$\Pr\{Z>\zeta\}=1-F(\zeta)=\mathrm{E}[N_\zeta/N_+]\approx\mathrm{E}[N_\zeta]/\mathrm{E}[N_+] \qquad (3.25)$$

where $F(\zeta)$ is the cumulative distribution function of Z, and $\mathrm{E}[\ ]$ stands for the expected value.

Rice (1944, 1945) and Middleton (1960) show that the expected (average) number of positive maxima above the level ζ per unit time can be written as

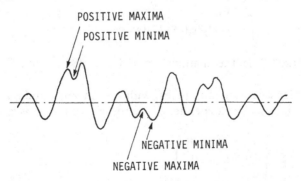

Fig. 3.7. Definition of local maxima and minima of non-narrow-band random waves.

$$E[N_\zeta] = \int_\zeta^\infty \int_{-\infty}^0 |\ddot{x}| f(x,0,\ddot{x}) \, d\ddot{x} \, dx \qquad (3.26)$$

where $f(x,0,\ddot{x})$ is the joint probability density function of displacement, velocity and acceleration of the Gaussian random process $x(t)$.

By letting $\zeta = 0$, we may write the expected number of positive maxima per unit time as

$$E[N_+] = \int_0^\infty \int_{-\infty}^0 |\ddot{x}| f(x,0,\ddot{x}) \, d\ddot{x} \, dx \qquad (3.27)$$

Then, the probability density function of the positive maxima $f(\zeta)$ can be obtained from Eqs. (3.25) through (3.27) as

$$f(\zeta) = \frac{d}{d\zeta}\left\{1 - \frac{E[N_\zeta]}{E[N_+]}\right\} = \frac{d}{d\zeta}\left\{1 - \frac{\displaystyle\int_\zeta^\infty \int_{-\infty}^0 |\ddot{x}| f(x,0,\ddot{x}) d\ddot{x} \, dx}{\displaystyle\int_0^\infty \int_{-\infty}^0 |\ddot{x}| f(x,0,\ddot{x}) d\ddot{x} \, dx}\right\}$$

$$= \frac{d}{d\zeta}\left\{1 - \frac{\displaystyle\int_{-\infty}^0 |\ddot{x}| F_x(\infty,0,\ddot{x}) d\ddot{x} - \int_{-\infty}^0 |\ddot{x}| F_x(\zeta,0,\ddot{x}) d\ddot{x}}{\displaystyle\int_0^\infty \int_{-\infty}^0 |\ddot{x}| f(x,0,\ddot{x}) d\ddot{x} \, dx}\right\}$$

$$= \frac{\displaystyle\int_{-\infty}^0 |\ddot{x}| f(\zeta,0,\ddot{x}) d\ddot{x}}{\displaystyle\int_0^\infty \int_{-\infty}^0 |\ddot{x}| f(x,0,\ddot{x}) d\ddot{x} \, dx} \qquad (3.28)$$

where $F_x(x,0,\ddot{x})$ is the marginal cumulative distribution function of $f(x,0,\ddot{x})$.

For a Gaussian random process with zero mean, x and $\dot{x}$ as well as $\dot{x}$ and $\ddot{x}$ are statistically independent. The covariance matrix is given by

$$\Sigma = \begin{pmatrix} m_0 & 0 & -m_2 \\ 0 & m_2 & 0 \\ -m_2 & 0 & m_4 \end{pmatrix} \qquad (3.29)$$

where $m_j = j$th moment of spectrum $= \int_0^\infty \omega^j S(\omega) d\omega$.

Hence, the joint probability density function $f(x,0,\ddot{x})$ becomes

$$f(x,0,\ddot{x}) = \frac{1}{(2\pi)^{3/2}\sqrt{m_2\Delta}}\, e^{-\frac{m_4x^2 + 2m_2x\ddot{x} + m_0\ddot{x}^2}{2\Delta}} \tag{3.30}$$

where $\Delta = m_0m_4 - m_2^2$. Then, the numerator of Eq. (3.28) becomes

$$\int_{-\infty}^{0} |\ddot{x}|\, f(\zeta,0,\ddot{x})\, d\ddot{x} = \frac{1}{2\pi\sqrt{m_0}}\, \sqrt{\frac{m_4}{m_2}}\left[\frac{\epsilon}{\sqrt{2\pi}}\exp\left\{-\frac{1}{2\epsilon^2}\left(\frac{\zeta}{\sqrt{m_0}}\right)^2\right\}\right.$$

$$\left. + \sqrt{1-\epsilon^2}\left(\frac{\zeta}{\sqrt{m_0}}\right)\exp\left\{-\frac{1}{2}\left(\frac{\zeta}{\sqrt{m_0}}\right)^2\right\}\Phi\left(\frac{\sqrt{1-\epsilon^2}}{\epsilon}\,\frac{\zeta}{\sqrt{m_0}}\right)\right] \tag{3.31}$$

where

$$\Phi(u) = \frac{1}{\sqrt{2\pi}}\int_{-\infty}^{u}\exp\{-u^2/2\}\, du$$

$$\epsilon = \sqrt{1 - m_2^2/m_0m_4} \tag{3.32}$$

The parameter ϵ is called the *bandwidth parameter* of the spectrum (Cartwright and Longuet-Higgins 1956), and $\epsilon=0$ represents a random process with a narrow-band spectrum. Although the parameter ϵ does not necessarily represent a measure of the energy spreading of a spectrum, it plays a convenient role in further development of the theory. Additional discussion on the bandwidth parameter is given at the end of this section.

In order to evaluate the denominator of Eq. (3.28), integration with respect to x is first carried out, and then integration with respect to $\ddot{x}$ is performed. This gives

$$\int_{0}^{\infty}\int_{-\infty}^{0} |\ddot{x}|\, f(x,0,\ddot{x})\, d\ddot{x}\, dx = \frac{1}{4\pi}\left(\frac{1+\sqrt{1-\epsilon^2}}{\sqrt{1-\epsilon^2}}\right)\sqrt{\frac{m_2}{m_0}} \tag{3.33}$$

It is noted that the inverse of the above quantity is equal to the average time interval between successive positive maxima. From Eqs. (3.31) and (3.33), the probability density function for the positive maxima given in Eq. (3.28) becomes

$$f(\zeta) = \frac{2/\sqrt{m_0}}{1+\sqrt{1-\epsilon^2}}\left[\frac{\epsilon}{\sqrt{2\pi}}\exp\left\{-\frac{1}{2\epsilon^2}\left(\frac{\zeta}{\sqrt{m_0}}\right)^2\right\}\right.$$

$$\left. + \sqrt{1-\epsilon^2}\left(\frac{\zeta}{\sqrt{m_0}}\right)\exp\left\{-\frac{1}{2}\left(\frac{\zeta}{\sqrt{m_0}}\right)^2\right\}\Phi\left(\frac{\sqrt{1-\epsilon^2}}{\epsilon}\,\frac{\zeta}{\sqrt{m_0}}\right)\right]$$

$$0 \leq \zeta < \infty \tag{3.34}$$

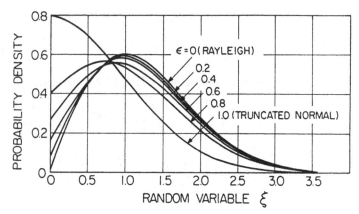

Fig. 3.8. Probability density function of maxima as a function of bandwidth
parameter ϵ.

We may write the above probability density function in dimensionless
form defining the random variable $\xi = \zeta/\sqrt{m_0}$. That is,

$$f(\xi) = \frac{2}{1+\sqrt{1-\epsilon^2}} \left[\frac{\epsilon}{\sqrt{2\pi}} e^{-\xi^2/2\epsilon^2} \right.$$
$$\left. + \sqrt{1-\epsilon^2}\, \xi e^{-\xi^2/2} \Phi\left(\frac{\sqrt{1-\epsilon^2}}{\epsilon} \xi \right) \right] \qquad (3.35)$$

As can be seen in Eq. (3.35), the probability density function is a function
of the bandwidth parameter ϵ, and $f(\xi)$ reduces to a Rayleigh probability
density function (in dimensionless form) for $\epsilon = 0$. On the other hand, for $\epsilon = 1$
which represents a wide-band random process consisting theoretically of an
infinite number of frequencies, the probability density function becomes

$$f(\xi) = \sqrt{\frac{2}{\pi}} e^{-\xi^2/2} \qquad 0 \leq \xi < \infty \qquad (3.36)$$

This is a truncated normal distribution (truncated at $\xi = 0$).

The probability density function $f(\xi)$ for various ϵ-values are shown in
Figure 3.8. As stated earlier, $f(\xi)$ can also be applied to the distribution of
negative minima.

The cumulative distribution function of the positive maxima ζ (dimensional) and ξ (dimensionless) can be derived by integrating Eqs. (3.34)
and (3.35) with respect to ζ and ξ, respectively. The results are:

$$F(\zeta) = \frac{2}{1+\sqrt{1-\epsilon^2}} \left[-\frac{1}{2}(1-\sqrt{1-\epsilon^2}) + \Phi\left(\frac{\zeta}{\epsilon\sqrt{m_0}} \right) \right.$$
$$\left. - \sqrt{1-\epsilon^2} \exp\left\{ -\frac{1}{2}\left(\frac{\zeta}{\sqrt{m_0}} \right)^2 \right\} \Phi\left(\frac{\sqrt{1-\epsilon^2}}{\epsilon} \frac{\zeta}{\sqrt{m_0}} \right) \right]$$
$$0 \leq \zeta < \infty \qquad (3.37)$$

and

$$F(\xi)=\frac{2}{1+\sqrt{1-\epsilon^2}}\left[-\frac{1}{2}(1-\sqrt{1-\epsilon^2})+\Phi\left(\frac{\xi}{\epsilon}\right)\right.$$

$$\left.-\sqrt{1-\epsilon^2}\,e^{-\xi^2/2}\Phi\left(\frac{\sqrt{1-\epsilon^2}}{\epsilon}\xi\right)\right]\qquad 0\leqslant\xi<\infty \qquad (3.38)$$

Next, the probability density function of maxima including both positive and negative maxima will be developed. In this case, the procedure for developing the density function is exactly the same as that of positive maxima except the sample space is extended into the negative range. Hence, the integration domain of the random variable x in Eq.(3.27) is from $-\infty$ to ∞. The probability density function in dimensionless form becomes (Cartwright and Loguet-Higgins 1956)

$$f(\xi)=\frac{\epsilon}{\sqrt{2\pi}}\,e^{-\xi^2/2\epsilon^2}+\sqrt{1-\epsilon^2}\,\xi e^{-\xi^2/2}\Phi\left(\frac{\sqrt{1-\epsilon^2}}{\epsilon}\xi\right)$$

$$-\infty<\xi<\infty \qquad (3.39)$$

The probability density function of $f(\xi)$ for various ϵ-values is shown in Figure 3.9. In this case, $f(\xi)$ reduces to a Rayleigh probability distribution

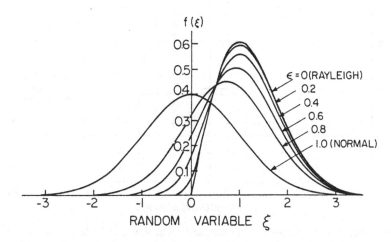

Fig. 3.9. Probability density function of maximum point process as a function of bandwidth parameter ϵ (Cartwright and Longuet-Higgins 1956).

(in dimensionless form) for $\epsilon=0$, and reduces to the standardized normal distribution for $\epsilon=1$.

As can be seen in the derivation of the probability density function of wave amplitude, the bandwidth parameter ϵ plays a significant role. As mentioned earlier, 'bandwidth' gives the impression that it represents the range of frequencies covered by the spectrum. However, this is not the case. Since the parameter ϵ depends on the magnitude of the 4th moment of the wave spectrum, the magnitude of ϵ depends to a great extent on the frequency range where dominant wave energy exists. The ϵ-value for ocean waves lies in the range of 0.40 to 0.80, while that for waves in the surf zone lies in a much higher range; of the order 0.70 to 0.95. Since wave spectra with ϵ less than 0.4 do not exist in reality, square-shaped spectra having ϵ less than 0.4 are artificially developed for research purposes as shown in Figure 3.10 (Ochi 1979b).

As a parameter representing reasonably well the degree of spectral energy spreading over the frequency range, Goda (1970) proposes the *spectral peakedness parameter* defined as

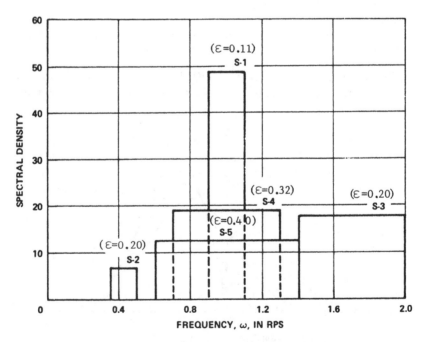

Fig. 3.10. Square-shape spectra with band-width parameter ϵ less than 0.4 (Ochi 1979b).

$$Q_p = \frac{\displaystyle\int_0^\infty \omega\{S(\omega)\}^2 d\omega}{\left(\displaystyle\int_0^\infty S(\omega)d\omega\right)^2} \tag{3.40}$$

On the other hand, Longuet-Higgins (1983) proposes the *spectral width parameter* defined by

$$\nu = \sqrt{(m_2 m_0/m_1^2) - 1} \tag{3.41}$$

and the wave spectrum is characterized as being narrow when $\nu \ll 1$.

3.4 JOINT DISTRIBUTION OF TWO WAVE AMPLITUDES

In this section, we consider the joint probability distribution of two wave amplitudes separated by time τ. For this, we may assume a random process with narrow-band spectrum and consider the joint probability distribution of two points of an envelope process $x(t)$ and $x(t+\tau)$ which are separated by time τ.

Let the wave profile $x(t)$ be presented in the form $\rho(t) \exp\{i\phi(t)\}$ given in Eq. (3.20), and write the two components $x_c(t)$ and $x_s(t)$ as

$$x_c(t) = \sum_n a_n \cos\{(\omega_n - \overline{\omega})t + \epsilon_n\}$$

$$x_s(t) = \sum_n a_n \sin\{(\omega_n - \overline{\omega})t + \epsilon_n\} \tag{3.42}$$

We may write the random variables x_{c1} and x_{s1} as the value of $x_c(t)$ and $x_s(t)$, respectively, at time t, and x_{c2}, x_{s2} as those at time $(t+\tau)$. Since a_n are normally distributed with zero mean, the random variables x_{c1}, x_{s1}, x_{c2} and x_{s2} are jointly normally distributed with the following properties:

(i) $\mathrm{E}[x_{c1}] = \mathrm{E}[x_{s1}] = \mathrm{E}[x_{c2}] = \mathrm{E}[x_{s2}] = 0$ $\hspace{2cm}$ (3.43)

(ii) $\mathrm{E}[x_{c1}x_{s1}] = \mathrm{E}[x_{c2}x_{s2}] = 0$ $\hspace{2.5cm}$ (3.44)

(iii) $\mathrm{E}[x_{c1}^2] = \mathrm{E}[x_{s1}^2] = \mathrm{E}[x_{c2}^2] = \mathrm{E}[x_{s2}^2] = \sigma^2$ $\hspace{1.3cm}$ (3.45)

(iv) $\mathrm{E}[x_{c1}x_{c2}] = \mathrm{E}[x_{s1}x_{s2}] = \displaystyle\int_0^\infty S(\omega)\cos(\omega - \overline{\omega})\tau\, d\omega = \rho$ $\hspace{0.5cm}$ (3.46)

(v) $\mathrm{E}[x_{c1}x_{s2}] = -\mathrm{E}[x_{s1}x_{c2}] = \displaystyle\int_0^\infty S(\omega)\sin(\omega - \overline{\omega})\tau\, d\omega = \lambda$ $\hspace{0.3cm}$ (3.47)

where $S(\omega)$ is the spectral density function of the random process.

Derivations of $E[x_{c1}x_{c2}]$ and $E[x_{c1}x_{s2}]$ are given in the following:

$$E[x_{c1}x_{c2}]=E\left[\sum_n a_n \cos\{(\omega_n-\overline{\omega})t+\epsilon_n\}\cdot\sum_n a_n \cos\{(\omega_n-\overline{\omega})(t+\tau)+\epsilon_n\}\right]$$

$$=\sum_n a_n^2 \overline{\cos^2\{(\omega_n-\omega)t+\epsilon_n\} \cos(\omega_n-\overline{\omega})\tau}$$

$$=\sum_n \frac{1}{2}a_n^2 \cos(\omega_n-\overline{\omega})\tau \qquad\qquad (3.48)$$

where the bar denotes the average value with respect to time t. The amplitudes are related with the spectral density function as

$$\frac{1}{2}a_n^2=S(\omega_n)\Delta\omega_n \qquad\qquad (3.49)$$

and thereby Eq. (3.46) can be derived as follows:

$$E[x_{c1}x_{c2}]=\sum_n S(\omega_n) \cos(\omega_n-\overline{\omega})\tau\Delta\omega_n$$

$$=\int_0^\infty S(\omega) \cos(\omega-\overline{\omega})\tau\,d\omega=\rho$$

Similarly, we have

$$E[x_{c1}x_{s2}]=E\left[\sum_n a_n \cos\{(\omega_n-\overline{\omega})t+\epsilon_n\}\cdot\sum_n a_n \sin\{(\omega_n-\overline{\omega})(t+\tau)+\epsilon_n\}\right]$$

$$=\sum_n a_n^2 \overline{\cos^2\{(\omega_n-\overline{\omega})t+\epsilon_n)\} \sin(\omega_n-\overline{\omega})\tau}$$

$$=\sum_n \frac{1}{2}a_n^2 \sin(\omega_n-\overline{\omega})\tau=\sum_n S(\omega_n) \sin(\omega_n-\overline{\omega})\tau\Delta\omega_n$$

$$=\int_0^\infty S(\omega) \sin(\omega-\overline{\omega})\tau\,d\omega$$

This is shown in Eq. (3.47) as λ.

Thus, from Eqs. (3.44) through (3.47), the covariance matrix of x_{c1}, x_{s1}, x_{c2} and x_{s2} can be written as

$$\Sigma=\begin{pmatrix} \sigma^2 & 0 & \rho & \lambda \\ 0 & \sigma^2 & -\lambda & \rho \\ \rho & -\lambda & \sigma^2 & 0 \\ \lambda & \rho & 0 & \sigma^2 \end{pmatrix} \qquad\qquad (3.50)$$

Consequently, the joint normal probability density function of the random variables $(x_{c1}, x_{s1}, x_{c2}, x_{s2})$ with zero means can be written as follows:

$$f(\mathbf{X}) = \frac{1}{(2\pi)^2 \sqrt{|\boldsymbol{\Sigma}|}} e^{-\frac{1}{2}\mathbf{X}^T \boldsymbol{\Sigma}^{-1} \mathbf{X}} \tag{3.51}$$

where,

$$\mathbf{X}^T = (x_{c1}, x_{s1}, x_{c2}, x_{s2})$$

$$\sqrt{|\boldsymbol{\Sigma}|} = \sigma^4 - \rho^2 - \lambda^2$$

and

$$\boldsymbol{\Sigma}^{-1} = \frac{1}{\sqrt{|\boldsymbol{\Sigma}|}} \begin{pmatrix} \sigma^2 & 0 & -\rho & -\lambda \\ 0 & \sigma^2 & \lambda & -\rho \\ -\rho & \lambda & \sigma^2 & 0 \\ -\lambda & -\rho & 0 & \sigma^2 \end{pmatrix}$$

Then, Eq. (3.51) yields

$$f(x_{c1}, x_{s1}, x_{c2}, x_{s2}) = \frac{1}{(2\pi)^2 \sqrt{|\boldsymbol{\Sigma}|}} \exp\left\{ -\frac{1}{2\sqrt{|\boldsymbol{\Sigma}|}} \left[\sigma^2(x_{c1}^2 + x_{s1}^2 + x_{c2}^2 + x_{s2}^2) \right.\right.$$
$$\left.\left. - 2\rho(x_{c1}x_{c2} + x_{s1}x_{s2}) - 2\lambda(x_{c1}x_{s2} - x_{s1}x_{c2}) \right] \right\} \tag{3.52}$$

We now transform the random variables $(x_{c1}, x_{s1}, x_{c2}, x_{s2})$ to random variables representing amplitudes and phases $(A_1, \epsilon_1, A_2, \epsilon_2)$ by using the functional relationships

$$\begin{aligned} x_{c1}(t) &= A_1(t)\cos\epsilon_1(t) \\ x_{s1}(t) &= A_1(t)\sin\epsilon_1(t) \\ x_{c2}(x+\tau) &= A_2(t+\tau)\cos\epsilon_2(t+\tau) \\ x_{s2}(x+\tau) &= A_2(t+\tau)\sin\epsilon_2(t+\tau) \end{aligned} \tag{3.53}$$

That is, by applying the transformation method, the joint probability density function of A_1, ϵ_1, A_2 and ϵ_2 becomes

$$f(A_1, \epsilon_1, A_2, \epsilon_2) = \frac{1}{4\pi^2 \sqrt{|\boldsymbol{\Sigma}|}} \exp\left\{ -\frac{1}{2\sqrt{|\boldsymbol{\Sigma}|}} \left[\sigma^2(A_1^2 + A_2^2) \right.\right.$$
$$\left.\left. - 2A_1 A_2 \{\rho\cos(\epsilon_2 - \epsilon_1) + \lambda\sin(\epsilon_2 - \epsilon_1)\} \right] \right\}$$

$$0 \leqslant A_1 < \infty \qquad 0 \leqslant A_2 < \infty \qquad 0 \leqslant \epsilon_1 \leqslant 2\pi \qquad 0 \leqslant \epsilon_2 \leqslant 2\pi \tag{3.54}$$

The joint probability distribution of amplitudes A_1 and A_2 can be obtained by integrating Eq. (3.54) with respect to ϵ_1 and ϵ_2. For this we may write

$$\rho \cos(\epsilon_2 - \epsilon_1) + \lambda \sin(\epsilon_2 - \epsilon_1)$$
$$= \sqrt{\rho^2 + \lambda^2} \cos(\epsilon_2 - \epsilon_1 - \tan^{-1}(\lambda/\rho)) \qquad (3.55)$$

Then, we have

$$f(A_1, A_2) = \frac{A_1 A_2}{4\pi^2 \sqrt{|\Sigma|}} \exp\left\{-\frac{1}{2\sqrt{|\Sigma|}} \sigma^2(A_1^2 + A_2^2)\right\}$$

$$\times \int_0^{2\pi} \int_0^{2\pi} \exp\left\{\frac{A_1 A_2}{\sqrt{|\Sigma|}} \sqrt{\rho^2 + \lambda^2} \cos(\epsilon_1 - \epsilon_2 - \tan^{-1}(\lambda/\rho))\right\} d\epsilon_1 \, d\epsilon_2 \quad (3.56)$$

In carrying out the double integration in the above equation, let us first consider transformation of the two random variables ϵ_1 and ϵ_2 to a single random variable $(\epsilon_1 - \epsilon_2)$. It can be easily shown that the double integral in Eq. (3.56) is equivalent to the constant 2π times a single integral with respect to $(\epsilon_1 - \epsilon_2)$. Furthermore, $\cos\{\epsilon_1 - \epsilon_2 - \tan^{-1}(\lambda/\rho)\}$ is periodic. Hence, Eq. (3.56) can be written as,

$$f(A_1, A_2) = \frac{A_1 A_2}{\sqrt{|\Sigma|}} \exp\left\{-\frac{1}{2\sqrt{|\Sigma|}} \sigma^2(A_1 + A_2)\right\}$$

$$\times \int_0^{2\pi} \frac{1}{2\pi} \exp\left\{\frac{\sqrt{\rho^2 + \lambda^2}}{\sqrt{|\Sigma|}} A_1 A_2 \cos \phi\right\} d\phi$$

$$= \frac{A_1 A_2}{\sqrt{|\Sigma|}} \exp\left\{-\frac{1}{2\sqrt{|\Sigma|}} \sigma^2(A_1^2 + A_2^2)\right\}$$

$$\times I_0\left(\frac{\sqrt{\rho^2 + \lambda^2}}{\sqrt{|\Sigma|}} A_1 A_2\right) \qquad (3.57)$$

where $\phi = \epsilon_2 - \epsilon_1 - \tan^{-1}(\lambda/\rho)$ and I_0 is the modified Bessel function of zero order.

This is the joint probability density function of two amplitudes A_1 and A_2, separated by the time interval τ, in a narrow-band Gaussian random process.

The joint probability density function $f(A_1, A_2)$ is often presented in another form. That is, by writing $\sigma^2 = m_0$ and by defining $k = \sqrt{\rho^2 + \lambda^2}/m_0$, we have $\sqrt{|\Sigma|} = m_0^2(1 - k^2)$. Then, the joint probability density function of A_1 and A_2 given in Eq. (3.57) can be written as

$$f(A_1, A_2) = \frac{1}{1 - k^2} \frac{A_1 A_2}{m_0^2} e^{-\frac{1}{2(1-k^2)} \frac{A_1^2 + A_2^2}{m_0}} I_0\left(\frac{1}{1 - k^2} \frac{A_1 A_2}{m_0}\right) \qquad (3.58)$$

If $k = 0$, the joint probability density becomes the product of two Rayleigh density functions; namely the random variables A_1 and A_2 are statistically independent and thereby they are uncorrelated. However, k is not a correlation coefficient. The correlation coefficient, denoted by γ, following its definition (see Appendix A) is given by Uhlenbeck (1943) as

$$\gamma=\left\{E(k)-\frac{1}{2}(1-k^2)K(k)-\frac{\pi}{4}\right\}\Big/\left(1-\frac{\pi}{4}\right) \tag{3.59}$$

where $K(k)$ and $E(k)$ are complete elliptic integrals of the 1st and 2nd kind, respectively. These are written as

$$K(k)=\int_0^{\pi/2}(1-k^2\sin^2\theta)^{-1/2}d\theta$$

$$E(k)=\int_0^{\pi/2}(1-k^2\sin^2\theta)^{1/2}d\theta \tag{3.60}$$

The values of the correlation coefficient γ for various k^2-values and its approximation are given by Longuet-Higgins (1984) (see Section 8.2.3).

By letting the separation time τ equal $2\pi/\overline{\omega}$, A_1 and A_2 may approximately represent the amplitudes of two successive waves.

The joint probability density function of any two amplitudes A_1 and A_2, without the restraint of the separation time, may be simply obtained by modifying the joint probability density function of the exponential distribution. That is, let the random variables X_1 and X_2 be two outcomes of the exponential probability distribution with parameter λ, and let ρ be the correlation coefficient between the two random variables. Then, the joint exponential probability distribution can be given by

$$f(x_1,x_2)=\frac{\lambda^2}{1-\rho}e^{-\frac{\lambda}{1-\rho}(x_1+x_2)}I_0\left(\frac{2\sqrt{\rho}}{1-\rho}\lambda\sqrt{x_1x_2}\right) \tag{3.61}$$

The random variables X_1 and X_2 are now transformed to A_1 and A_2 by letting the functional relationship between them be as follows:

$$X_1=A_1^2/\lambda R \quad \text{and} \quad X_2=A_2^2/\lambda R \tag{3.62}$$

Then, the joint probability density function of A_1 and A_2 yields the following joint Rayleigh probability density function:

$$f(A_1,A_2)=\frac{4}{1-\rho}\frac{A_1A_2}{R^2}e^{-\frac{1}{1-\rho}\frac{1}{R}(A_1^2+A_2^2)}I_0\left(\frac{2\sqrt{\rho}}{1-\rho}\frac{A_1A_2}{R}\right) \tag{3.63}$$

The joint probability density functions derived in Eqs. (3.57), (3.58) and (3.63) can be applied to solve many interesting properties of random waves. Examples of their application to the probability distribution of wave height will be presented in the next section, and their application to group waves will be discussed in Section 8.2.

3.5 PROBABILITY DISTRIBUTION OF PEAK-TO-TROUGH EXCURSIONS (WAVE HEIGHT)

The probability distribution of peak-to-tough excursions may be simply obtained if we assume that the magnitude of excursions is twice the amplitude. This assumption is commonly made for random waves in deep water. That is, the probability distribution of wave height, denoted by H, for a narrow-band random process can be obtained from Eq. (3.9) by transforming the random variable from A to $H=2A$. This results in the density function of wave height being the Rayleigh probability distribution given by

$$f(H)=\frac{H}{4\sigma^2}\exp\{H^2/8\sigma^2\}\qquad 0\leqslant H<\infty \qquad (3.64)$$

If the density function is written in the form of Eq. (3.10), then the parameter R of the Rayleigh distribution is equal to $8\sigma^2$.

For the probability density function of waves with a non-narrow-band spectrum, it may not be appropriate to use the same simple concept considered in the derivation of Eq. (3.64). This is because twice the magnitude of the positive maxima does not represent the wave height, in general. Nevertheless, it is common practice to consider the probability distribution of wave height to be of the same form as given in Eq. (3.34) but with the random variable ζ doubled.

For the probability distribution of wave height, it may be more appropriate to consider either the sum of two statistically independent amplitudes or, more precisely, the sum of two amplitudes separated by one-half of the average period. In order to apply the first concept for the derivation of wave height distribution, we may consider two Rayleigh probability distributions given by Eq. (3.9) with independent amplitudes A_1 and A_2. The probability density function of the sum of these two random variables H can be obtained by the following convolution integral:

$$f(H)=\int_0^H f(A_1)f(H-A_1)\mathrm{d}A_1$$

$$=\int_0^H \frac{A_1}{\sigma^2}\exp\left\{-\frac{A_1^2}{2\sigma^2}\right\}\cdot\frac{H-A_1}{\sigma^2}\exp\left\{-\frac{(H-A_1)^2}{2\sigma^2}\right\}\mathrm{d}A_1 \quad (3.65)$$

Tayfun (1981) derives the wave height distribution by considering the sum of amplitudes A_1 and A_2 that are separated by one-half the period. That is, by writing the wave height as $H=A_1+A_2$, the probability density function of wave height may be written by applying Eq. (3.57) in the following form:

$$f(H|\tau=T/2) = \int_0^H f(H-A_2, A_2; \tau=T/2)\, dA_2 \qquad (3.66)$$

Taking into account all values of the period, the density function of H can be obtained as

$$f(H) = \int_0^\infty \int_0^H f(H-A_2, A_2; \tau=T/2)f(T)\, dA_2\, dT \qquad (3.67)$$

where $f(T)$ is the probability density function of the period. Since the computation of the density function given in Eq. (3.67) is extremely complicated, Tayfun assumes that the period is a constant, and is the mean period given by $T=2\pi/\overline{\omega}$ where $\overline{\omega}$ is the mean frequency. Then, the probability density function of wave height is given by

$$f(H) = \int_0^H f(H-A_2, A_2; \tau=\pi/\overline{\omega})\, dA_2 \qquad (3.68)$$

In evaluating $f(H)$, Tayfun non-dimensionalizes A_1 and A_2 by dividing by $\sqrt{R}$.

Figure 3.11 shows an example of a comparison between the probability distribution given by Eq. (3.68), Rayleigh probability distribution, and the histogram constructed from the measured data. It can be seen that the peak-to-trough distribution computed by Eq. (3.68) yields a larger probability density around the mean value than the Rayleigh distribution and that it agrees well with the observed data.

For the probability distribution function applicable to peak-to-trough excursions, Naess (1985a) considers the following probability which is an extension of the concept given in Eq. (3.25). That is, assuming a narrow-band random process and assuming the wave height is equal to twice the amplitude and, in addition, taking into consideration the correlation between peak and trough which are separated by a time equal to one-half of the dominant wave period, we may write

$$\begin{aligned} \text{Pr \{wave amplitude} > \zeta\} &= \text{Pr \{peak} > \zeta,\ \text{trough} < -\zeta\} \\ &= E[N_{\zeta,-\zeta}]/E[N_{+,-}] \end{aligned} \qquad (3.69)$$

where $E[N_{\zeta,-\zeta}]$ = expected number of simultaneous occurrences (per unit time) of an up-crossing of the level ζ followed by a down-crossing of the level $-\zeta$ at a time $T/2$ later

$E[N_{+,-}]$ = expected number of upward zero-crossing (per unit time) followed by downward zero-crossing at a time $T/2$ later.

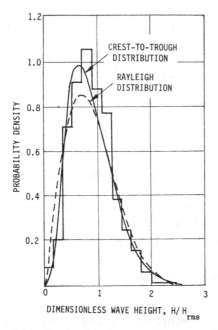

Fig. 3.11. Comparison between histogram of a peak-to-trough wave excursion and Rayleigh distribution and the probability density function given in Eq.(3.68) (Tayfun 1981).

Naess shows that the expected number of crossings $E[N_{\zeta,-\zeta}]$ with time difference $\tau = T/2$ between peaks and troughs for Gaussian waves may be evaluated by extending a procedure given in Eq. (3.13) to four random variables $(x_1, \dot{x}_1, x_2$ and $\dot{x}_2)$ where the subscript 1 stands for the up-crossing displacement and velocity, while the subscript 2 are those for the down-crossing. That is,

$$E[N_{\zeta,-\zeta}] = \int_{-\infty}^{0} \int_{0}^{\infty} \dot{x}_1 \dot{x}_2 f(\zeta, \dot{x}_1, -\zeta, \dot{x}_2) d\dot{x}_1 \, d\dot{x}_2$$

$$= \frac{1}{(2\pi)^2 |\boldsymbol{\Sigma}|^{1/2}} \int_{-\infty}^{0} \int_{0}^{\infty} x_1 x_2 \exp\left\{ -\frac{1}{2} \mathbf{x}^T \boldsymbol{\Sigma}^{-1} \mathbf{x} \right\} d\dot{x}_1 d\dot{x}_2 \quad (3.70)$$

where $\mathbf{x} = (\zeta, \dot{x}_1, -\zeta, \dot{x}_2)^T$

$$\boldsymbol{\Sigma} = \begin{pmatrix} R(0) & -\dot{R}(0) & R(\tau) & \dot{R}(\tau) \\ \dot{R}(0) & -\ddot{R}(0) & -\dot{R}(\tau) & -\ddot{R}(\tau) \\ R(\tau) & -\dot{R}(\tau) & R(0) & -\dot{R}(0) \\ \dot{R}(\tau) & -\ddot{R}(\tau) & R(0) & -\ddot{R}(0) \end{pmatrix}$$

$$R(\tau) = \int_{0}^{\infty} S(\omega) \cos \omega\tau \, d\omega \qquad \text{with } \tau = T/2$$

By evaluating Eq. (3.70), Eq. (3.69) yields

$$\Pr\{\text{wave amplitude}>\zeta\}=\exp\left\{-\frac{\zeta^2}{\sigma^2\{1-r(T/2)\}}\right\} \tag{3.71}$$

where σ^2 is the variance of $x(t)$ and $r(T/2)=R(T/2)/\sigma^2$.

From Eq. (3.71), the probability density function of wave amplitude becomes

$$f(\zeta)=\frac{2\zeta}{\sigma^2\{1-r(T/2)\}}\exp\left\{-\frac{\zeta^2}{\sigma^2\{1-r(T/2)\}}\right\}$$

$$0<\zeta<\infty \tag{3.72}$$

and thereby the probability density function for the peak-to-trough excursion (wave height H) can be derived as follows:

$$f(H)=\frac{H}{2\sigma^2\{1-r(T/2)\}}\exp\left\{-\frac{H^2}{4\sigma^2\{1-r(T/2)\}}\right\}$$

$$0<H<\infty \tag{3.73}$$

Equation (3.73) is the Rayleigh probability distribution with a parameter which is a function of the dominant wave period T. When the bandwidth of a spectral density function is zero (namely, a strictly narrow-band random process), $r(T/2)$ becomes -1, and hence $f(H)$ is the same as that given in Eq. (3.64).

3.6 SIGNIFICANT WAVE HEIGHT

As stated in Section 3.1, significant wave height (introduced by Sverdrop and Munk (1947)) is the measure most commonly used for representing the severity of sea conditions. It is defined as the average of the one-third highest observed or measured wave heights. However, it is rarely evaluated following the definition; instead, it is commonly evaluated by using the variance computed from a spectrum, or by applying the statistical inference theory discussed in Section 3.9. The principle underlying the evaluation of significant wave height based on the Rayleigh probability density function is as follows.

By assuming a narow-band wave spectrum, wave height follows the Rayleigh probability distribution. For convenience, we may write the Rayleigh distribution as

$$f(x)=\frac{2x}{R}\exp\{-x^2/R\} \qquad 0\leqslant x<\infty \tag{3.74}$$

where $R=8m_0$ if x represents wave height and m_0 is the area under the spectral density function.

Let us write the lower limit of the highest one-third of the probability density function as x_*. Since significant wave height deals with the highest one-third wave heights, we can write the probability of exceeding x_* is 1/3 as follows (see Figure 3.12):

$$\Pr\{X \geq x_*\} = \int_{x_*}^{\infty} \frac{2x}{R} \exp\{-x^2/R\}\ dx = \frac{1}{3} \qquad (3.75)$$

which yields

$$x_* = \sqrt{R}(\ln 3) = 1.048\sqrt{R} \qquad (3.76)$$

By evaluating the moment about the origin, we have

$$\int_{x_*}^{\infty} xf(x)dx = x_* \exp\{-x_*^2/R)\} + \sqrt{\pi R}\left\{1 - \Phi\left(\sqrt{\frac{2}{R}}x_*\right)\right\}$$

$$= \frac{1}{3}H_s \qquad (3.77)$$

where H_s = significant wave height.

Then, from Eqs. (3.76) and (3.77), the significant wave height can be obtained as

$$H_s = [\sqrt{\ln 3} + 3\sqrt{\pi}\{1 - \Phi(\sqrt{2\ln 3})\}]\sqrt{R} \sim 1.42\sqrt{R} \qquad (3.78)$$

Since the parameter R of the Rayleigh distribution for peak-to-trough excursions is eight times the variance, the significant wave height becomes

$$H_s = 1.42\sqrt{8\,m_0} = 4.01\sqrt{m_0} \qquad (3.79)$$

Thus, the significant wave height is equal to four times the square-root of the area under the spectral density function with the narrow-band

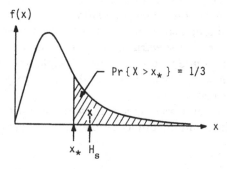

Fig. 3.12. Computation of significant wave height from probability density function.

random process assumption. If this assumpton is removed, significant wave height can be approximately evaluated through the same procedure but by applying the probability density function of positive maxima given in Eq. (3.34). It is necessary, however, to double the evaluated value in this case since Eq. (3.34) is applicable to the positive side only.

The results of computations of significant wave height for various bandwidth parameters ϵ are shown in Figure 3.13. The range of ϵ for waves observed in the ocean is on the order of 0.4 to 0.8; hence, it appears that the significant wave height evaluated on the basis of the narrow-band assumption yields an overestimation of approximately 1.5–8 percent.

The formula for evaluating significant wave height given in Eq. (3.78) can be generalized to evaluate the average of the highest $1/n$ observations, denoted by $x_{1/n}$, for the Rayleigh distribution. That is,

$$x_{1/n} = [\sqrt{\ln n} + n\sqrt{\pi}\{1 - \Phi(\sqrt{2\ln n})\}]\sqrt{R} \qquad (3.80)$$

For large n, the second term in Eq. (3.80) becomes negligibly small in comparison with the first term, and hence, Eq. (3.80) may be approximated by

$$x_{1/n} = \sqrt{\ln n}\,\sqrt{R} = 2\sqrt{2\ln n}\,\sqrt{m_0} \qquad (3.81)$$

It is noted that the formula given in Eq. (3.81) agrees with the probable extreme wave height expected in n observations (where n is large), which will be derived in Chapter 6.

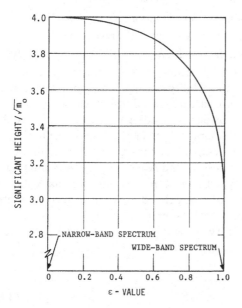

Fig. 3.13. Significant wave height as a function of bandwidth parameter ϵ.

3.7 PROBABILITY DISTRIBUTION OF HALF-CYCLE EXCURSIONS

The definition of half-cycle excursions is given with reference to Figure 3.2; the four different local peaks and troughs associated with the half-cycle excursions are defined as the positive maxima, positive minima, negative maxima and negative minima as illustrated in Figure 3.7. Statistical analysis of displacement (vertical distance) between successive local peak to local trough excursions or vice versa is called *half-cycle excursion analysis*.

Half-cycle excursions can be categorized into two groups; those crossing the zero-line and those located above or below the zero-line. We may define these half-cycle excursions as *Type I* and *Type II excursions*, respectively. Although there are ascending and descending excursions for each case, the statistical properties remain the same.

The probability density function of half-cycle excursions includes the probability density functions of both Type I and Type II excursions taking into consideration the frequency of occurrence of each type (Ochi and Eckhoff 1984). For this, we may write the dimensionless probability density function of the half-cycle excursion as

$$h(\zeta) = q_1 h_1(\zeta) + q_2 h_2(\zeta) \tag{3.82}$$

where $\qquad\zeta$ = dimensionless half-cycle excursion

$\qquad\qquad$ = (half-cycle excursion)$/\sqrt{m_0}$

$\qquad m_0$ = area under the spectral density function

$h_1(\zeta), h_2(\zeta)$ = probability density function of Type I and

$\qquad\qquad$ Type II half-cycle excursions, respectively

$\qquad q_1, q_2$ = weighting factors for $h_1(\zeta)$ and $h_2(\zeta)$, respectively.

It may be thought that Type I excursions are twice the magnitude of the positive maxima whose probability density function is given in Eq. (3.34). However, this is not the case for the following reason.

For convenience, let us consider descending excursions $A_I C$ and $A_{II} B$ shown in Figure 3.14. The former is a Type I excursion, while the latter is a Type II excursion. Nevertheless, the probability distribution of the positive maxima is applicable to both A_I and A_{II}. Therefore in evaluating the probability distribution of Type I excursions $A_I C$, the maxima A_{II} belonging to the Type II excursions must be deleted by taking into consideration the frequency of occurrence of each excursion.

The expected numbers of Type I and Type II half-cycles, denoted by $\overline{N}_{AC}$ and $\overline{N}_{AB}$, respectively, are given by

$$\overline{N}_{AC} = \overline{N}_A - \overline{N}_B$$

$$\overline{N}_{AB} = \overline{N}_B \tag{3.83}$$

where $\overline{N}_A$ and $\overline{N}_B$ are the expected numbers of positive maxima and positive minima, respectively. $\overline{N}_A$ and $\overline{N}_B$ are given by Cartwright and Longuet-Higgins (1956) as

$$\overline{N}_A = \frac{1}{4\pi}\left(\frac{1+\sqrt{1-\epsilon^2}}{\sqrt{1-\epsilon^2}}\right)\sqrt{m_2/m_0}$$

$$\overline{N}_B = \frac{1}{4\pi}\left(\frac{1-\sqrt{1-\epsilon^2}}{\sqrt{1-\epsilon^2}}\right)\sqrt{m_2/m_0} \qquad (3.84)$$

Thus, the probability of occurrence of Type I and Type II descending excursions on the positive side, denoted by p_1 and p_2 can be evaluated from Eqs. (3.83) and (3.84) as

$$p_1 = \frac{\overline{N}_{AC}}{\overline{N}_A} = \frac{2\sqrt{1-\epsilon^2}}{1+\sqrt{1-\epsilon^2}}$$

$$p_2 = \frac{\overline{N}_{AB}}{\overline{N}_A} = \frac{1-\sqrt{1-\epsilon^2}}{1+\sqrt{1-\epsilon^2}} \qquad (3.85)$$

The conditional probability density function of descending Type I excursions, given that positive maxima have occurred, can be written as $p_1 h_1(\zeta)$. Since the probability of occurrence of positive maxima is p_1, the (unconditional) probability density function of descending Type I excursions becomes $p_1^2 h_1(\zeta)$. Here $h_1(\zeta)$ is twice the magnitude of the positive maxima. By applying the probability density function given in Eq. (3.35), we may write the probability density function of descending Type I half-cycle excursions (in dimensionless form) as

$$p_1^2 h_1(\zeta) = \frac{p_1^2}{2}[f(\xi)]_{\xi=\zeta/2} \qquad (3.86)$$

where $f(\zeta)$ is the probability density function of positive maxima in dimensionless form given in Eq. (3.35).

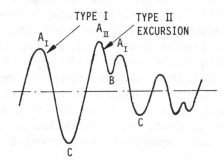

Fig. 3.14. Type I and Type II excursions.

The probability density function of ascending Type I excursion is the same as that given in Eq. (3.86) because the statistical properties of a Gaussian random process on the positive and negative sides are identical.

For the probability density function of Type II half-cycle excursions, we must first consider the conditional probability density function of the positive maxima, given that the positive maxima are greater than positive minima. Under this condition, the joint probability density function of positive maxima and positive minima is given in dimensionless form as follows:

$$f(\xi,\nu) = \frac{f(\xi)}{1 - [F(\xi)]_{\xi=\nu}} f(\nu) \qquad (3.87)$$

where $F(\xi)$ is the cumulative distribution function of positive maxima in dimensionless form given in Eq. (3.38), and $f(\nu)$ is the probability density function of positive minima in dimensionless form given by

$$f(\nu) = \frac{2}{1 - \sqrt{1-\epsilon^2}} \left[\frac{\epsilon}{\sqrt{2\pi}} \exp\left\{ -\frac{1}{2}\left(\frac{\nu}{\epsilon}\right)^2 \right\} \right.$$

$$\left. - \sqrt{1-\epsilon^2}\, \nu \exp\left\{ -\frac{\nu^2}{2} \right\} \Phi\left(\frac{\sqrt{1-\epsilon^2}}{\epsilon} \nu \right) \right] \qquad (3.88)$$

The probability density function of Type II half-cycle excursions can be derived from Eq. (3.87) by applying the technique change of random variables. Since Type II half-cycle excursions are the difference in the magnitude of positive maxima and positive minima, we may write $\zeta = \xi - \nu$. Then, the probability density function $h_2(\zeta)$ can be obtained by replacing the random variable ξ in the numerator of Eq. (3.87) by $(\zeta+\nu)$, and by integrating with respect to ν. That is,

$$h_2(\zeta) = \int_0^\infty \frac{[f(\xi)]_{\xi=\zeta+\nu}}{1 - [F(\xi)]_{\xi=\nu}} f(\nu)\, d\nu \qquad (3.89)$$

The probability of occurrence of Type II descending half-cycle excursions p_2 is given in Eq. (3.85). Since positive minima yield both ascending and descending half-cycle excursions on the positive side, the probability density function of Type II excursions becomes $2p_2 h_2(\zeta)$ on the positive side. The same probability density function can be applied to Type II half-cycle excursions on the negative side for a Gaussian random process.

The desired probability density function for half-cycle excursions is obtained by combining the two probability density functions given in Eqs. (3.86) and (3.89). By taking into consideration the condition required for the sum of two probability density functions, the result is given as follows:

$$h(\zeta)=\frac{1}{p_1^2+2p_2}\{p_1^2h_1(\zeta)+2p_2h_2(\zeta)\}$$

$$=q_1h_1(\zeta)+q_2h_2(\zeta) \tag{3.90}$$

where

$$q_1=\frac{p_1^2}{p_1^2+2p_2}=\text{weighting factor for } h_1(\zeta)$$

$$q_2=\frac{2p_2}{p_1^2+2p_2}=\text{weighting factor for } h_2(\zeta) \tag{3.91}$$

As shown in Eq. (3.85), p_1 and p_2 are functions of the bandwidth parameter ϵ; hence, it can be seen from Eq. (3.90) that the probability density function of half-cycle excursions (in dimensionless form) depends on only the bandwidth parameter ϵ.

As an example of the probability density function of half-cycle excursions, Figure 3.15 shows the density function computed from wave data obtained during hurricane CAMILLE. The data used in the computation were obtained during the growing stage of the hurricane in a significant wave height of 12.3 m. There is a total of 525 half-cycle excursions in the record and this agrees well with the predicted number of 524 excursions; twice the sum of the expected number of Type I and Type II half-cycle excursions computed by Eqs. (3.83) and (3.84). On the other hand, the total number of excursions computed by the narrow-band random process assumption is 360; a significant underprediction of the measured value. The difference can be attributed to the consideration of Type II excursions in the analysis.

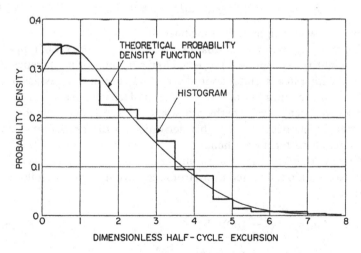

Fig. 3.15. Comparison between histogram of half-cycle excursions (dimensionless) and theoretical probability density function.

It is noted that half-cycle excursion analysis considers every excursion of a random process irrespective of its magnitude; therefore, the probability density function given in Eq. (3.90) can be used for stochastic analysis of the properties of various random processes; for example, for analysis of fatigue loads of marine systems (Ochi 1985).

3.8 LONG-TERM WAVE HEIGHT DISTRIBUTION

The statistical properties of waves presented so far are those in a steady-state sea condition, which is often referred to as the short-term wave statistics. It is of interest to examine the statistical properties of waves accumulated over a long time period. Long-term wave statistics plays an important role for the design of marine systems, since the accumulation of responses of a marine system in each short-term sea state over its lifetime provides information vital for evaluating fatigue loads of the system. The methods for estimating long-term individual wave statistics and a marine system's long-term response are essentially the same. For the latter case, however, the effects of additional conditions such as loading, heading to waves, speed, etc., have to be considered.

The long-term statistics of individual wave height is an accumulation of the statistics for all short-term sea conditions, taking into account the frequency of occurrence of each short-term sea state. Here, it is assumed that the wave height in the short term follows the Rayleigh probability law. By letting H_s and $\overline{T}_0$ be the significant wave height and average zero-crossing period, respectively, the probability that the magnitude of wave height in the long term will exceed a specified value 'a' may be written, in principle, as follows:

$$\Pr\{H_L > a\} = \Sigma \, \Pr\{H_L > a | H_s, \overline{T}_0\} \Pr\{H_s, \overline{T}_0\} \qquad (3.92)$$

where H_L = wave height in the long term.

The above equation for evaluating the statistical property of long-term wave height is conceptually correct. However, long-term statistics does not deal with a steady-state random process; hence, it is necessary to take into account the number of waves for each piecewise steady-state short-term sea in order to evaluate the probability. This was first brought to our attention by Battjes (1972) in his derivation of the probability density function of long-term wave height.

We may compute the average number of zero-crossings for each short-term sea condition, denoted by n_*, from the wave spectrum as follows (see Section 4.4):

$$n_* = \frac{1}{2\pi} \sqrt{\frac{m_2}{m_0}} \, (60)^2 \, T \qquad (3.93)$$

where m_0, m_2 = zeroth and second moment, respectively, of the
short-term wave spectrum

T=duration of short-term sea in hours.

By incorporating the number of waves in each short-term sea, Eq. (3.92) may be modified as

$$\Pr\{H_{\mathrm{L}}>a\}=\frac{\displaystyle\sum_{H_s}\sum_{T_0}n_\star\,\Pr\{H_{\mathrm{L}}>a\,|\,H_s,\overline{T}_0\}\,\Pr\{\overline{T}_0\,|\,H_s\}\,\Pr\{H_s\}}{\displaystyle\sum_{H_s}\sum_{T_0}n_\star\,\Pr\{\overline{T}_0\,|\,H_s\}\,\Pr\{H_s\}}\qquad(3.94)$$

The total number of waves in the long term can be evaluated from the denominator of Eq. (3.94) as follows:

$$\binom{\text{total number of waves}}{\text{in the long term}}=\sum_{H_s}\sum_{T_0}n_\star\,\Pr\{T_0|H_s\}\,\Pr\{H_s\}\qquad(3.95)$$

Since estimation of the duration of each short-term sea involved in Eq. (3.93) is difficult, the average duration is considered in practice.

The long-term probability density function of wave height is developed by Battjes (1972) using the joint probability density function of significant wave height and average zero-crossing period as follows:

$$f(x)=\frac{\displaystyle\int\!\!\int f_\star(x)\,t_0^{-1}\,f(H_s,\overline{T}_0)\,\mathrm{d}H_s\,\mathrm{d}\overline{T}_0}{\displaystyle\int\!\!\int t_0^{-1}\,f(H_s,\overline{T}_0)\,\mathrm{d}H_s\,\mathrm{d}\overline{T}_0}\qquad(3.96)$$

where $f_\star(x)$=probability density function for the short-term
wave height
t_0^{-1}=short-term average number of waves per unit time.

An example of the cumulative distribution of long-term individual wave heights obtained at three locations in the North Atlantic and the North Sea is shown in Figure 3.16. The data at each location consist of records of 12 minutes duration taken every 3 hours for one year. As seen in the figure, the cumulative distributions plotted on Weibull probability paper show that the long-term wave heights appear to follow the Weibull distribution with an exponent close to unity; that is, the distributions are close to the exponential probability distribution.

It may be of interest to note the total number of waves a marine structure might expect to encounter in its lifetime. An example of computations made for a 50-year lifetime of a semi-submersible-type offshore structure in various sea states in the North Sea up to a significant wave height of 13 m, taking into consideration various wave spectral shapes and headings to the seas, indicates this may be on the order of 3.58×10^8 by applying half-cycle excursion analysis, and 3.05×10^8 by applying the narrow-band random process assumption (Ochi 1987). The number

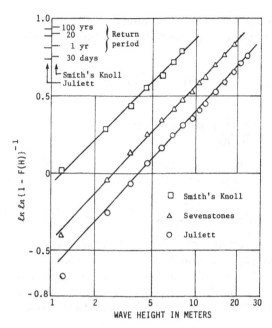

Fig. 3.16. Example of long-term probability density function of wave height (Battjes 1972).

will certainly be larger for marine vehicles, depending on their design speed.

The long-term wave statistics discussed in this section is the accumulation of all wave statistics in a given sea state (short-term wave statistics), with the sequence of the occurrence of sea states ignored. This information is sufficient, in general, for the marine systems design considerations; however, a more accurate estimation of long-term wave statistics may be achieved by taking the sea state history into consideration. Athanassoulis et al. (1992) and Athanassoulis and Soukissian (1993) developed a method for evaluating long-term wave statistics reflecting the sequence of occurrence of sea state at a given site. Waves in the long-term are considered to consist of two-level stochastic processes; one the slow-time component representing sea state (significant wave height) variation with time, the other the fast-time component associated with individual waves in a given sea state. They consider the number of maximum of the sea-surface elevation above a specified level occurring during a long time period [0, T], and they apply renewal theory. Although their approach is excellent, the application of renewal theory requires information regarding the time history of sea state for a sufficiently long period of time at a specified site. This may cause some difficulty in the application of the method at the present time.

3.9 STATISTICAL ANALYSIS OF AMPLITUDE AND HEIGHT FROM WAVE RECORDS

3.9.1 Introduction

Analyses and predictions of probabilistic properties of waves presented in this chapter so far are based on stochastic analysis of the wave spectrum. It is often important, however, to evaluate statistical properties of waves from measured data without carrying out spectral analysis, but instead, through analysis of random samples of a wave record by applying statistical estimation theory. There are two ways to take random samples: one is to read wave heights $(x_1, x_2, ..., x_n)$ as shown in Figure 3.17(a), the other is by reading wave deviations by drawing the mean line, as shown in Figure 3.17(b). These readings are not necessarily in sequence nor at equal intervals for wave deviation sampling.

Most commonly, we assume that waves are a narrow-band, Gaussian random process, and thereby wave heights obey the following Rayleigh probability law. The parameter R of the distribution will be estimated through statistical analysis of the random sample.

$$f(x) = \frac{2x}{R} e^{-x^2/R} \qquad 0 < x < \infty \tag{3.97}$$

The Rayleigh probability density function is derived in Section 3.2 following Rice's method developed in communication engineering. The probability distribution, however, was originally developed by Lord Rayleigh in 1880 in connection with the analysis of the resultant intensity of a large number of independent sounds. The distribution was introduced by Longuet-Higgins for predicting the magnitude of random waves not through spectral analysis but instead based on a concept similar to that considered by Lord Rayleigh. That is, assuming that a wave envelope exists and wave energy is being received from a large number of different sources whose phases are random, Longuet-Higgins shows that the probability distribution of such a sum (envelope) can be given in the form of Eq. (3.97).

The parameter R of the Rayleigh distribution is estimated based on a function composed of a set of random sample of size n. This is called the *estimator* of the parameter R. The accuracy of the estimation certainly depends on the sample size n; the larger the sample size, the more reliable

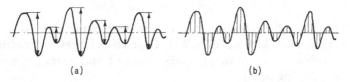

(a)   (b)

Fig. 3.17. Random sampling as applied to wave records.

the value can be acquired. For a large number of n (theoretically, infinite), the value of the estimator can be considered as the parameter. But, if the number is small, it is only possible to estimate the upper and lower limit values (confidence bands) of the parameter for a specified probability of assurance.

Prior to presenting the methodology for estimation, we outline the general requirements for determining the estimator. As will be shown later, the estimator itself is a random variable and it has its own probability distribution. It is commonly known that the best estimator should satisfy the following two conditions:

(i) The expected value of the estimator should be equal to the parameter. For the present case, $E[\hat{R}] = R$. Then, the estimator $\hat{R}$ is called an *unbiased estimator*.

(ii) The variance of the estimator is the smallest of all estimators; it is then called the *minimum variance estimator*.

A very useful technique for obtaining the best estimator is the maximum likelihood method. Although the maximum likelihood estimator does not always satisfy the above two conditions, it does satisfy the conditions for the Rayleigh probability distribution.

3.9.2 Maximum likelihood estimation

The method for finding the maximum likelihood estimator of the Rayleigh probability distribution from a random sample of wave heights is the following.

Each element of the random sample $(x_1, x_2, ..., x_n)$ is statistically independent, and is an outcome from a population having the Rayleigh probability distribution. Therefore, a joint probability distribution function of $(x_1, x_2, ..., x_n)$ is given by the product of the n Rayleigh distributions which is called the *likelihood function* denoted by L. We have

$$L = \prod_{i=1}^{n} \frac{2x_i}{R} \exp\left\{ -\frac{x_i^2}{R} \right\} = \left(\frac{2}{R}\right)^n \left(\prod_{i=1}^{n} x_i\right) \exp\left\{ -\frac{\sum_{i=1}^{n} x_i^2}{R} \right\} \quad (3.98)$$

The maximum likelihood estimator, $\hat{R}$ is defined such that the likelihood function L is larger than any other likelihood functions evaluated using other possible estimator R'. That is

$$L(x_1, x_2, ..., x_n | \hat{R}) > L(x_1, x_2, ..., x_n | R') \quad (3.99)$$

Therefore, the desired estimation can be determined by maximizing L; namely, as a solution of $dL/dR = 0$. In practice, however, the estimator $\hat{R}$ is evaluated by maximizing the logarithm of L and thereby as a solution of the following equation:

$$\frac{d}{dR}\ln L = -\frac{n}{R} + \sum_{i=1}^{n} x_i^2/R^2 = 0 \qquad (3.100)$$

From Eq. (3.100), we have

$$\hat{R} = \frac{1}{n}\sum_{i=1}^{n} x_i^2 \qquad (3.101)$$

where x_i are random samples of wave height.

As can be seen in the above equation, the estimator $\hat{R}$ is the average value of the sum of the squares of the sample elements x_i. The significant wave height of the sea from which the random sample is taken can be evaluated by Eq. (3.78) with the estimator $\hat{R}$. That is,

$$H_s = 1.42\sqrt{\hat{R}} \qquad (3.102)$$

Here, $\sqrt{\hat{R}}$ is called the *root-mean-square (rms) value*.

As can be seen in Eq. (3.101), the accuracy of the estimator $\hat{R}$ depends on the sample size n. It is common practice to consider a random sample size (wave height readings in the present case) of 120 to 150 to be sufficient for estimating a reliable R-value.

Figure 3.18 is an example of a comparison between the Rayleigh probability distribution with parameter R estimated from Eq. (3.101) and the wave height histogram constructed from the record. The estimated R-value from the sample of 125 readings is 33.9 m² as compared with 35.6 m² evaluated from spectral analysis.

Another way to estimate the parameter R of the Rayleigh distribution is to use the estimator evaluated through random sampling of the deviation from the mean value. As shown in Eq. (3.64), the parameter of the

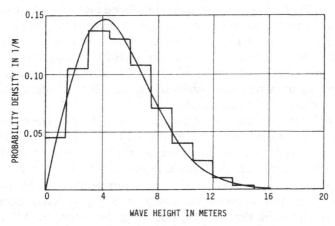

Fig. 3.18. Comparison between histogram of wave height and Rayleigh probability density function computed with estimator $\hat{R}$.

Rayleigh distribution applicable to wave height is equal to eight times the variance of a Gaussian random process. Therefore, estimation of the parameter can be achieved by estimating the variance σ^2 of the normal probability distribution; namely, we may take a random sample of deviations from the mean value $(x_1, x_2, ..., x_n)$ as shown in Figure 3.17(b). The likelihood function in this case becomes

$$L=\prod_{i=1}^{n} \frac{1}{\sqrt{2\pi}\sigma} \exp\left\{-\frac{x_i^2}{2\sigma^2}\right\}=\left(\frac{1}{2\pi\sigma^2}\right)^{n/2} \exp\left\{-\frac{1}{2\sigma^2}\sum_{i=1}^{n} x_i^2\right\} \quad (3.103)$$

By differentiating $\ln L$ with respect to σ^2, the estimator of the variance can be obtained as

$$\hat{\sigma}^2=\frac{1}{n}\sum_{i=1}^{n} x_i^2 \quad (3.104)$$

where x_i are random samples of wave profile (displacement).

It can be proved theoretically that the expected value of the estimated variance, $E[\hat{\sigma}^2]$, is equal to $(n-1)\sigma^2/n$, and hence the estimator $\hat{\sigma}^2$ is biased. Therefore, Eq. (3.104) may be used for large n. For small sample size, it is recommended that the following formula be used for estimating σ^2:

$$\hat{\sigma}^2=\frac{1}{n-1}\sum_{i=1}^{n} x_i^2 \quad (3.105)$$

The significant wave height can be estimated from Eq. (3.102) with $\hat{R}=8\hat{\sigma}^2$. That is,

$$H_s=4.01\sqrt{\hat{\sigma}^2} \quad (3.106)$$

As seen in Eq. (3.104), the maximum likelihood estimator of the variance, $\hat{\sigma}^2$ is the average value of the sum of the squares of the sample elements x_i. Hence, $\sqrt{\hat{\sigma}^2}$ in Eq. (3.106) is also called the *root-mean-square (rms) value*, the same name as $\sqrt{\hat{R}}$ in Eq. (3.102), although these two estimators are entirely different. Table 3.1 summarizes the formulae for estimating the parameter of the distribution applicable to significant wave height from observed random samples.

It is noted that estimation of the parameter of the Rayleigh probability distribution can be made either by $\hat{R}$ (Eq. 3.101) or by $\hat{\sigma}^2$ (Eq. 3.104) from the measured wave record. However, the latter method yields a more accurate result than the former, in general. This is because the sample size of the latter is much larger than the former. Figure 3.19 shows the same comparison between the histogram and probability density function as shown in Figure 3.18, but using the estimator $\hat{\sigma}^2$. The number of readings x_i in this case is 1120 compared to 125 for the estimator $\hat{R}$. This results in an estimator $\hat{\sigma}^2$ of 35.2 m^2, which is closer to the variance evaluated through spectral analysis.

Table 3.1. *Estimation of significant wave height from observed random sample.*

Random sample, x	RMS-value	Estimation of significant wave height
Wave height	$\sqrt{\hat{R}} = \dfrac{1}{n}\sum\limits_{i=1}^{n} x_i^2$	$1.42\sqrt{\hat{R}}$
Wave displacement	$\sqrt{\hat{\sigma}^2} = \dfrac{1}{n}\sum\limits_{i=1}^{n} x_i^2$	$4.01\sqrt{\hat{\sigma}}$

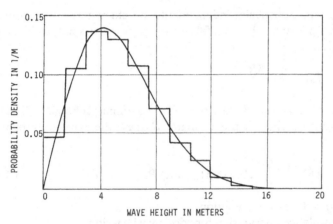

Fig. 3.19. Comparison between histogram of wave height and Rayleigh probability density function computed with estimator $\hat{\sigma}^2$.

3.9.3 Estimation of Rayleigh distribution parameter from a small number of observations

The methods of estimating the parameter R of the Rayleigh probability distribution in the preceding section are based on the condition that the sample size, n, must be large, on the order of 120 waves or greater. If the sample size is small, a reliable specific value of the parameter can no longer be estimated; instead an interval, called the *confidence interval* in which the true value of the unknown parameter R lies can be estimated. The estimation is made with a measure of assurance, $1 - \alpha$, called the *confidence coefficient*. Here, the parameter α, called the *level of significance*, may be interpreted as the probability of committing a possible error in the estimation.

The confidence interval can be estimated based on the probability function of the estimator. It is first necessary, therefore, to derive the probability density function of the estimator $\hat{R}$.

The estimator $\hat{R}$ is defined in Eq. (3.101) in which the set of x_i is assumed to be a random sample from the Rayleigh distribution with the parameter R. Hence, the probability density function of x_i^2 (denoted by y_i) can be obtained by applying the change of random variable technique (see Appendix A) as

$$f(y_i) = \frac{1}{R} \exp\{-y_i/R\} \tag{3.107}$$

This is the exponential probability distribution, and its characteristic function can be written as

$$\phi_{y_i}(t) = \int_0^\infty e^{ity_i} \frac{1}{R} e^{-y_i/R} dy_i = (1 - iRt)^{-1} \tag{3.108}$$

Since y_i are statistically independent, the characteristic function of $\hat{R} = (1/n)\Sigma_{i=1}^{n} y_i$ can be obtained following the properties of the characteristic function

$$\phi_{\hat{R}}(t) = \left(1 - \frac{iRt}{n}\right)^{-n} \tag{3.109}$$

which is the characteristic function of the gamma probability distribution. We thus obtain that the estimator $\hat{R}$ obeys the gamma probability law given by

$$f(\hat{R}) = \frac{(n/R)^n}{\Gamma(n)} \hat{R}^{n-1} \exp\{-(n/R)\hat{R}\} \qquad 0 \leqslant \hat{R} < \infty \tag{3.110}$$

where $\Gamma(n)$ is the gamma function.

The cumulative distribution function may be written as

$$F(R) = \frac{\Gamma\left(n, \dfrac{n}{R}\hat{R}\right)}{\Gamma(n)} \tag{3.111}$$

where

$$\Gamma\left(n, \frac{n}{R}\hat{R}\right)$$

is the incomplete gamma function.

In determining the confidence interval of the estimator $\hat{R}$ for a specified confidence coefficient $1 - \alpha$, it is common practice to choose the confidence interval such that the possibility of committing an error may occur at the lower and upper portions of the distribution with an equal probability of error $\alpha/2$, as shown in Figure 3.20.

Thus, the confidence interval of the estimator $\hat{R}$ can be determined from the following probability:

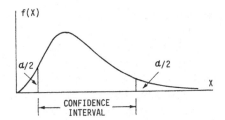

Fig. 3.20. Sketch indicating the confidence interval for the confidence coefficient $1-\alpha$.

$$\Pr\{R_L < \hat{R} < R_U\} = 1-\alpha \qquad (3.112)$$

where R_L and R_U satisfy the cumulative distribution function of the gamma probability distribution given by

$$F(R_L) = \alpha/2 \quad \text{and} \quad F(R_U) = 1 - (\alpha/2) \qquad (3.113)$$

Hence, for a specified α, the upper and lower bounds of $\hat{R}$ can be obtained from Eqs. (3.111) through (3.113) as

$$n\frac{R_L}{R} < \hat{R} < n\frac{R_U}{R} \qquad (3.114)$$

which results in the confidence interval of the parameter R with confidence coefficient $1-\alpha$ being

$$n\frac{R_L}{\hat{R}} < R < n\frac{R_U}{\hat{R}} \qquad (3.115)$$

It is possible to evaluate the confidence interval given in Eq. (3.115) through the χ^2 distribution. That is, it can be proved that if $\hat{R}$ has the gamma distribution with parameters n and n/R as given in Eq. (3.110), then the random variable $2(n/R)\hat{R}$ obeys the χ^2 distribution with $2n$ degrees of freedom. Therefore, we write the probability representing the confidence interval given in Eq. (3.112) as follows:

$$\Pr\left\{\chi^2_{(2n)L} < \frac{2n}{R}\hat{R} < \chi^2_{(2n)U}\right\} = 1-\alpha \qquad (3.116)$$

where $\chi^2_{(2n)L}$ and $\chi^2_{(2n)U}$ are the values of the χ^2 distribution with $2n$ degrees of freedom for which the cumulative distribution is $\alpha/2$ and $1-(\alpha/2)$, respectively. These values may easily be found in the χ^2 distribution table.

Thus, the confidence interval of the parameter R with confidence coefficient $1-\alpha$ can be determined from Eq. (3.116) as follows:

$$\frac{2n\hat{R}}{\chi^2_{(2n)U}} < R < \frac{2n\hat{R}}{\chi^2_{(2n)L}} \qquad (3.117)$$

The above formula to evaluate the confidence interval is valid for any number of samples of size n. However, the χ^2 distribution is asymptotically equal to the normal distribution with mean n and variance $2n$ for large n, say $n=30$ or greater. Hence, for the present case, the random variable $2(n/R)\hat{R}$ is asymptotically normally distributed with mean $2n$ and variance $4n$. Then, by standardizing the normal distribution, the confidence interval given in Eq. (3.116) can be written as

$$\Pr\left\{U_{\alpha/2}<\frac{(2n/R)\,\hat{R}-2n}{\sqrt{4n}}<|U_{\alpha/2}|\right\}=1-\alpha \qquad (3.118)$$

where $U_{\alpha/2}$ is the value of the standard normal distribution for which the cumulative distribution is $\alpha/2$ ($U_{\alpha/2}$ is negative).

Thus, for a sample size n greater than 30, the confidence interval for the parameter R with confidence coefficient $1-\alpha$ becomes as follows:

$$\frac{\hat{R}}{1+(|U_{\alpha/2}|/\sqrt{n})}<R<\frac{\hat{R}}{1+(U_{\alpha/2}/\sqrt{n})} \qquad (3.119)$$

where $U_{\alpha/2}<0$.

For estimating the parameter R of the Rayleigh distribution through a random sample of displacement from the mean value of the wave record, it was shown in Eq. (3.104) that the random variable $\hat{\sigma}^2=\Sigma_{i=1}^{n}\,x_i^2/n$ is the estimator for σ^2. In this case, the random sample x_i is considered to be an outcome from a population which is normally distributed with zero mean and variance σ^2. It can be proved, in general, that if a random variable x obeys the normal probability law with zero mean and variance σ^2, then $\Sigma_{i=1}^{n}\,x_i^2/\sigma^2$ follows the χ^2 distribution with n degrees of freedom. Hence, for the present case, the random variable $n\hat{\sigma}^2/\sigma^2$ has the χ^2 distribution with n degrees of freedom. The mean and variance of the distribution are n and $2n$, respectively. Since the sample size n is usually larger than 30, we may estimate the confidence interval based on the normal distribution by applying the same method used for the derivation of Eq. (3.118). That is, by standardizing the random variable $n\hat{\sigma}^2/\sigma^2$, the confidence interval should satisfy the following probability:

$$\Pr\left\{U_{\alpha/2}<\frac{(n\hat{\sigma}^2/\sigma^2)-n}{\sqrt{2n}}<|U_{\alpha/2}|\right\}=1-\alpha \qquad (3.120)$$

where $U_{\alpha/2}$ is defined in connection with Eq. (3.118).

Then, we can obtain the confidence interval of σ^2 as

$$\frac{\hat{\sigma}^2}{1+\sqrt{\dfrac{2}{n}}|U_{\alpha/2}|}<\sigma^2<\frac{\hat{\sigma}^2}{1+\sqrt{\dfrac{2}{n}}\,U_{\alpha/2}} \qquad (3.121)$$

and thereby the confidence interval of the parameter R of the Rayleigh distribution applicable for wave height with the confidence interval $1-\alpha$ is given as follows

$$\frac{8\,\hat{\sigma}^2}{1+\sqrt{\frac{2}{n}}|U_{\alpha/2}|}<R<\frac{8\,\hat{\sigma}^2}{1+\sqrt{\frac{2}{n}}\,U_{\alpha/2}} \tag{3.122}$$

where $U_{\alpha/2}<0$.

3.9.4 Goodness-of-fit tests

In analyzing wave data obtained in the ocean or in the laboratory, it is sometimes necessary to examine whether or not the observed data can be considered as a sample from a population having a presumed probability distribution. For example, waves in finite water depths are assumed to be a non-Gaussian random process in general, but they may quite likely be a Gaussian random process if the sea severity is mild. Therefore, it is necessary to examine the data sample with a statistically based assurance. This can be done by carrying out so called *goodness-of-fit tests* in statistical inference theory.

Although several goodness-of-fit tests are available, each having its advantage and disadvantage, only typical tests suitable for analysis of wave data are discussed here. One is the χ^2 test, which has been used extensively in various fields of applied statistics, the other is the Kolmogorov–Smirnov's test, which has the advantage of being applicable even for a sample consisting of a small number of observations.

(i) Chi-square (χ^2) test

This test establishes the confidence with which a sample of observed frequencies of occurrence of particular events can be assumed to belong to a hypothesized distribution. The test is valid only for a large number of observations (at least 120 or so is necessary). The test procedure is as follows.

Classify the observed data into k mutually exclusive and exhaustive divisions, even though the widths of divisions may not necessarily be equal. It is recommended that the number of divisions, k, should be greater than eight. Let n_i be the number of observed outcomes of the event which fall into the ith division ($i=1, 2, ..., n$), and let $\Sigma_{i=1}^{k}\, n_i=N$.

Next, let p_i be the probability of the hypothesized theoretical distribution in the ith division. Here, $\Sigma_{i=1}^{k}\, p_i=1$ and the number of samples in the ith division is theoretically p_iN. It is highly desirable that p_iN should be at least five. If it is difficult to meet this condition, it is common practice to combine or rearrange the classification of data until the p_iN reaches a satisfactory size. Then, we have the statistical criterion that

$$U=\sum_{i=1}^{k}\frac{(n_{i}-p_{i}N)^{2}}{p_{i}N} \tag{3.123}$$

is approximately the χ^2 distribution with $k-1$ degrees of freedom provided N is large. When the parameter of the theoretical distribution (hypothesis) is unknown (this is the usual case), then the unknown parameter must be estimated from the data. If there are more than one unknown parameters, then the degrees of freedom are reduced by one for every unknown parameter to be estimated.

If the observed value of n is exactly equal to p_iN, then $U=0$. This implies that the observed data follow the hypothesized probability law. It can be said therefore that a small value of U provides more assurance that the observed sample obeys the hypothesized probability distribution. Hence, we may reject the hypothesis that the sample is from the population having the presumed probability distribution if

$$U>\chi^2_{k-1;\alpha} \tag{3.124}$$

where $\chi^2_{k-1;\alpha}$ stands for a value of the χ^2 distribution with $k-1$ degrees of freedom and level of significance α. In other words, the observed data are considered to be significant at level α, and they do not substantiate the hypothesized probability distribution. The level of significance α is usually taken as 0.05; namely, χ^2 test is conducted with 95 percent confidence, or with a 5 percent possible error.

An example of the use of the χ^2 test to examine whether or not waves observed at a water depth of 8.7 m during a storm can be considered to be a Gaussian random process follows. The data are taken for 17 minutes in which 140 waves are recorded. Spectral analysis of the data is performed from which the variance of this wave data is obtained as 0.398 m^2. A total of 1140 wave displacements from the zero-line are read from the record, and they are classified into 13 divisions as shown in Figure 3.21. In the

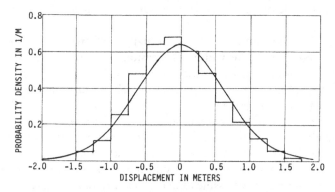

Fig. 3.21. Example of χ^2 test on waves observed in finite water depth.

computation, two divisions at the positive ends of the data are combined to meet the requirement that $p_r N$ in each division be greater than five. Hence, the test is carried out for $k=12$ divisions.

Included also in Figure 3.21 is the hypothesized normal probability distribution with zero mean and variance 0.398 m^2. The U-value computed from Eq. (3.123) is 25.32 as compared to 19.68 from the χ^2 distribution with $k-1=11$ degrees of freedom for level of significance, $\alpha=0.05$. Since the U-value is greater than the hypothesized χ^2 value with 11 degrees of freedom for $\alpha=0.05$, it may be said that the measured waves cannot be considered to be a Gaussian random process at the 5 percent level of significance.

(ii) Kolmogorov–Smirnov test

This test does not consider the difference between the observed frequency of occurrence and the probability of the hypothesized distribution in a classified division; instead, it is concerned with the difference in the cumulative distribution between observed and the hypothesized probability law. The observed sample, therefore, is not classified into groups, but is rearranged in an ordered sequence for this test.

Let $x_1, x_2, ..., x_n$ be the ordered sample of size n, and let its cumulative distribution be $F_n(x)$ given by

$$F_n(x) = \begin{cases} 0 & \text{for } x=x_1 \\ r/n & \text{for } x_r \leqslant x \leqslant x_{r+1} \qquad r=1, 2, ..., (n-1) \\ 1 & \text{for } x=x_n \end{cases} \tag{3.125}$$

$F_n(x)$ is a step function. In the case where several values are observed to be the same for $x=x_j$, $F_n(x_j)$ jumps significantly. Let $F(x)$ be the cumulative distribution function of the hypothesized probability distribution. The Kolmogorov–Smirnov test is based on the statistics

$$D_n = \sup |F_n(x) - F(x)| \tag{3.126}$$

where sup implies the maximum value over the entire range of the sample domain. Kolmogorov shows that if a sample of size n is large (on the order of 35 or greater) the probability of D_n exceeding the value ϵ is given approximately by

$$\Pr\{D_n > \epsilon\} = 2 \sum_{r=1}^{\infty} (-1)^{r-1} \exp\{-2r^2 n\epsilon^2\} \tag{3.127}$$

Since this series converges rapidly, we may take the first term only, namely $2 \exp\{-2n\epsilon^2\}$. By letting this value be equal to the level of significance α, we have

$$\epsilon = \sqrt{-\ln(\alpha/2)/2n} = \begin{cases} 1.63/\sqrt{n} & \text{for } \alpha=0.01 \\ 1.48/\sqrt{n} & \text{for } \alpha=0.025 \\ 1.36/\sqrt{n} & \text{for } \alpha=0.05 \end{cases} \tag{3.128}$$

The test is then carried out by which we may reject the hypothesis with level of significance α if the statistics D_n exceeds ϵ.

It can also be shown that $4nD_n^2$ is approximately the χ^2 distribution with two degrees of freedom, denoted by $\chi_{2,\alpha}^2$ (see Kendall and Stuart 1961). Hence, the assumption that the observed data come from the hypothesized distribution should be rejected with level of significance α if

$$4nD_n \geqslant \chi_{2,\alpha}^2 \tag{3.129}$$

The Kolmogorov–Smirnov test has the feature that it can be applied to data from a small sample size provided that the parameter of the hypothesized distribution is either assumed or known in advance.

As an example of the Kolmogorov–Smirnov test, the same wave data for which the χ^2 test was made (see Figure 3.21) are used, but this time the test is carried out by reading 50 samples of wave height instead of wave displacement. We may hypothesize that wave height follows the Rayleigh probability distribution with the parameter eight times the variance ($\sigma^2 = 0.398$) which is obtained from the spectral analysis.

The largest discrepancy between the cumulative distribution of the ascending ordered sample and the hypothesized Rayleigh distribution is observed at $r = 12$ where $x = 1.45$ m in a total of 50 ordered samples. The cumulative distribution of the Rayleigh distribution with $R = 8\sigma^2$ is 0.483 for $x = 1.45$ m, while the cumulative distribution for $r = 12$ in a total of 50 ordered sample becomes $F_n(x) = 12/50 = 0.240$ and thereby we have $D_n = 0.243$. Since D_n exceeds the ϵ-value 0.192 for $n = 50$ with $\alpha = 0.05$ computed by Eq. (3.128), the hypothesized probability distribution is rejected with a level of significance $\alpha = 0.05$.

If we test by applying the formula given in Eq. (3.129), $4nD_n^2$ is equal to 6.77 which is greater than 5.99, the computed χ^2 value with two degrees of freedom for level of significance $\alpha = 0.05$. Hence, we may conclude that wave height obtained at a location where the water depth is 8.7 m during this storm cannot be considered to obey the Rayleigh probability law. This supports the results of the χ^2 test computed by Eq. (3.124) which concluded that the waves could not be considered a Gaussian random process.

4 WAVE HEIGHT AND ASSOCIATED PERIOD

4.1 INTRODUCTION

For a complete description of wind-generated random waves, it is necessary to consider wave height and period as well as the direction of travel. In particular, serious consideration must be given to the combined effect of height, period and direction, if any correlation exists. Wave data measured in the ocean show that period dispersion for very large wave height is not widely spread; as is also the case of height dispersion for large wave periods. In other words, height and period of incident waves are not statistically independent. Hence, the joint probability distribution of wave height and period plays a significant role in predicting statistical properties of waves such as the frequency of occurrence of wave breaking in a seaway.

Wave breaking takes place when wave height and period cannot maintain the equilibrium condition needed for stability. Therefore, for estimating the possibility of the occurrence of wave breaking in a given sea condition, knowledge of statistics on wave height and associated period, namely, the joint probability distribution, is necessary.

Further, the joint probability distribution of wave height and period is of the utmost importance for the design of floating marine systems. This is because one of the most important considerations for the design of a floating marine system lies in estimation of the possible occurrence of resonant motion which may occur when wave periods are close to the natural motion period of the system. The latter depends on the size and underwater configuration of the system, determined by the designer.

If the wave period is sufficiently long or sufficiently short in comparison with the natural period of a system's motion, the system may be in no danger even though wave height is large. On the other hand, the magnitude of motion will reach a level critical for the system if wave periods are close to the system's natural period. In designing a floating marine system, it is therefore extremely important to know the statistical information concerning wave height with periods that are close to the system's natural period in areas where the structure will be operated.

Several probability distributions of combined wave height (or amplitude or maxima) and wave period (or time interval between peaks) have been developed. Longuet-Higgins (1975, 1983) developed the joint probability distribution of wave height (peak-to-trough excursions) and associated zero-crossing period, while a group of researchers at the Centre National pour l'Exploitation (CNEXO) developed a joint distribution of positive maxima and the time interval between them (Cavanié *et al.* 1976, Arhan *et al.* 1976). On the other hand, Lindgren (1970, 1972), and Lindgren and Rychlik (1982) deal with the joint distribution of local peak-to-trough excursion and associated time interval (half-period). Lindgren's distribution is mathematically elaborate but the distribution is not given in closed form, and the computation is too complex to use in practice. A critical comparison of these three joint probability distributions along with measured data is given by Srokosz and Challenor (1987). Based on the results of their comparison, Longuet-Higgins' and CNEXO's joint distributions appear to be more appropriate in ocean engineering; hence, these will be presented in Sections 4.2 and 4.3, respectively.

As to the statistical properties of wave period, the probability distribution derived as the marginal distribution of the joint wave height and period distribution has been usually considered. Although the probability distribution of the period of a random process was derived by Rice (1944) in connection with the level-crossing problem, Rice's distribution has not been applied to the analysis of ocean wave period since his distribution appears to be appropriate for a random process having relatively high frequencies. The probability distribution of wave period derived from the joint distribution of wave height and period will be presented in Section 4.4.

Statistical information on wave height and the direction of travel of incident random waves is developed by Isobe (1988) by evaluating the ratio of two rectangular components of the horizontal water particle velocity. The joint probability distribution of wave amplitude and direction thus developed will be presented in Section 4.5.

4.2 JOINT PROBABILITY DISTRIBUTION OF WAVE HEIGHT AND PERIOD

The joint probability density function applicable to wave height and period was first developed by Longuet-Higgins (1975), and this joint distribution is superseded by his revised distribution developed in 1983. Here, the latter distribution is summarized below.

Assuming waves to be a Gaussian random process, we may write the wave profile as

$$x(t) = \sum_{n=1}^{\infty} a_n \cos(\omega_n t + \epsilon_n)$$

$$= \sum a_n \cos\{(\omega_n - \overline{\omega})t + \overline{\omega}t + \epsilon\}$$

$$= x_c(t) \cos \overline{\omega}t - x_s(t) \sin \overline{\omega}t \qquad (4.1)$$

where

$$x_c(t) = \sum_{n=1}^{\infty} a_n \cos\{(\omega_n - \overline{\omega})t + \epsilon_n\}$$

$$x_s(t) = \sum_{n=1}^{\infty} a_n \sin\{(\omega_n - \overline{\omega})t + \epsilon_n\}$$

$$\overline{\omega} = \text{mean frequency defined in Eq. (2.74).} \qquad (4.2)$$

This is the same presentation as given in Eq. (3.42).

It is further assumed that waves are a narrow-band random process whose spectrum is concentrated in the vicinity of the mean frequency $\overline{\omega}$. By employing the mean frequency $\overline{\omega}$, a significant advantage will be seen in the development of the joint distribution of wave profile $x(t)$ and its velocity $\dot{x}(t)$. Consider the joint distribution of x_c, $\dot{x}_c$, x_s and $\dot{x}_s$. It can be proved that these random variables are statistically independent and normally distributed with the mean

$$E[x_c] = E[\dot{x}_c] = E[x_s] = E[\dot{x}_s] = 0 \qquad (4.3)$$

and the covariance matrix given by

$$\Sigma = \begin{pmatrix} \mu_0 & 0 & 0 & 0 \\ 0 & \mu_2 & 0 & 0 \\ 0 & 0 & \mu_0 & 0 \\ 0 & 0 & 0 & \mu_2 \end{pmatrix} \qquad (4.4)$$

where

$$\mu_0 = \int_0^{\infty} S(\omega)\, d\omega = m_0$$

$$\mu_2 = \int_0^{\infty} (\omega - \overline{\omega})^2 S(\omega)\, d\omega = (m_0 m_2 - m_1^2)/m_0$$

$m_j = j$th moment of the spectrum.

Hence, the joint probability density function of x_c, $\dot{x}_c$, x_s and $\dot{x}_s$ is obtained as

$$f(x_c, \dot{x}_c, x_s, \dot{x}_s) = \frac{1}{(2\pi)^2 \mu_0 \mu_2} \exp\left\{-\frac{x_c^2 + x_s^2}{2\mu_0}\right\} \cdot \exp\left\{-\frac{\dot{x}_c^2 + \dot{x}_s^2}{2\mu_2}\right\}. \qquad (4.5)$$

Since waves are assumed to be narrow-banded in the vicinity of the mean frequency $\bar{\omega}$, $x(t)$ may be written as given in Eq. (3.20). That is,

$$x(t) = \mathrm{Re}\{\rho(t)\,e^{i\phi(t)}e^{i\bar{\omega}t}\} \tag{4.6}$$

where $\rho(t)$ = amplitude, $\phi(t)$ = phase.

Note that we have the following relationship:

$$\rho(t)e^{i\phi(t)} = \sum_{n=1}^{\infty} a_n \exp\{i(\omega_n - \bar{\omega})t + \epsilon_n\} \tag{4.7}$$

We now change the random variables from $(x_c, \dot{x}_c, x_s, \dot{x}_s)$ to $(\rho, \phi, \dot{\rho}, \dot{\phi})$ by letting

$$\begin{aligned} x_c &= \rho \cos \phi \\ x_s &= \rho \sin \phi \end{aligned} \tag{4.8}$$

The joint probability density function $(\rho, \phi, \dot{\rho}, \dot{\phi})$ then can be obtained as

$$(\rho, \phi, \dot{\rho}, \dot{\phi}) = \frac{\rho^2}{(2\pi)^2 \mu_0 \mu_2} \exp\left\{-\frac{\rho^2}{2\mu_0}\right\} \exp\left\{-\frac{\dot{\rho}^2 + \rho^2\dot{\phi}^2}{2\mu_2}\right\}$$

$$0 \leqslant \rho < \infty \qquad -\infty < \dot{\rho} < \infty \qquad 0 \leqslant \phi \leqslant 2\pi \qquad -\infty < \dot{\phi} < \infty \tag{4.9}$$

By integrating Eq. (4.9) with respect to $\dot{\rho}$ and ϕ, we have the joint probability density function of ρ and $\dot{\phi}$ as

$$f(\rho, \dot{\phi}) = \frac{\rho^2}{\sqrt{2\pi}\,\mu_0\sqrt{\mu_2}} \exp\left\{-\frac{\rho^2}{2\mu_0}\right\} \exp\left\{-\frac{\rho^2\dot{\phi}^2}{2\mu_2}\right\}$$

$$0 \leqslant \rho < \infty \qquad -\infty < \dot{\phi} < \infty \tag{4.10}$$

Next, it will be shown that the phase velocity $\dot{\phi}$ may be expressed in terms of period. For this, consider the *spectral width parameter* ν defined in Eq. (3.41). That is,

$$\nu = \left(\frac{m_0 m_2}{m_1^2} - 1\right)^{1/2} = \sqrt{\mu_2/\mu_0}\,(1/\bar{\omega}) \tag{4.11}$$

Here, ν can be expressed as a function of the average zero-crossing frequency, $\bar{\omega}_0$, and the mean frequency, $\bar{\omega}$; and hence, for narrow-banded random waves ν is much less than unity. That is,

$$\nu = ((\bar{\omega}_0/\bar{\omega})^2 - 1)^{1/2} \ll 1 \tag{4.12}$$

From Eqs. (4.11) and (4.12), we have

$$\sqrt{\mu_2/\mu_0} \ll \bar{\omega} \tag{4.13}$$

On the other hand, the marginal probability density function of the phase velocity, $\dot{\phi}$ can be obtained from Eq. (4.10) as

$$f(\dot{\phi}) = \frac{\sqrt{\mu_0/\mu_2}}{2\{1+(\mu_0/\mu_2)\dot{\phi}^2\}^{3/2}} \qquad -\infty < \dot{\phi} < \infty \qquad (4.14)$$

By letting $(\mu_0/\mu_2)\dot{\phi}=z$, it can be seen that the probability density function of z is sharply concentrated around $z=0$. Hence, by assuming it is a point probability distribution, we may write

$$z = \sqrt{\mu_2/\mu_0}\,\dot{\phi} \approx 1 \qquad (4.15)$$

and thereby $\sqrt{\mu_0/\mu_2}$ and $\dot{\phi}$ are in the same order of magnitude. Hence, from Eq. (4.13), we have

$$\dot{\phi} \ll \overline{\omega} \qquad (4.16)$$

If this is the case, the period of $x(t)$ given in Eq. (4.6) may be approximated as

$$T = \frac{2\pi}{\overline{\omega}+\dot{\phi}} \qquad (4.17)$$

Hence, we have

$$\dot{\phi} = (2\pi/T) - \overline{\omega} \qquad (4.18)$$

By using the relationship given in Eq. (4.18), the joint probability density function $f(\rho,\dot{\phi})$ can be converted to that of amplitude and period $f(\rho,T)$. For dimensionless presentation of the joint probability density function, let us define

$$\xi = \rho/\sqrt{m_0} = \text{dimensionless amplitude}$$

$$\eta = T/\overline{T} = \text{dimensionless period} \qquad (4.19)$$

where m_0 = area under the spectral density function

$\overline{T}$ = mean period = $2\pi/\overline{\omega} = 2\pi(m_0/m_1)$.

Then, the joint probability density function of ξ and η becomes

$$f(\xi,\eta) = L\frac{1}{\sqrt{2\pi}\nu}(\xi/\eta)^2\exp\left\{-\frac{\xi^2}{2}\left[1+\left(1-\frac{1}{\eta}\right)^2\frac{1}{\nu^2}\right]\right\} \qquad (4.20)$$

where L is a normalization factor required so that the joint probability density function satisfies the conditions that the integration of the density function over the entire sample space becomes unity. Longuet-Higgins obtains L as

$$L = \left(1+\frac{\nu^2}{4}\right) \qquad (4.21)$$

Thus, following Longuet-Higgins' derivation, the joint probability density function of wave height and period in dimensionless form is given by

$$f(\xi,\eta)=\frac{1}{\sqrt{2\pi}\nu}\left(1+\frac{\nu^2}{4}\right)(\xi/\eta)^2\exp\left\{-\frac{\xi^2}{2}\left[1+\left(1-\frac{1}{\eta}\right)^2\frac{1}{\nu^2}\right]\right\}$$

$$0<\xi<\infty \qquad 0<\eta<\infty \qquad\qquad\qquad (4.22)$$

It is noted that the above dimensionless density function differs slightly from the original density function given by Longuet-Higgins shown below in that his definition of dimensionless wave amplitude is given by $\rho\sqrt{2m_0}$ as compared with $\rho/\sqrt{m_0}$ defined here. By defining the dimensionless wave amplitude as $r=\rho/\sqrt{2m_0}$, the joint probability density function becomes as follows:

$$f(r,\eta)=\frac{2}{\sqrt{\pi}\nu}\left(1+\frac{\nu^2}{4}\right)(r/\eta)^2\exp\left\{-r^2\left[1+\left(1-\frac{1}{\eta}\right)^2\frac{1}{\nu^2}\right]\right\}$$

$$0<r<\infty \qquad 0<\eta<\infty \qquad\qquad\qquad (4.23)$$

The location of the mode where the joint probability density function $f(\xi,\eta)$ peaks is obtained from the condition $\partial f/\partial\xi=0$ and $\partial f/\partial\eta=0$. That is, the mode is located at

$$\xi=\sqrt{2}/\sqrt{1+\nu^2} \qquad \eta=1/(1+\nu^2) \qquad\qquad (4.24)$$

and the value of $f(\xi,\eta)$ at this point becomes

$$f_{\max}(\xi,\eta)=\sqrt{2/\pi}\,(1/e)\left(1+\frac{\nu^2}{4}\right)\left(\frac{1+\nu^2}{\nu}\right) \qquad (4.25)$$

Figure 4.1 shows the contour curves of the joint probability density function given in Eq. (4.23) for $\nu=0.3$ and 0.4.

In applying the joint density function given in Eq. (4.23), Shum and Melville (1984) define the amplitude as a continuous function of time for a given wave record. Their approach to apply the Hilbert transform technique demonstrates a significant contribution toward avoiding the ambiguity in defining wave amplitude and period. They also develop a method for a more accurate resolution of wave period from the record. Figure 4.2 shows a comparison of Longuet-Higgins' theoretical joint probability density function and data obtained during hurricane CAMILLE in which 3109 combinations of amplitude and period are accumulated.

The marginal probability density function of wave amplitude can be obtained from Eq. (4.22) as

$$f(\xi) = \int_0^\infty f(\xi,\eta)\,\mathrm{d}\eta = \xi \exp\{-\xi^2/2\}\left(1+\frac{\nu^2}{4}\right)\Phi(\xi/\nu)$$

$$0<\xi<\infty \tag{4.26}$$

The above equation reduces to the Rayleigh probability distribution (in dimensionless form) for $\nu=0$. Even though $\nu\neq0$, the value of $(1+(\nu^2/4))\,\Phi(\xi/\nu)$ is very close to unity for $\nu<0.6$; hence it may safely be assumed in practice that the marginal probability distribution becomes the Rayleigh distribution.

The joint probability density function of wave height, H, and period in dimensionless form can be obtained from Eq. (4.22) by letting $\zeta=H/\sqrt{m_0}$ as

$$f(\zeta,\eta) = \frac{1}{8\sqrt{2\pi}\nu}\left(1+\frac{\nu^2}{4}\right)\left(\frac{\zeta}{\eta}\right)^2 \exp\left\{-\frac{\zeta^2}{8}\left[1+\left(1-\frac{1}{\eta}\right)^2\frac{1}{\nu^2}\right]\right\}$$

$$0<\zeta<\infty \qquad 0<\eta<\infty \tag{4.27}$$

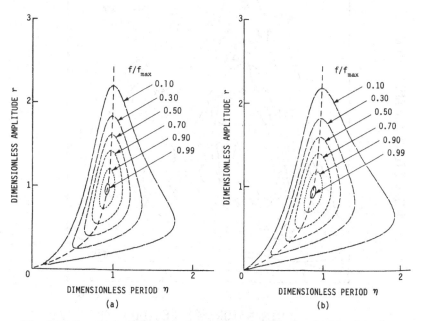

Fig. 4.1. Contours of dimensionless joint probability density function of wave amplitude and period for (a) $\nu=0.3$ and (b) $\nu=0.4$ (Longuet-Higgins 1983).

4.3 JOINT PROBABILITY DISTRIBUTION OF POSITIVE MAXIMA AND TIME INTERVAL

For non-narrow-band random waves, the amplitude as well as period cannot be defined in the strict sense, since there are multiple crests (positive maxima). Therefore, the joint probability density for non-narrow-band random waves developed by Cavanié et al. (1976) is that referred to positive maxima, Ξ, and the time interval, T, between them, shown in Figure 4.3.

The basic principle of the derivation is to first obtain the joint probability distribution of the maxima and associated acceleration, and then convert it to the joint probability distribution of the maxima and period. The former joint distribution can be derived by the same approach as presented in Section 3.3; but the acceleration term is retained in the present case. The probability that the positive maxima, Ξ, exceeds a specified level ζ with acceleration $\ddot{x}$ is given by

$$\Pr\{\Xi > \zeta \text{ with acceleration } \ddot{x}\} = \overline{N}_{\zeta,\ddot{x}}/\overline{N}_+ \tag{4.28}$$

where $\overline{N}_{\zeta,\ddot{x}}$ = average number of positive maxima above a specified level ζ with acceleration $\ddot{x}$ per unit time

$\overline{N}_+$ = average number of positive maxima per unit time.

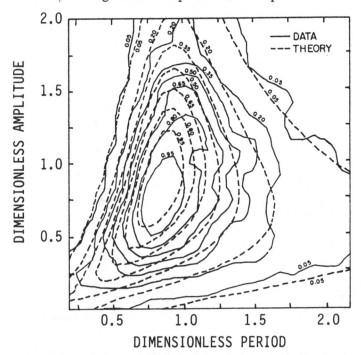

Fig. 4.2. Comparison of theoretical joint probability density function of dimensionless wave amplitude and period and data obtained during Hurricane CAMILLE (Shum and Melville 1984).

By letting $f(x,\dot{x},\ddot{x})$ be the joint probability density function of wave displacement, velocity and acceleration, $\overline{N}_{\zeta,\ddot{x}}$ and $\overline{N}_+$ can be written by

$$\overline{N}_{\zeta,\ddot{x}} = \int_\zeta^\infty |\ddot{x}| f(x,0,\ddot{x})\,dx$$

$$\overline{N}_+ = \int_0^\infty \int_{-\infty}^0 |\ddot{x}| f(x,0,\ddot{x})\,d\ddot{x}\,dx \qquad (4.29)$$

By the same procedure as shown in Eq. (3.28), the joint probability density function of the positive maxima and acceleration can be written

$$f(\zeta,\ddot{x}) = \frac{d}{d\zeta}\left(1 - \frac{\overline{N}_{\zeta,\ddot{x}}}{\overline{N}_+}\right)$$

$$= \frac{|\ddot{x}| f(\zeta,0,\ddot{x})}{\displaystyle\int_0^\infty \int_{-\infty}^0 |\ddot{x}| f(x,0,\ddot{x})\,d\ddot{x}\,dx} \qquad (4.30)$$

Here, the probability density function $f(x,0,\ddot{x})$ is given in Eq. (3.30). By performing integration of the above equation, the joint probability density function of maxima and associated acceleration can be derived as

$$f(\zeta,\ddot{x}) = \frac{2|\ddot{x}|}{\sqrt{2\pi}\,\sqrt{\Delta m_4}\,(1+\sqrt{1-\epsilon^2})} \exp\left\{-\frac{m_0\ddot{x}+2m_2\zeta\ddot{x}+m_4\zeta^2}{2\Delta}\right\}$$

$$0 \leqslant \zeta < \infty \qquad -\infty < \ddot{x} \ll 0 \qquad (4.31)$$

where $m_j = j$th moment of the spectrum

$$\Delta = m_0 m_4 - m_2^2$$

$$\epsilon = \sqrt{1 - m_2^2/m_0 m_4} \text{ (see Eq. 3.32)}.$$

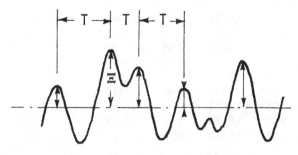

Fig. 4.3. Definition of positive maxima and time interval between them.

We next transform the joint distribution of maxima and acceleration to that of maxima and time interval by using the functional relationship given by

$$\ddot{x} = -\omega^2 \zeta = -(2\pi/T)^2 \zeta \tag{4.32}$$

By applying the change of random variable technique, the joint probability density function of the maxima and time interval can be derived. In order to present the joint density function in dimensionless form, let us write

$$\xi = \zeta/\sqrt{m_0} \quad \tau = T/\overline{T}_\text{m} \tag{4.33}$$

where m_0 = area under the spectrum
$\overline{T}_\text{m}$ = average time interval between successive positive maxima

$$= 1/\overline{N}_+ = 4\pi \left(\frac{\sqrt{1-\epsilon^2}}{1+\sqrt{1-\epsilon^2}} \right) \sqrt{\frac{m_0}{m_2}}$$

Then, the dimensionless joint density function may be presented as

$$f(\xi,\tau) = \sqrt{\frac{2}{\pi}} \frac{\alpha^3 \xi^2}{\epsilon(1-\epsilon^2)\tau^5} \exp\left\{ -\frac{\xi^2}{2\epsilon^2\tau^4} \left[(\tau^2-\alpha^2)^2 + \alpha^4\beta^2 \right] \right\} \tag{4.34}$$

where

$$\alpha = \frac{1}{2}(1+\sqrt{1-\epsilon^2})$$

$$\beta = \epsilon/\sqrt{1-\epsilon^2}$$

Table 4.1. *Average time interval $\bar{\tau}$ as a function of ϵ (Arhan et al. 1976).*

ϵ	$\bar{\tau}(\epsilon)$	ϵ	$\bar{\tau}(\epsilon)$
0.05	0.9988	0.55	0.9500
0.10	0.9963	0.60	0.9445
0.15	0.9928	0.65	0.9396
0.20	0.9886	0.70	0.9355
0.25	0.9838	0.75	0.9331
0.30	0.9787	0.80	0.9335
0.35	0.9732	0.85	0.9393
0.40	0.9675	0.90	0.9573
0.45	0.9617	0.95	1.0133
0.50	0.9558		

The joint probability density function given in Eq. (4.34) needs to be modified since the average time interval computed by

$$\bar{\tau}=\int_0^\infty \int_0^\infty \tau f(\xi,\tau)\,d\xi\,d\tau \tag{4.35}$$

is not unity; instead, $\bar{\tau}$ depends on ϵ-value as shown in Table 4.1. Hence, by reading $\bar{\tau}(\epsilon)$ for a given ϵ from the table and defining the dimensionless time interval as

$$\tau_* = \frac{\tau}{\bar{\tau}(\epsilon)} = \frac{T}{T_{\mathrm{m}}}\frac{1}{\bar{\tau}(\epsilon)} \tag{4.36}$$

the joint probability density of the positive maxima and its time interval in dimensionless form is given by

$$f(\xi,\tau_*) = \sqrt{\frac{2}{\pi}}\frac{\alpha^3\xi^2}{\epsilon(1-\epsilon^2)\tau_*^5\bar{\tau}^4}$$

$$\times \exp\left\{-\frac{\xi^2}{2\epsilon^2\tau_*^4\bar{\tau}^4}\left[(\tau_*^2\bar{\tau}^2-\alpha^2)^2+\alpha^4\beta^2\right]\right\} \tag{4.37}$$

By assuming twice the positive maxima as wave height, Arhan *et al.* (1976) give the dimensionless joint probability density function of wave height and period as

$$f(h,\tau_*) = \frac{1}{4\sqrt{2\pi}}\frac{\alpha^3 h^2}{\epsilon(1-\epsilon^2)\tau_*^5\bar{\tau}^4}$$

$$\times \exp\left\{-\frac{h^2}{8\epsilon^2\tau_*^4\bar{\tau}^4}\left[(\tau_*^2\bar{\tau}^2-\alpha^2)^2+\alpha^4\beta^2\right]\right\} \tag{4.38}$$

where $h=2\xi=2\zeta/\sqrt{m_0}$.

Figure 4.4 shows the dimensionless density function $f(h,\tau_*)$ for the bandwidth parameter $\epsilon=0.6$ and 0.8.

It is noted that the theoretical joint probability density functions of wave height and period presented in Sections 4.2 and 4.3 are developed under the condition of a steady-state, Gaussian and narrow-banded random process. It appears that these joint probability density functions, in particular the period distributions, are sensitive to the shape of the wave spectrum as pointed out by Sobey (1992). In general, these theoretical joint density functions may be compared with data obtained in ordinary wind-generated seas in deep water areas.

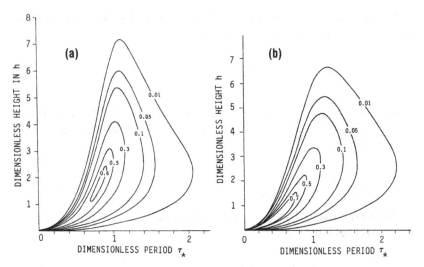

Fig. 4.4. Contours of dimensionless joint probability density function of wave height and period for (a) $\epsilon = 0.6$ and (b) $\epsilon = 0.8$ (Arhan *et al.* 1976).

4.4 PROBABILITY DISTRIBUTION OF WAVE PERIOD

As stated in Section 4.1, the probability distribution function of wave period, more precisely the zero-crossing period, may be derived as the marginal distribution of the joint probability distribution of wave height and period discussed in Section 4.2. The average value of the zero-crossing period which provides information necessary for estimating the long-term probability distribution of wave height (see Section 3.8) may be independently derived.

Let us consider the expected number of wave crossings of the level x with velocity $\dot{x}$ per unit time. This is analogous to the concept of evaluating the expected number of positive maxima per unit time discussed in Section 3.3. As was shown in Figure 3.4, the required time to cross the distance dx with velocity $\dot{x}$ is given by $dx/|\dot{x}|$. On the other hand, the probability that the displacement is in $(x, x+dx)$ with velocity $(\dot{x}, \dot{x}+d\dot{x})$ can be written in terms of the joint probability density function as $f(x, \dot{x})dx\, d\dot{x}$. This is interpreted as the amount of time the wave spends in $(x, x+dx)$. Hence, the number of crossings per unit time with velocity $\dot{x}$ becomes

$$f(x, \dot{x})dx\, d\dot{x}/(dx/|\dot{x}|) = |\dot{x}| f(x, \dot{x})d\dot{x} \qquad (4.39)$$

and thereby the expected number of crossings of the level x irrespective of the velocity magnitude can be obtained as

$$\overline{N}_x = \int_{-\infty}^{0} |\dot{x}| f(x, \dot{x})d\dot{x} \qquad (4.40)$$

We may consider only the up-crossings, and by letting the level $x=0$, the average number of zero-crossing with positive slope can be written as

$$\overline{N}_{0+}=\frac{1}{2}\int_{-\infty}^{0}|\dot{x}|f(0,\dot{x})\,\mathrm{d}\dot{x} \qquad (4.41)$$

For a Gaussian random process with the covariance matrix given in Eq. (3.29), Eq. (4.41) becomes

$$\overline{N}_{0+}=\frac{1}{2\pi\sqrt{m_0 m_2}}\int_{0}^{\infty}\dot{x}\exp\{-\dot{x}^2/2m_2\}\,\mathrm{d}\dot{x}=\frac{1}{2\pi}\sqrt{m_2/m_0} \qquad (4.42)$$

Thus, the inverse of $\overline{N}_{0+}$ which represents the average zero-crossing period, $\overline{T}_0$, is given by

$$\overline{T}_0=2\pi\sqrt{m_0/m_2} \qquad (4.43)$$

It is noted that the moments are evaluated for the spectral density function given in terms of the frequency ω. For the spectral density function given in terms of the frequency f, the average number of crossings as well as the average zero-crossing period are given by

$$\overline{N}_0=\sqrt{m_{2f}/m_{0f}}$$
$$\overline{T}_0=\sqrt{m_{0f}/m_{2f}} \qquad (4.44)$$

where m_{0f}, m_{2f} are the moments of the spectral density function $S(f)$.

Next, the probability density function of wave period will be derived from the joint probability density function of wave height and period. By integrating the joint probability density function given in Eq. (4.22) with respect to ξ, the density function of period in dimensionless form can be derived as follows (Longuet-Higgins 1983):

$$f(\eta)=\left(1+\frac{\nu^2}{4}\right)\frac{1}{2\nu\eta^2}\left\{1+\left(1-\frac{1}{\eta}\right)^2\frac{1}{\nu^2}\right\}^{-3/2} \qquad 0<\eta<\infty \qquad (4.45)$$

where

$\eta=T/\overline{T}$
$\overline{T}=$ mean period $=2\pi(m_0/m_1)$
$\nu=$ spectral width parameter given in Eq. (4.11).

The probability density function $f(\eta)$ is shown in Figure 4.5 as a function of the parameter ν. The mean value of the distribution computed from Eq. (4.45) is infinite, hence the variance of the distribution does not exist theoretically. As to the mean period, Longuet-Higgins suggests the use of the average zero-crossing period, $\overline{T}_0$. The dimensionless average period, therefore, may be written as

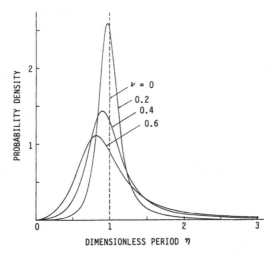

Fig. 4.5. Dimensionless probability density function of wave period as a function of parameter ν (Longuet-Higgins 1983).

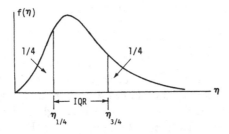

Fig. 4.6. Definition of interquartile range (IQR).

$$\eta_{\mathrm{ave}} = \overline{T}_0/\overline{T} = 1/\sqrt{1+\nu^2} \qquad (4.46)$$

The spectral width parameter ν defined in Eq. (4.11) can generally be evaluated from knowledge of the moments of the wave spectrum. Longuet-Higgins, however, shows that the parameter can be evaluated from observed data by using the *interquartile range* (IQR). The IQR is defined as the distance between two points $\eta_{1/4}$ and $\eta_{3/4}$ of the probability density function $f(\eta)$ for which the probability less than these values is 1/4 and 3/4, respectively (see Figure 4.6). In other words, the IQR covers 50 percent of the distribution. It can be evaluated from the probability density function given in Eq. (4.45) as

$$\mathrm{IQR} = \eta_{3/4} - \eta_{1/4} = \frac{1}{1-\nu\beta_3} - \frac{1}{1-\nu\beta_1} \qquad (4.47)$$

where

$$\beta_3 = \left(\frac{3}{2L}-1\right)\bigg/\sqrt{1-\left(\frac{3}{2L}-1\right)^2}$$

$$\beta_1 = \left(\frac{1}{2L}-1\right)\bigg/\sqrt{1-\left(\frac{1}{2L}-1\right)^2}$$

$$L = 1 + \nu^2/4$$

Figure 4.7 shows IQR and η_{ave} computed by Eqs. (4.47) and (4.46), respectively, as a function of ν. By evaluating $\eta_{1/4}$ and $\eta_{3/4}$ as well as IQR from the cumulative distribution function constructed from wave period data, the parameter ν can be obtained from Figure 4.7.

Information on statistical properties of successive wave periods is useful for estimating a possible critical resonant condition of a marine system with waves. At present, two joint probability density functions applicable for successive wave periods are available; one is the two-dimensional Weibull distribution developed by Kimura (1980), the other is the two-dimensional probability distribution developed by Myrhaug and Rue (1993) which is based on Bretschneider's assumption (1959) that the square of the wave period obeys the Rayleigh probability law. These probability distributions are empirically derived and hence more justification of their validity through comparison with measured data is highly desirable.

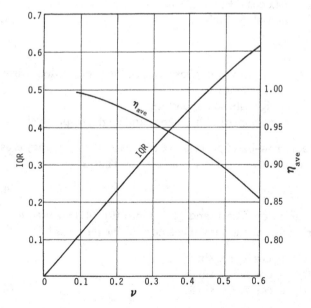

Fig. 4.7. Interquartile range (IQR) and dimensionless average period as a function of parameter ν (Longuet-Higgins 1983).

4.5 JOINT PROBABILITY DISTRIBUTION OF WAVE HEIGHT AND DIRECTION OF WAVE ENERGY TRAVEL

An interesting method for estimating the probabilistic relationship between individual wave height and the direction in which the energy is traveling is developed by Isobe (1988). The basic concept of his method is to consider the joint probability distribution of envelope processes of wave displacement and the two rectangular components of the horizontal water particle velocities; the ratio of the latter two components yields the direction of traveling energy. His method is outlined below.

Consider waves in a finite water depth h and let $\eta(t)$, $u(t)$ and $v(t)$ be wave displacement and the two rectangular components of the horizontal particle velocity, respectively. We may write

$$\eta(t) = \sum_{i=1}^{\infty} \sum_{j=1}^{\infty} a_{ij} \cos \phi_{ij}$$

$$u(t) = \sum_{i=1}^{\infty} \sum_{j=1}^{\infty} H_i \cos \theta_j \cdot a_{ij} \cos \phi_{ij}$$

$$v(t) = \sum_{i=1}^{\infty} \sum_{j=1}^{\infty} H_i \sin \theta_j \cdot a_{ij} \cos \phi_{ij} \tag{4.48}$$

$$\phi_{ij} = k_i(x \cos \theta_j + y \sin \theta_j) - \omega_i t - \epsilon_{ij} \tag{4.49}$$
$$H_i = \omega_i \cosh k_i z / \sinh k_i h \tag{4.50}$$

where

a_{ij}, ϵ_{ij} = amplitude and phase, respectively, of incident wave
h = water depth
z = height above the bottom
i,j = component of frequency and angle, respectively.

Assuming a narrow-band process, we may write the frequency ω as the sum of mean frequency $\overline{\omega}$ and its deviation $\Delta\omega$. That is

$$\omega_i = \overline{\omega} + \Delta\omega_i \tag{4.51}$$

By substituting Eqs. (4.50) and (4.51) into Eq. (4.48), $\eta(t)$, $u(t)$ and $v(t)$ can be presented as the sum of cosine and sine terms as follows:

$$\eta(t) = \eta_c \cos \overline{\omega} t + \eta_s \sin \overline{\omega} t$$
$$u(t) = u_c \cos \overline{\omega} t + u_s \sin \overline{\omega} t$$
$$v(t) = v_c \cos \overline{\omega} t + v_s \sin \overline{\omega} t \tag{4.52}$$

where η_c, η_s, u_c, u_s, v_c, and v_s are all slowly varying envelope functions which may be written as

$$\binom{\eta_c}{\eta_s}=\sum_{i=1}^{\infty}\sum_{j=1}^{\infty}a_{ij}\binom{\cos\phi'_{ij}}{\sin\phi'_{ij}}$$

$$\binom{u_c}{u_s}=\sum_{i=1}^{\infty}\sum_{j=1}^{\infty}H_i\cos\theta_j\cdot a_{ij}\binom{\cos\phi'_{ij}}{\sin\phi'_{ij}}$$

$$\binom{v_c}{v_s}=\sum_{i=1}^{\infty}\sum_{j=1}^{\infty}H_i\sin\theta_j\cdot a_{ij}\binom{\cos\phi'_{ij}}{\sin\phi'_{ij}} \tag{4.53}$$

where

$$\phi'_{ij}=k_i(x\cos\theta_j+y\sin\theta_j)-\Delta\omega_i t-\epsilon_{ij} \tag{4.54}$$

Six random variables involved in Eq. (4.53) are all normally distributed with zero mean. Here, the random variables with subscript c and those with subscript s are statistically independent and their covariance matrix is given by

$$\Sigma=\begin{pmatrix}\Sigma_0 & 0\\ 0 & \Sigma_0\end{pmatrix} \tag{4.55}$$

where

$$\Sigma_0=\begin{pmatrix} m_{00} & m_{10} & m_{01}\\ m_{10} & m_{20} & m_{11}\\ m_{01} & m_{11} & m_{02}\end{pmatrix} \tag{4.56}$$

and

$$\begin{aligned}
E[\eta_c^2] &=E[\eta_s^2] &=E[\eta^2] &=m_{00}\\
E[u_c^2] &=E[u_s^2] &=E[u^2] &=m_{20}\\
E[v_c^2] &=E[v_s^2] &=E[v^2] &=m_{02}\\
E[\eta_c u_c] &=E[\eta_s u_s] &=E[\eta u] &=m_{10}\\
E[\eta_c v_c] &=E[\eta_s v_s] &=E[\eta v] &=m_{01}\\
E[u_c v_c] &=E[u_s v_s] &=E[uv] &=m_{11}
\end{aligned} \tag{4.57}$$

Next, by taking the x-axis as the principal direction, we have $m_{11}=0$. Furthermore, the random variables are non-dimensionalized as follows:

$$N_c=\eta_c/\sqrt{m_{00}} \qquad U_c=u_c/\sqrt{m_{20}} \qquad V_c=v_c/\sqrt{m_{02}}$$

$$N_s=\eta_s/\sqrt{m_{00}} \qquad U_s=u_s/\sqrt{m_{20}} \qquad V_s=v_s/\sqrt{m_{02}} \tag{4.58}$$

The amplitude and phase of the displacement as well as the surface particle velocities of an individual wave can be determined from the cosine and sine components of the envelope function. For this, let N_p be the dimensionless wave amplitude and let δ be the phase of the surface displacement. Then, N_c and N_s become

$$N_c = N_p \cos \delta$$
$$N_s = N_p \sin \delta \tag{4.59}$$

Furthermore, by using δ as a reference phase, we may write

$$U_c = U_p \cos \delta - U_q \sin \delta$$
$$U_s = U_p \sin \delta + U_q \cos \delta$$
$$V_c = V_p \cos \delta - V_q \sin \delta \tag{4.60}$$
$$V_s = V_p \sin \delta + V_q \cos \delta$$

where the components with subscript p are in phase with the water surface displacement, while those with subscript q are 90 degrees out of phase. Thus, by using the functional relationship given in Eqs. (4.58) through (4.60), the joint normal probability density function of six random variables η_c, η_s, u_c, u_s, v_c, and v_s is transformed to the joint probability density function of N_p, δ, U_p, U_q, V_p and V_q. After some mathematical manipulation, the following joint probability density function of N_p, U_p and V_p can be derived.

$$f(N_p, U_p, V_p) = \frac{N_p}{2\pi\sqrt{\Delta}} \exp\left\{-\frac{1}{2\Delta}\left[N_p^2 + (1 - r_{01}^2)U_p^2 + (1 - r_{10}^2)V_p^2\right.\right.$$
$$\left.\left. -2r_{10}N_pU_p - 2r_{01}N_pV_p + 2r_{10}r_{01}U_pV_p\right]\right\} \tag{4.61}$$

where

$$\Delta = 1 - r_{10}^2 - r_{01}^2 + 2r_{10}r_{01}r_{11} - r_{11}^2$$
$$r_{10} = m_{10}/\sqrt{m_{00}m_{20}}$$
$$r_{01} = m_{01}/\sqrt{m_{00}m_{02}}$$
$$r_{11} = m_{11}/\sqrt{m_{20}m_{02}}$$

The direction of wave energy travel is given by $\alpha = \tan^{-1}(v_p/u_p)$. Hence, in order to obtain the joint probability density function of wave amplitude N_p and direction α, the random variables U_p and V_p are expressed in terms of the polar coordinate (W, α). That is,

$$U_p = W \cos \alpha$$
$$V_p = (W/\gamma) \sin \alpha \tag{4.62}$$

where γ is a factor associated with the non-dimensionalization shown in Eq. (4.58), and it can be obtained as

$$\gamma = \sqrt{m_{02}/m_{20}} \tag{4.63}$$

By applying the functional relationship given in Eq. (4.62), the joint probability density function $f(N_p, U_p, V_p)$ given in Eq. (4.61) is transformed to the joint probability density function $f(N_p, W, \alpha)$. Then, by integrating this density function with respect to W from 0 to ∞, the desired joint density function of wave amplitude (in dimensionless form N_p) and direction of energy traveling α can be derived as follows:

$$f(N_p,\alpha)=\frac{1}{2\pi\gamma}\left[\frac{\sqrt{\Delta}}{a}N_p\exp\{-N_p^2/(2\Delta)\}\right.$$

$$\left.+\frac{b}{a^{3/2}}N_p^2\exp\{-cN_p^2/(2a)\}\sqrt{\pi/2}\,(1+\mathrm{erf}\{bN_p/\sqrt{\Delta a}\})\right] \qquad (4.64)$$

where

$$a=(1-r_{01}^2)\cos^2\alpha+2r_{10}r_{01}\cos\alpha(\sin\alpha/\gamma)$$

$$+(1-r_{10}^2)(\sin\alpha/\gamma)^2$$

$$b=r_{10}\cos\alpha+r_{01}(\sin\alpha/\gamma)$$

$$c=\cos^2\alpha+(\sin\alpha/\gamma)^2$$

$$\mathrm{erf}(x)=\sqrt{2/\pi}\int_0^x\exp\{-t^2/2\}\,dt$$

As can be seen in Eq. (4.64) the joint probability density function of wave amplitude and direction of energy travel can be expressed as a function of r_{10}, r_{01} and γ. In other words, the distribution can be evaluated from knowledge of variance and covariance of wave displacement, velocities u and v given in Eq. (4.57).

In particular, in case the directional distribution of wave energy is narrow and symmetric, we have $r_{01}=0$ and $r_{10}=1$. Hence, Eq. (4.64) can be simplified as

$$f(N_p,\alpha)=\frac{1}{\sqrt{2\pi}\,\gamma}\frac{N_p^2}{\cos^2\alpha}\exp\left\{-\frac{1}{2}\left(1+\frac{\tan^2\alpha}{\gamma^2}\right)N_p^2\right\} \qquad (4.65)$$

where $\cos\alpha>0$.

An example of the joint probability density function of wave amplitude and direction, $f(N_p,\alpha)$ for $r_{10}=0.9$, $r_{01}=0.2$ and $\gamma=0.4$ is shown in Figure 4.8.

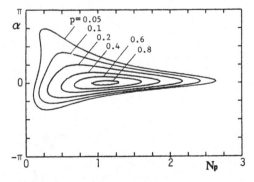

Fig. 4.8. Example of joint probability density function of dimensionless wave amplitude and direction of energy travel ($r_{10}=0.9$, $r_{01}=0.2$ and $\gamma=0.4$) (Isobe 1988).

Kwon and Deguchi (1994) further develop the joint probability distribution of wave height, period and direction of energy travel following the same procedure as developed by Isobe. In this case, the density function is initiated for eight random variables including $\dot{\eta}_c$ and $\dot{\eta}_s$ in addition to the six random variables shown in Eq. (4.53).

5 SEA SEVERITY

5.1 STATISTICAL PRESENTATION OF SEA SEVERITY

5.1.1 Probability distribution of significant wave height

Statistical presentation of sea severity provides information vital for the design and operation of marine systems. For the design of marine systems, information is necessary not only on the severest sea condition expected to occur during the system's lifetime (50 years for example), but also on the frequency of occurrence of all sea conditions, the latter being especially necessary for evaluating fatigue loadings.

The most commonly available information on sea severity is the statistical tabulation of significant wave height constructed from data accumulated over several years. Of course, the greater the number of accumulations, the more reliable the data. As to the time interval between data sampling, it is highly desirable that data be obtained at least at 3-hour intervals so that a relatively fast change in sea condition will not be missed. During a storm, sampling at no more than one hour intervals is strongly recommended (see Section 5.2.1).

Table 5.1 is a tabulation of 5412 significant wave heights obtained over a 3-year period in the North Sea (Bouws 1978). The data indicate that the measurements were made, on average, five times per day throughout the 3 years. In order to estimate the extreme significant wave height expected to occur in 50 years (for example) at this location, it is first required to find a probability distribution which accurately represents the data.

In general, the sea severity as evaluated from wave measurements depends to a great extent on the geographical location where the data are obtained, since the crucial factors for sea severity are frequency of occurrence of storms, water depth, wind direction, etc. In addition, sea severity depends on the stage of growth and decay of a storm, even though wind speed is the same. Thus, there is no scientific basis for selecting a specific probability distribution function to represent the statistical properties of sea state (significant wave height). Because of this, various probability distribution functions have been proposed which appear to best fit

Table 5.1. *Significant wave height data obtained from measurements in the North Sea (Bouws 1978).*

Significant wave height (m)	Number of observations
0–0.5	1280
0.5–1.0	1549
1.0–1.5	1088
1.5–2.0	628
2.0–2.5	402
2.5–3.0	192
3.0–3.5	115
3.5–4.0	63
4.0–4.5	38
4.5–5.0	18
5.0–5.5	21
5.5–6.0	7
6.0–6.5	8
6.5–7.0	2
7.0–7.5	1
	Total 5412 in 3 years

particular sets of observed data. These include: (a) log-normal distribution (Ochi 1978b); (b) modified log-normal distribution (Fang and Hogben 1982); (c) three-parameter Weibull distribution (Burrows and Salih 1986, Mathisen and Bitner-Gregersen 1990); (d) combined exponential and power of significant wave height (Ochi and Whalen 1980); and (e) modified exponential distribution (Thompson and Harris 1972).

It is a general trend that the greater part of significant wave height data is well represented by the log-normal probability distribution; however, the data diverge from the log-normal distribution for large significant wave heights, which are critical for estimating extreme values. As an example, Figure 5.1 shows the cumulative probability distribution of significant wave height data given in Table 5.1 plotted on log-normal probability paper. Included also in the figure is a straight line which represents the cumulative probability distribution of the following log-normal probability distribution:

$$f(x) = \frac{1}{\sqrt{2\pi}\,\sigma\,x} \exp\left\{ -\frac{1}{2}\left(\frac{\ln x - \mu}{\sigma}\right)^2 \right\} \qquad 0 \leqslant x < \infty \qquad (5.1)$$

Another general trend observed on the statistical distribution of significant wave height is that the cumulative distribution function of large significant wave heights can be well represented by the Weibull probability distribution, but approximately 30 percent of the lower portion of data fail to follow. As an example, Figure 5.2 shows a comparison between the cumulative distribution function of the same data shown in Figure 5.1 plotted on Weibull probability paper. The straight line in the figure is the Weibull distribution given by

$$f(x) = c\lambda^c x^{c-1} \exp\{-(\lambda x)^c\} \qquad a \leqslant x < \infty \qquad (5.2)$$

As seen in Figure 5.2, the lower portion of the data deviates from the Weibull distribution to a great extent. In order to improve this situation, the following three-parameter Weibull distribution is often considered:

$$f(x) = c\lambda^c (x-a)^{c-1} \exp\{-(\lambda(x-a))^c\} \qquad 9 \leqslant x < \infty \qquad (5.3)$$

Note that the three-parameter Weibull distribution carries a minimum non-zero value 'a' as one of its parameters, and it is difficult to explain the physical meaning of this minimum significant wave height in the distribution. The sample space of significant wave height has to be

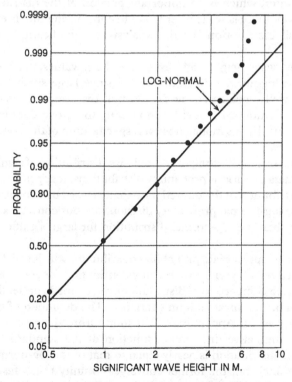

Fig. 5.1. Comparison between cumulative distribution function of significant wave height data given in Table 5.1 and the log-normal probability distribution.

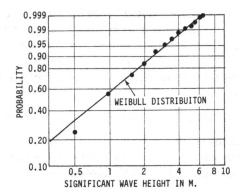

Fig. 5.2. Comparison between cumulative distribution function of significant
wave height data given in Table 5.1 and that of Weibull probability
distribution.

chosen between zero and ∞, where a significant wave height of zero
represents calm water, which is an important portion of the distribution. For this reason, there is some reservation in considering the three-
parameter Weibull distribution for the analysis of significant wave
height data.

Since significant wave height data, except for large values, are well
represented by the log-normal distribution, and since large significant
wave heights are well represented by the Weibull distribution, Haver
(1985) proposes an empirical model by combining these two distributions at a certain value. In this model, however, specification of the transition point is arbitrary.

Judging from the results of fitting the cumulative distribution of significant wave height data by various probability distributions, it appears that
the cumulative distribution of the desired probability law is, by and large,
close to that of the log-normal probability distribution, but converges to
unity much faster than the log-normal distribution for large significant
wave heights.

In order to find an appropriate probability distribution which satisfies
the condition stated above, various probability distribution functions are
standardized so that a comparison of distributions can be made under the
uniform condition of zero mean and unit variance. The definition of the
standardization is given in Appendix A. It is found from the results of
analysis that the cumulative distribution function of the standardized
generalized gamma distribution is nearly equal to that of the log-normal
distribution up to 0.90, but the former converges to unity much faster
than the latter above 0.90. This feature of the generalized gamma distribution is considered to make its use advantageous in the statistical analysis
of significant wave height data (Ochi 1992).

The probability density function, $f(x)$, and cumulative distribution function, $F(x)$, of the generalized gamma distribution (non-standardized form) are as follows:

$$f(x) = \frac{c}{\Gamma(m)} \lambda^{cm} x^{cm-1} \exp\{-(\lambda x)^c\} \qquad 0<x<\infty \qquad (5.4)$$

$$F(x) = \Gamma\{m, (\lambda x)^c\}/\Gamma(m) \qquad (5.5)$$

where $\Gamma(m)$ is a gamma function, and the numerator of $F(x)$ is an incomplete gamma function.

A comparison between the cumulative distribution function of the same data shown in Figures 5.1 and 5.2 and the generalized gamma distribution is given in Figure 5.3. The parameter values of the generalized gamma distribution in this example are $m=1.60$, $c=0.98$ and $\lambda=1.37$. From a comparison of Figures 5.1 through 5.3, it can be seen that the generalized gamma distribution accurately represents the data in the domains where the log-normal and the Weibull distributions fail to satisfactorily agree with the measured data.

Other examples of comparisons between the generalized gamma distribution and measured data are shown in Figures 5.4 and 5.5. The data

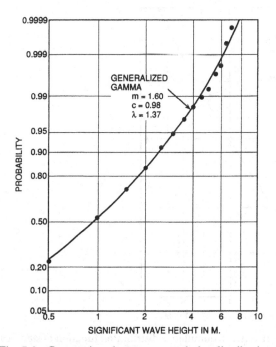

Fig. 5.3. Comparison between cumulative distribution function of significant wave height data given in Table 5.1 and the generalized gamma probability distribution (Ochi 1992).

presented in Figure 5.4 were obtained in the North Sea off the Norwegian coast (Mathisen and Bitner-Gregersen 1990), while those presented in Figure 5.5 were obtained in the North Pacific off Canada (Teng *et al.* 1993).

The values of the three parameters, m, c and λ involved in the generalized gamma distribution may be simply estimated by equating the sample moments to theoretical moments, since the sample size of significant wave height data is usually very large, on the order of several thousand or greater. The theoretical jth moment of the generalized gamma distribution is given by

$$E[x^j] = \frac{1}{\lambda^j} \frac{\Gamma\left(m + \dfrac{j}{c}\right)}{\Gamma(m)} \tag{5.6}$$

From a set of three equations for $j=2$, 3 and 4 in Eq. (5.6), we can derive the following two equations by eliminating the parameter λ:

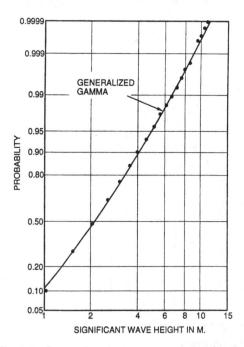

Fig. 5.4. Comparison between cumulative distribution function of significant wave height data obtained in the North Sea off the Norwegian coast and the generalized gamma probability distribution (data from Mathisen and Bitner-Gregersen 1990).

$$\frac{\{\Gamma(m)\}^{1/2}\,\Gamma\!\left(m+\dfrac{3}{c}\right)}{\left\{\Gamma\!\left(m+\dfrac{2}{c}\right)\right\}^{3/2}}=\frac{E[x^3]}{\{E[x^2]\}^{3/2}}$$

$$\frac{\Gamma(m)\,\Gamma\!\left(m+\dfrac{4}{c}\right)}{\left\{\Gamma\!\left(m+\dfrac{2}{c}\right)\right\}^{2}}=\frac{E[x^4]}{\{E[x^2]\}^{2}} \tag{5.7}$$

The parameters m and c are determined from the above equations, and λ can be obtained from Eq. (5.6) by letting $j=2$ or 3.

As stated earlier, the prime purpose of representing the significant wave height data by an appropriate probability distribution is for the estimation of (a) the extreme sea state expected to occur in a long time period, say 50 or 100 years, and (b) the frequency of occurrence of all sea states, and from this the long-term statistics of individual waves needed

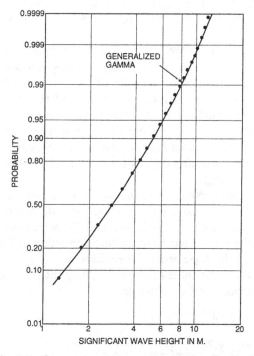

Fig. 5.5. Comparison between cumulative distribution function of significant wave height data obtained in the North Pacific off Canada and the generalized gamma probability distribution (data from Teng *et al.* 1993).

for evaluating fatigue loads on marine systems. The estimation of extreme sea state will be discussed in Section 6.1, while the estimation of long-term wave statistics was presented in Section 3.8.

5.1.2 Joint probability distribution of significant wave height and period

Information on long-term sea severity is broadened to a great extent by including, in addition to wave height, wave period, usually the average value of the zero-crossing wave periods in a specified sea, denoted by $\overline{T}_0$. The combined significant wave height and average zero-crossing period information is presented in tabular form, often called the *contingency table* or the *scatter diagram*.

Table 5.2 shows an example of the contingency table of significant wave height and zero-crossing period obtained from NOAA Buoy 46001 located in the North Pacific. It comprises a tabulation of 13 years of data obtained at one hour intervals. The statistical information comprising the contingency table may be obtained through finding the appropriate joint probability distribution which best represents the data. For example, the joint log-normal probability distribution has been proposed (Ochi 1978b), but this joint distribution does not represent the severe sea states data very well.

Since it is somewhat difficult to derive the joint probability distribution of significant wave height, H_s, and average zero-crossing period, $\overline{T}_0$, directly from data, a method commonly considered for the derivation of $f(H_s, \overline{T}_0)$ is to find the marginal probability density function $f(H_s)$ and conditional probability density function $f(\overline{T}_0|H_s)$. The parameters involved in the latter distribution function are evaluated from analysis of each conditional distribution of $\overline{T}_0$ for a specified H_s, and presented them as a function of H_s. Then, the joint probability density function can be derived as the product of $f(H_s)$ and $f(T_0|H_s)$.

Burrows and Salih (1986) present the results of two combinations for the marginal Weibull distribution; one a conditional Weibull, the other a conditional log-normal distribution. Mathisen and Bitner-Gregersen (1990) demonstrate that the combination of the three-parameter Weibull distribution for $f(H_s)$ and the log-normal distribution for $f(T_0|H_s)$ best represents data obtained off the coast of Norway. The joint probability density function of this model can be written as

$$f(H_s, \overline{T}_0) = f(H_s)f(\overline{T}_0|H_s)$$

$$= \frac{\beta(H_s - \gamma)^{\beta-1}}{\alpha^\beta} \exp\left\{-\left(\frac{H_s - \gamma}{\alpha}\right)^\beta\right\}$$

$$\times \frac{1}{\sqrt{2\pi}\,\sigma(H_s)\overline{T}_0} \exp\left\{-\frac{[\ln\overline{T}_0 - \mu(H_s)]^2}{2\{\sigma(H_s)\}^2}\right\} \qquad (5.8)$$

Table 5.2 Contingency table of significant wave height and average zero-crossing period obtained from NOAA Buoy 64001 data (Teng et al. 1993).

| T_{avg}(s) → | 0 | 2.0 | 3.0 | 4.0 | 5.0 | 6.0 | 7.0 | 8.0 | 9.0 | 10.0 | 11.0 | 12.0 | 13.0 | 14.0 | 15.0 | 16.0 | Sum | % |
H_s(m) ↓	–2.0	–3.0	–4.0	–5.0	–6.0	–7.0	–8.0	–9.0	–10.0	–11.0	–12.0	–13.0	–14.0	–15.0	–16.0	>		
0.0–0.5	0	12	76	141	74	24	0	0	0	0	0	0	0	0	0	0	327	0.31
0.5–1.0	0	5	345	1990	2157	966	238	41	11	0	0	0	0	0	0	0	5753	5.54
1.0–1.5	0	0	131	3834	5942	2840	594	108	31	1	0	0	0	0	0	0	13481	12.98
1.5–2.0	0	0	6	2033	7979	5119	1539	272	44	0	0	0	0	0	0	0	16992	16.36
2.0–2.5	0	0	1	485	6411	6301	2687	635	109	10	0	0	0	0	0	0	16639	16.02
2.5–3.0	0	0	0	22	2719	6513	3383	950	138	16	0	0	0	0	0	0	13741	13.23
3.0–3.5	0	0	0	0	751	5132	3659	1046	219	19	1	0	0	0	0	0	10827	10.42
3.5–4.0	0	0	0	0	85	2827	3469	1323	268	41	5	0	0	0	0	0	8018	7.72
4.0–4.5	0	0	0	0	2	1073	3083	1392	325	48	5	3	0	0	0	0	5931	5.71
4.5–5.0	0	0	0	0	0	252	2149	1300	386	67	11	4	1	0	0	0	4170	4.01
5.0–5.5	0	0	0	0	0	29	1256	1138	432	79	7	0	0	0	0	0	2941	2.83
5.5–6.0	0	0	0	0	0	0	408	946	346	68	10	0	0	0	0	0	1778	1.71
6.0–6.5	0	0	0	0	0	0	87	728	293	75	14	0	0	0	0	0	1197	1.15
6.5–7.0	0	0	0	0	0	0	12	454	275	76	15	0	0	0	0	0	832	0.80
7.0–7.5	0	0	0	0	0	0	0	199	227	54	15	1	0	0	0	0	496	0.48
7.5–8.0	0	0	0	0	0	0	0	68	184	52	12	3	0	0	0	0	319	0.31
8.0–8.5	0	0	0	0	0	0	0	20	122	47	11	0	0	0	0	0	200	0.19
8.5–9.0	0	0	0	0	0	0	0	1	73	32	12	3	0	0	0	0	121	0.12
9.0–9.5	0	0	0	0	0	0	0	0	22	29	9	3	1	0	0	0	64	0.06
9.5–10.0	0	0	0	0	0	0	0	0	1	18	4	4	0	0	0	0	27	0.03
10.0–10.5	0	0	0	0	0	0	0	0	1	6	4	6	0	0	0	0	17	0.02
10.5–11.0	0	0	0	0	0	0	0	0	0	3	2	1	0	0	0	0	6	0.01
11.0–11.5	0	0	0	0	0	0	0	0	0	2	1	0	2	0	0	0	5	0.00
11.5–12.0	0	0	0	0	0	0	0	0	0	0	1	1	0	0	0	0	2	0.00
12.0–12.5	0	0	0	0	0	0	0	0	0	0	1	0	0	0	0	0	1	0.00
>12.5	0	0	0	0	0	0	0	0	0	0	1	0	0	0	0	0	1	0.00
Sum	0	17	559	8505	26120	31076	22564	10621	3507	743	141	29	4	0	0	0	103886	
%	0.00	0.02	0.54	8.19	25.14	29.91	21.72	10.22	3.38	0.72	0.14	0.03	0.00	0.00	0.00	0.00		

Here, the parameters of the conditional log-normal distribution are given as a function of significant wave height H_s as follows:

$$\mu(H_s) = a_1 + a_2 H_s^{a_3}$$
$$\sigma(H_s) = b_1 + b_2 \exp\{b_3 H_s\} \tag{5.9}$$

where a_1, a_2, a_3, b_1, b_2, and b_3 are constants determined from data.

In general, the approach for deriving the joint probability density function of significant wave height and the average zero-crossing period through a combination of the marginal distribution $f(H_s)$ and the conditional distribution $f(\overline{T}_0|H_s)$ appears quite promising. However, as stated in the previous section, use of the Weibull probability distribution with three parameters distorts the sample space of significant wave height. On the other hand, the generalized gamma distribution accurately represents the marginal distribution, $f(H_s)$, over its entire domain. Taking this into consideration, the joint probability density function $f(H_s, \overline{T}_s)$ is developed by combining the generalized gamma distribution for the marginal distribution $f(H_s)$, and the log-normal distribution for the conditional distribution $f(\overline{T}_0|H_s)$ (Ochi et al. 1996). In this case, the joint probability distribution may be written as

$$f(H_s, \overline{T}_0) = \frac{c}{\Gamma(m)} \lambda^{cm} H_s^{cm-1} \exp\{-(\lambda H_s)^c\}$$
$$\times \frac{1}{\sqrt{2\pi}\, \sigma(H_s)\overline{T}_0} \exp\left\{-\frac{[\ln \overline{T}_0 - \mu(H_s)]^2}{2\{\sigma(H_s)\}^2}\right\} \tag{5.10}$$

The parameters $\mu(H_s)$ and $\sigma(H_s)$ are those given in Eq. (5.9), which are developed by Mathisen and Bitner-Gregersen (1990).

As an example, the joint probability density function given in Eq. (5.10) is applied to the data given in Table 5.2, and a comparison between the data and contour curves of the density function is shown in Figure 5.6. As can be seen, the joint probability density function accurately represents the contingency table. It is of interest to observe in the figure that the contour curve of 1×10^{-6} covers almost all of the data points, and that this indicates the domain of H_s and $\overline{T}_0$ to be considered for the design of marine systems operating in the site.

Another interesting method to construct the joint probability distribution of two random variables from a knowledge of the individual marginal distribution function has been developed by several statisticians; Morgenstern (1956), Plackett (1965), Kimeldorf and Sampson (1975), among others. Athanassoulis et al. (1994) first introduced this approach for analysis of sea state data as will be shown later. Here, Plackett's method is outlined below.

Let $F(x)$ and $F(y)$ be the cumulative distribution functions of two random variables X and Y, respectively, and let $F(x,y)$ be the joint cumulative distribution function. Here, $F(x,y)$ must satisfy the Frécht's inequality given by

$$\max\{F(x)+F(y)-1,0\}\leqslant F(x,y)\leqslant\min\{F(x),F(y)\} \qquad (5.11)$$

Next, evaluate the probabilities of data in four quadrants separated by drawing two lines parallel to the axes at an arbitrary point (x,y) in the sample space. By letting these probabilities be

$$\begin{aligned}
p_1 &= \Pr\{X<x, Y<y\} = F(x,y)\\
p_2 &= \Pr\{X<x, Y>y\} = F(x)-F(x,y)\\
p_3 &= \Pr\{X>x, Y<y\} = F(y)-F(x,y)\\
p_4 &= \Pr\{X>x, Y>y\} = 1-F(x)-F(y)+F(x,y)
\end{aligned} \qquad (5.12)$$

and defining the *coefficient of contingency*, denoted by ψ as

$$\psi = \frac{p_1 p_4}{p_2 p_3} \qquad (5.13)$$

Plackett (1965) constructs a bivariate distribution $F(x,y)$ for given marginal cumulative distribution functions $F(x)$ and $F(y)$ as a solution of the following equation derived from Eqs. (5.12) and (5.13):

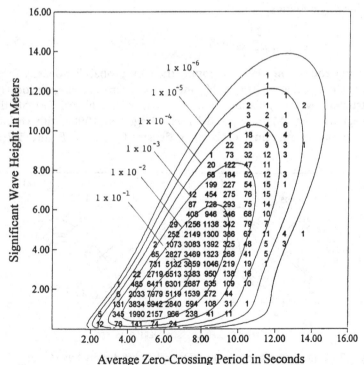

Fig. 5.6. Comparison between joint probability density function and data given in Table 5.2.

$$(\psi-1)\{F(x,y)\}^2 - [1+\{F(x)+F(y)\}(\psi-1)F(x,y)] \\ +\psi F(x)F(y)=0 \qquad (5.14)$$

As a solution of Eq. (5.14), Mardia (1967) shows that the joint cumulative distribution function is given by

$$F(x,y)=[S-\{S^2-4\psi(\psi-1)F(x)F(y)\}^{1/2}]/\{2(\psi-1)\} \qquad (5.15)$$

where

$$S=1+\{F(x)+F(y)\}(\psi-1)$$

The joint probability density function $f(x,y)$ then becomes

$$f(x,y)=\psi f(x)f(y)\frac{(\psi-1)\{F(x)+F(y)-2F(x)F(y)\}+1}{\{S^2-4\psi(\psi-1)F(x)F(y)\}^{3/2}} \qquad (5.16)$$

where $f(x)$ and $f(y)$ are the probability density functions of x and y, respectively.

The coefficient of the contingency, ψ, may be evaluated numerically through the maximum likelihood method but it is extremely complicated. One way to evaluate it is to use the sample correlation coefficient, ρ, by the following equation

$$\rho=\frac{\psi+1}{\psi-1}-\frac{2\psi}{(\psi-1)^2}\ln\psi \qquad (5.17)$$

Athanassoulis *et al.* (1994) compare the joint probability density function of significant wave height, H_s, and average zero-crossing period, $\overline{T}_0$, constructed by interpolating data obtained in the North Sea off the Norwegian coast (Mathisen and Bitner-Gregersen 1990) with various combinations of marginal probability density functions. Figure 5.7(a)

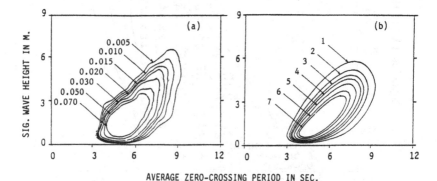

Fig. 5.7. Comparison between the joint probability density function of significant wave height, H_s, and average zero-crossing period, $\overline{T}_0$ (a) obtained from data and (b) computed by applying Plackett's method (Athanassoulis *et al.* 1994).

shows the joint density function obtained from the data, and Figure 5.7(b) shows the computed joint density function which best represents the data. In this example, the gamma probability function is chosen for the marginal density function of H_s and the log-normal probability function for $\bar{T}_0$. The seven contour curves in Figures 5.7(a) and (b) are equivalent in sequence; namely, curve 1 in (b) is equivalent to the outermost probability density curve 0.005 in (a).

5.1.3 Time series analysis of sea state data

Another approach to acquire long-term sea state information is to analyze a long-term time series of significant wave height data. For this, data measured continuously or for short time intervals over a long period of time are required. For a series of records with small intervals between recordings, the series may safely be assumed to be that of a continuous random process. However, it is a non-stationary random process. Hence, the general characteristics of the process such as (a) effect of sampling time interval, (b) existence of piece-wise stationarity and (c) statistical independence of successive sea states, etc., must be carefully examined.

Regarding the effect of sampling time interval on the statistical characteristics of significant wave height, it is necessary to consider the time duration of sea state (Labeyrie 1990). Athanassoulis and Stefanakos (1995) review available data on time duration of sea states and conclude that a period of 3–6 hours between samplings can be considered appropriate for the time series analysis of significant wave height data.

From analysis of data obtained for 6 years by a wave-rider buoy located off the coast of Oregon, Medina et al. (1991) show seasonal characteristic effects on the long-term analysis of significant wave height, H_s, and significant wave period, T_s. They claim that both H_s and T_s follow the log-normal probability law with annually periodic parameters. As an example, Figure 5.8 shows the variation of the parameter μ in the log-normal probability distribution over a one year period. A1 and A2 in the figure are the μ-values for significant wave height, H_s, and significant wave period, T_s, respectively (H_s is given in cm and T_s in s). The figure clearly shows the seasonal trend of the parametric values. In developing a numerical model for synthesizing the time series of significant wave height data, Medina et al. standardized the time series seasonally.

Athanassoulis and Stefanakos (1995) analyze a twenty-year time series of hindcast significant wave height data at five locations in the North Atlantic. In their analysis, assuming a yearly statistical periodicity, the annual mean value is subtracted from each set of annual significant wave height data. Then the average value and standard deviation of seasonal fluctuation are evaluated as a function of time. These are given by

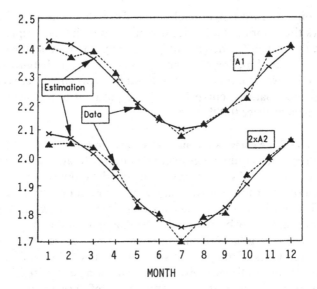

Fig. 5.8. Estimation of parameter μ of log-normal distribution. (A1 is μ-value for significant wave height in cm, A2 is that for significant period in s.) (Medina *et al.* 1991.)

$$m(\tau) = \frac{1}{N} \sum_{j=1}^{N} Y_j(\tau)$$

$$s(\tau) = \frac{1}{N} \sum_{j=1}^{N} \{Y_j(\tau) - m(\tau)\} \tag{5.18}$$

where

$Y_j(\tau)$ = seasonal significant wave height fluctuation in the *j*th year = $X_j(\tau) - \overline{X}_j$

$X_j(\tau)$ = annual time series significant wave height data in the *j*th year

$\overline{X}_j$ = annual mean value of significant wave height in the *j*th year.

An example of the seasonal fluctuation averaged over 19 years of data is shown in Figure 5.9. Included also in the figure is the first-order Fourier representation of the data. Athanassoulis and Stefanakos examine the effect of time length of data acquisition on the seasonal fluctuation of significant wave height, and find that 10 years of data appear to be sufficient for evaluating the seasonal variability.

By letting $\mu(\tau)$ and $\sigma(\tau)$ be the smoothed curves of the low-order Fourier series of $m(\tau)$ and $s(\tau)$, respectively, the seasonal significant wave height fluctuation in the *j*th year can be standardized. That is,

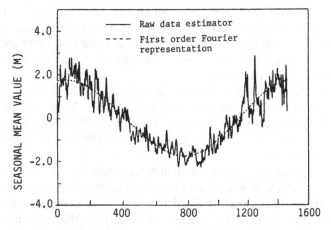

Fig. 5.9. Example of seasonal variation of significant wave height (Athanassoulis and Stefanakos 1995).

$$W_j(\tau) = \frac{Y_j(\tau) - \mu(\tau)}{\sigma(\tau)} \qquad (5.19)$$

From a comparison of the residual time series $W_j(\tau)$ for segments of different seasons, it is found that $W_j(\tau)$ can be considered as a stationary random process, $W(\tau)$. Hence, it can be concluded that the seasonal significant wave height fluctuation is a periodically correlated stochastic process, written as

$$Y(\tau) = \mu(\tau) + \sigma(\tau)W(\tau) \qquad (5.20)$$

and based on this relationship, a simulation model can be developed.

The methodology of how simulation models are developed for a time series of significant wave height is a subject beyond the scope of this text. It may suffice to state that besides the two references stated above, interesting and different techniques are available, Scheffner and Borgman (1992), for example.

5.2 HURRICANE-ASSOCIATED SEAS

5.2.1 Introduction

Wind-generated random seas defined in this text are those generated by ordinary winds blowing continuously for several hours with almost constant speed. The situation is different for random seas associated with tropical cyclones/hurricanes. Although a hurricane is defined as an intense tropical cyclone with mean wind speed greater than 75 miles per hour (65.2 knots, 33.5 m/s), this definition is not strictly observed in the discussion in this section. A significant feature of random seas associated with hurricanes is that the input source of energy generating waves is advancing at a speed of 5 to 12 knots. Therefore, the rate of change of wind speed at a location in the path of a hurricane is much greater and thus the time duration of a given wind speed is much shorter than that observed during an ordinary storm.

Many studies have been carried out on hurricane-generated seas, primarily through hindcasting and forecasting techniques. These include Cardone *et al.* (1976), Bretschneider and Tamaye (1976), Ross and Cardone (1978), Ross (1979), Young and Sobey (1981), and Young (1988), among others.

In hindcasting and forecasting mathematical models for sea severity generated by hurricanes, it is common to consider several parameters of a hurricane such as maximum sustained wind speed, central pressure, storm forward velocity, radius of maximum wind, etc. Although these hindcasting and forecasting models provide valuable information for individual hurricanes, it is difficult to draw general conclusions from them regarding the severity of the sea and the shape of wave spectra during hurricanes.

For estimating the severest sea at a specific location in the ocean under hurricane conditions, Donoso *et al.* (1987) consider the hurricane track relative to the site, in addition to the hurricane parameters. By using the historical record of hurricane parameters for storms that have passed within 100 nautical miles (185 km) of a selected location in the Gulf of Mexico, they estimate the extreme significant wave height expected to occur at the site. Although their approach for estimating the sea severity associated with hurricanes is interesting and useful, the magnitude of estimated extreme sea states depends entirely on the mathematical model employed.

On the other hand, the results of analysis of measured wind speed and sea severity during hurricanes indicate that there exists a relatively simple relationship between the sea severity and mean wind speed as far as deep water is concerned. In fact, there are two different categories: one is the growing state of hurricane-generated seas in which the wind speed is increasing at an extremely high rate and hence the sea severity is difficult to follow, the other is the sea condition resulting from winds of relatively mild severity blowing continuously for one week or longer and then followed by a tropical cyclone. The sea severity of these two cases are presented in the following section.

5.2.2 Sea severity measured during hurricanes

It is of interest to observe the sea condition at a specific location in the ocean as a hurricane approaches and examine the relationship between wind speed and significant wave height. As an example, Figure 5.10 shows the relationship between the measured mean wind speed at 10 m height and significant wave height obtained by NOAA Buoy EB10 in hurricane

ELOISE. The buoy was located at 88.0°W, 27.5°N. in the Gulf of Mexico, approximately 330 km off the Florida coast. The hurricane traveled about 660 km in open sea in the Gulf before it reached the buoy, and its center passed within 16 km of the buoy. The open circles in the figure indicate the wind and wave relationship during the growing stage of the hurricane.

It can be seen in Figure 5.10 that the sea severity increases almost linearly with increase in wind speed during the growing stage, and becomes severest (significant wave height 8.8 m) when the wind speed becomes maximum (35.2 m/s) as the hurricane leading edge approaches, and is then followed by a transition stage. That is, the wind speed significantly reduces in magnitude (from 35.2 m/s to 8 m/s) during the two hours following the hurricane eye passing near the buoy. The sea state follows the change in energy source by reducing in severity from a significant wave height of 8.8 m to 5.0 m. Then, during the following two hours, the hurricane trailing edge passes and the wind speed and sea state come back to nearly the same levels as they were in the growing stage. After that, the sea severity reduces almost linearly with reduction in wind speed. Thus, it can be seen that sea state severity is highly dependent on wind speed during the hurricane.

The wind-speed–sea-severity relationship during the growing stage of hurricanes is obtained for an additional seven hurricanes and the results are summarized in Figure 5.11 (Ochi 1993). Wave data for seven hurricanes were measured by NOAA buoys in deep water (NOAA Buoy Office

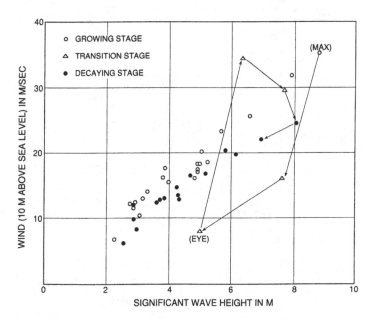

Fig. 5.10. Relationship between mean wind speed and significant wave height observed during hurricane ELOISE (Ochi 1993).

1975, 1978, 1981, and 1986), and data for hurricane CAMILLE were obtained by a resistance-type wave gage attached to a tower located in a water depth of 104 m (Patterson 1974). Some wind speeds measured at 5 m above the buoy are converted to 10 m height. Some wind speeds included in the figure were below hurricane level (33.5 m/s) when they passed over the buoys, but the wind severity reached hurricane level later.

As can be seen in Figure 5.11, sea severity increases almost linearly with increase in wind speed during the growing stage of the hurricanes. By establishing an upper-bound by drawing a straight line which includes the majority of the points in the figure, the significant wave height, H_s, can be simply obtained as a function of the mean wind speed at 10 m, $\overline{U}_{10}$, as

$$H_s = 0.24\overline{U}_{10} \qquad (5.21)$$

where H_s is in meters and $\overline{U}_{10}$ in meters per second.

Included in the figure is the functional relationship between wind speed and significant wave height for fully developed seas obtained from the Pierson–Moskowitz spectrum (1964). The original Pierson–Moskowitz spectral formulation for fully developed seas is given as a function of the mean wind speed referred to 19.6 m height above sea level, however, the

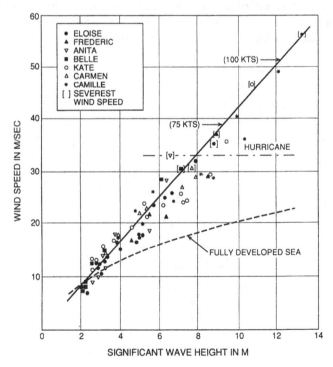

Fig. 5.11. Mean wind speed and significant wave height during the growing stage of hurricanes (Ochi 1993).

wind speed is converted to 10 m height in the figure. As expected, the figure shows that the sea severity during the growing stage of a hurricane for a given wind speed is much less than that for fully developed seas. This is because the time duration of a given wind speed is extremely short during hurricanes in comparison with that required for fully developed seas.

Sometimes we observe a sea condition resulting from winds of mild severity blowing continuously for one week or longer and then followed by a storm, usually a tropical cyclone. In this case, the sea becomes severe. For example, before tropical cyclone GLORIA passed near NOAA Buoy 41002 (32.3°N, 75.3°W) in 1985, the sea condition (significant wave height) in that area was 2.5–4.0 m for 10 days with consistently blowing winds of 7–11 m/s. That is, the sea had the potential for being easily augmented in severity when GLORIA came to the area. Another example of this situation can be observed in the sea state during a storm in the North Atlantic (Snider and Chakrabarti 1973). In this case, winds of 6–14 m/s were blowing for some time before a storm came to Station J (53°N, 18°W). When the storm approached the area, the significant wave height increased from 5.5 m to 16.8 m in 21 hours at an almost constant rate.

Figure 5.12 shows the relationship between the mean wind speed and

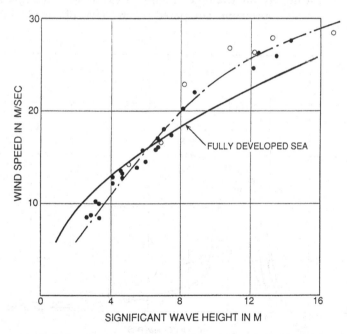

Fig. 5.12. Relationship between mean wind speed and significant wave height (solid circles are Hurricane GLORIA, open circles are North Atlantic storm). (North Atlantic data from Sneider and Chakrabarti 1973.)

significant wave height measured by the NOAA buoy during tropical cyclone GLORIA (solid circles) and that measured by a Tucker meter installed on the Ocean Weather Ship during a storm in the North Atlantic (open circles). Included in the figure is the functional relationship applicable for fully developed seas. As can be seen, the relationship between the average wind velocity and sea severity during the tropical cyclone and the storm with similar weather background is almost the same even though the geographical locations are far apart. The relationship between the average wind speed and significant wave height is close to that applicable for fully developed seas. This does not imply that, for a given significant wave height, the shape of the wave spectrum is close to that of a fully developed sea, since the sea severity at issue is not associated with the long sustained strong wind speed required for fully developed seas.

5.2.3 Wave spectra and wave heights in hurricane-generated seas

The shape of wave spectra obtained from data taken during hurricane-generated seas is different from that obtained in seas associated with ordinary storms. As an example, Figure 5.13 shows a comparison between a wave spectrum computed from data obtained during

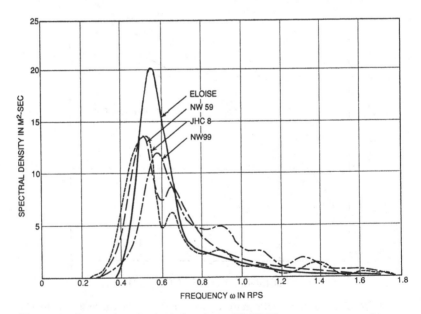

Fig. 5.13. Comparison between hurricane ELOISE wave spectrum and wave spectra in ordinary wind-generated seas having the same significant wave height 8.8 m.

hurricane ELOISE and various spectra with the same severity (significant wave height of 8.8 m) obtained from data taken in the North Atlantic. As can be seen, the energy density is concentrated primarily in the neighbourhood of the peak frequency of the spectrum for the hurricane-generated seas, contrasting with the energy being spread over a wide frequency range, including double peaks, for wave spectra obtained during ordinary storms.

Ross (1979) presents the results of a least square fit of the hurricane wave data to the JONSWAP spectral formulation. Although these spectral shapes are well represented by the JONSWAP formulation (see Eq. 2.83), values of the parameters of the formulation are quite different from those originally specified. Through regression analysis of data, he derives the parameters of the formulation as a function of the radial distance from the eye of the hurricane as follows:

$$\alpha = 0.035 \bar{f}_m^{0.82}$$

$$\gamma = 4.70 \bar{X}_r^{-0.13}$$

$$\bar{f}_m = 0.97 \bar{X}_r^{-0.21} \tag{5.22}$$

where

$\bar{f}_m$ = dimensionless modal frequency = $f_m U/g$

$\bar{X}_r$ = dimensionless radial distance = $X_r g/U^2$

f_m = modal frequency

U = wind speed at 10 m height above the sea surface

X_r = radial distance from the eye of hurricane.

Foster (1982) represents approximately 400 wave spectra obtained during hurricanes by the JONSWAP spectral formulation. From statistical analysis of these data, he presents the parameter α of the JONSWAP formulation as a function of significant wave height and modal frequency as follows:

$$\alpha = 4.5 H_s^2 f_m^4 \tag{5.23}$$

It is also found from Foster's analysis that there is a strong correlation between the peak energy density, $S(f_m)$, and significant wave height, H_s, for hurricane-generated seas, as shown in Figure 5.14, which can be presented as

$$S(f_m) = 0.75 H_s^{2.34} \tag{5.24}$$

On the other hand, the peak energy density can be obtained from Eq. (2.83) as

$$S(f_m) = \alpha \frac{g^2}{(2\pi)^4 f_m^5} e^{-1.25} \gamma \qquad (5.25)$$

Thus, from Eqs. (5.23) through (5.25), the peak-shape parameter, γ, can be expressed as a function of significant wave height and modal frequency as follows:

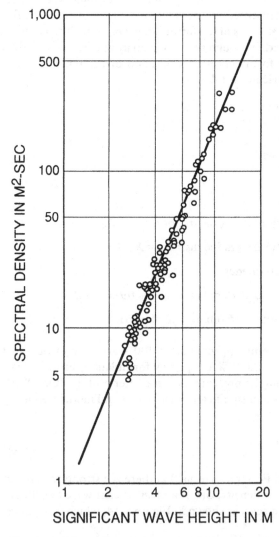

Fig. 5.14. Wave energy density at modal frequency versus significant wave height (Foster 1982).

$$\gamma = 9.5 H_s^{0.34} f_m \tag{5.26}$$

It is noted that the value of the peak-shape parameter, γ, for hurricane-generated seas ranges from 0.6 to 4.0, and its mean values is much smaller than 3.30 as originally given. A histogram of γ constructed from data obtained during hurricane-generated seas and its probability density function is shown in Figure 5.15. By using these relationships, a wave spectral formulation specifically applicable for hurricane-generated seas in the form of the JONSWAP formulation presented as a function of significant wave height and modal frequency is as follows:

$$S(f) = \frac{4.5}{(2\pi)^4}(H_s g)^2 (f_m^4/f^5)\exp\{-1.25(f_m/f)^4\}$$
$$\times (9.5 H_s^{0.34} f_m)^{\exp\{-(f-f_m)^2/2(\sigma f_m)^2\}} \tag{5.27}$$

where the units are meters and seconds. The above formula can also be written as a function of frequency ω as follows:

$$S(\omega) = \frac{4.5}{(2\pi)^4}(H_s g)^2 (\omega_m^4/\omega^5)\exp\{-1.25(\omega_m/\omega)^4\}$$
$$\times \left(\frac{9.5}{2\pi} H_s^{0.34}\omega_m\right)^{\exp\{-(\omega-\omega_m)^2/2(\sigma\omega_m)^2\}} \tag{5.28}$$

Figures 5.16(a) and (b) show examples of comparisons between wave spectra obtained from data during hurricanes ELOISE and KATE, respectively, and those computed by Eq. (5.28).

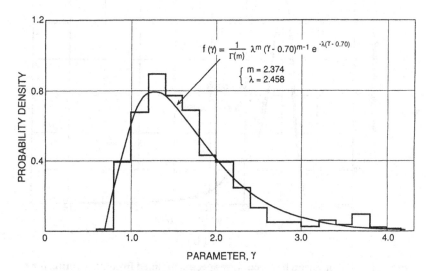

Fig. 5.15. Histogram of parameter γ constructed from data obtained during hurricane-generated seas (Foster 1982).

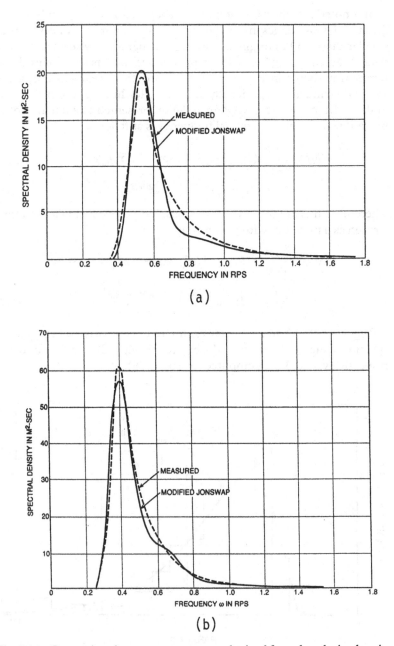

Fig. 5.16. Comparison between wave spectra obtained from data during hurricanes and modified JONSWAP spectra: (a) hurricane ELOISE, H_s=8.8 m; (b) hurricane KATE, H_s=10.0 m (Ochi 1993).

It has been shown that the shape of wave spectra during hurricanes is different from that observed in ordinary wind-generated seas. Statistical properties (such as extreme height) of waves during hurricanes, however, appear to be the same as those observed in ordinary storms. Borgman (1973) shows in his extensive study on the prediction of the largest wave height during hurricanes that the largest waves can be predicted based on extreme value statistics associated with the Rayleigh distribution.

Some results of analysis of wave data obtained during hurricanes show that the Rayleigh probability distribution overestimates the magnitude of the highest waves (Forristall 1978, 1984; Haring *et al.* 1976, among others). However, the overestimation of high waves during hurricanes by applying the Rayleigh distribution may not be attributed to sea severity *per se*, but due to a combination of sea severity and finite water depth. As will be presented in Chapter 9, the statistical distribution of wave profiles evaluated from data obtained in finite water depths deviates from the Gaussian distribution when the sea state is severe, and the Rayleigh distribution overestimates high waves under this situation.

Conversely, when wave records obtained in hurricane-generated seas show the Gaussian property, the probabilistic prediction of wave height, including extreme height, may be made based on the Rayleigh probability distribution. As an example, Figure 5.17 shows a comparison between the Rayleigh distribution and a histogram of wave height (dimensionless) constructed from wave data obtained during hurricane CAMILLE in the Gulf of Mexico (29.0°N, 88.7°W) where the water depth is reported as 103.7 m. This water depth may be increased by several meters due to storm surge during the hurricane. The significant wave height evaluated from the spectrum is 12.3 m. Thirty-two (32) waves in 180 measured waves have double peaks but do not show any noticeable excess of high

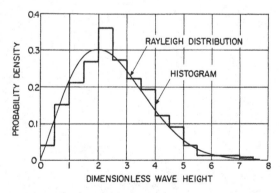

Fig. 5.17. Comparison between histogram of wave height (dimensionless) obtained during hurricane CAMILLE and Rayleigh probability density function.

crests nor round troughs; that is, waves can be considered as a Gaussian random process in this case.

Although some discrepancy between the histogram and the Rayleigh distribution can be seen in Figure 5.17, the data pass the χ^2-test for a level of significance $\alpha = 0.05$. The extreme wave height measured in this sea is 22.6 m as compared with a probable extreme wave height of 19.8 computed based on the Rayleigh distribution. The measured extreme height exceeds the estimated probable extreme value, but this is not surprising since the probability that the extreme value will exceed the probable extreme value is theoretically 0.63 (see Section 6.2). The sea state given in this example is not the severest sea measured during hurricane CAMILLE, but this sea severity (significant wave height 12.3 m) appears to be the maximum which can be considered as a Gaussian random process for the water depth at this site.

Estimation of extreme wave height during a hurricane, including various sea conditions in the growing and decaying stages, will be discussed in Section 6.4.

6 ESTIMATION OF EXTREME WAVE HEIGHT AND SEA STATE

6.1 BASIC CONCEPT OF EXTREME VALUES

This section presents the theoretical background for predicting extreme values (extreme wave height, extreme sea state, etc.) which provide invaluable information for the design and operation of marine systems. The *extreme value* is defined as the largest value of a random variable expected to occur in a specified number of observations. Note that the extreme value is defined as a function of the number of samples. In the naval and ocean engineering area, however, it is highly desirable to estimate the largest wave height expected to occur in one hour, or the severest sea state expected to be encountered in 50 years, for example. This information can be obtained by estimating the number of waves (or sea states) per unit time, and thereby the number of samples necessary for evaluating the extreme value is converted to time.

The concept supporting the estimation of extreme values is *order statistics* which is outlined in the following. Let us consider a sample set consisting of wave heights taken in the sequence of observations $(x_1, x_2, ..., x_n)$. Each element of the random sample x_i is assumed to be statistically independent having the same probability density function $f(x)$. In the case of wave height observations, each x_i is considered to obey the Rayleigh probability law. Next, let us rearrange the elements of this random sample in ascending order of magnitude such that y_1 is the smallest wave height, and y_n is the largest wave height in the set; namely, $y_1 y_2 ... y_n$. Then, $(y_1, y_2, ..., y_n)$ is called the *ordered sample*.

Once the sample wave heights are ordered, the sample elements $y_1, y_2, ..., y_n$ are statistically independent random variables and each has its own probability density function. Note that the largest extreme wave height y_n is a random variable. This can be easily understood because every time we take a sample of n waves, presumably the magnitude of the largest height will be different.

The cumulative distribution function of y_n, denoted by $G(y_n)$ is given by

$$G(y_n) = [\{F(x)\}^n]_{x=y_n} \tag{6.1}$$

and thereby the probability density function, denoted by $g(y_n)$ becomes

$$g(y_n)=n[f(x)\{F(x)\}^{n-1}]_{x=y_n} \qquad (6.2)$$

Here, $f(x)$ and $F(x)$ are the probability density function and cumulative distribution function, respectively, of wave height, and they are called the *initial probability density function* and *distribution function*, respectively, in order statistics. For the derivation of Eqs. (6.1) and (6.2) the reader is referred to extreme value statistics (Ochi 1990a, for example).

As an example, Figure 6.1 shows the extreme value probability density functions $g(y_n)$ for $n=50$, 100 and 200 for waves having a significant height of 5.66 m, assuming wave height follows the Rayleigh probability distribution. It can be seen that the shape of the probability density function $g(y_n)$ is much more sharply concentrated about its modal value than the initial probability density function, and that the modal value shifts to higher wave height with increase in number n. It is easily understood, therefore, that the probability of extreme wave heights (or amplitudes) exceeding a specified value depends on the number of waves. Since the largest value in a specified number of waves, y_n, is a random variable and obeys its own probability law, the question arises as to the use of the extreme value distribution in the design considerations of marine systems. This subject will be discussed in detail in Section 6.2.

Note that Eqs. (6.1) and (6.2) show that the extreme values cannot be evaluated without knowledge of the initial distribution, $f(x)$ or $F(x)$, which may not always be known in practice. Estimation of extreme values

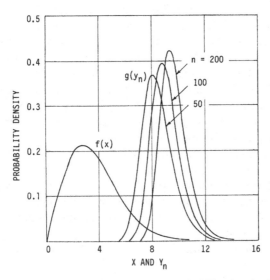

Fig. 6.1. Initial and extreme value probability density functions of wave height with significant wave height of 5.66 m.

under this situation, however, can be made by assuming a specific form of the initial cumulative distribution function. This subject will be discussed in Section 6.3.

6.2 PROBABLE AND DESIGN EXTREME WAVE HEIGHT

As stated in the previous section, the extreme value probability density function $g(y_n)$ is sharply concentrated around its modal value. Hence, it is reasonable to assume that the modal value may be most likely to occur. This modal value is denoted by $\overline{Y}_n$ as shown in Figure 6.2, and is called the *probable extreme value* or *characteristic value*. The magnitude of $\overline{y}_n$ can be obtained as the solution of the following equation:

$$\frac{d}{dy_n} g(y_n) = 0 \tag{6.3}$$

which yields from Eq. (6.2)

$$f'(y_n)F(y_n) + (n-1)\{f(y_n)\}^2 = 0 \tag{6.4}$$

Assuming wave height (or amplitude) obeys the Rayleigh probability law, we may write the initial probability density function as

$$f(x) = \frac{2x}{R} e^{-x^2/R} \tag{6.5}$$

where

$$R = \begin{cases} 2m_0 & \text{for } x \text{ being wave amplitude} \\ 8m_0 & \text{for } x \text{ being wave height} \end{cases}$$

$m_0 = $ area under wave spectrum.

Then, Eq. (6.4) becomes

$$\left\{\frac{2}{R} - \left(\frac{2y_n}{R}\right)^2\right\}\left(1 - \exp\left\{-\frac{y_n^2}{R}\right\}\right) + (n-1)\left(\frac{2y_n}{R}\right)^2 \exp\left\{-\frac{y_n^2}{R}\right\} = 0 \tag{6.6}$$

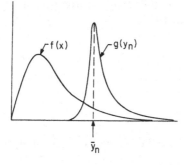

$$\overline{y}_n$$

Fig. 6.2. Explanatory sketch of initial probability density function $f(x)$ and extreme value probability density function $g(y_n)$.

Since y_n is large, $(2y_n/R)^2$ in the first term is much greater than $(2/R)$, and thereby $(2/R)$ can be discarded. Furthermore, n is supposed to be very large. Then, the solution of Eq. (6.6), $\bar{y}_n$, can be simply obtained as

$$\bar{y}_n = \sqrt{\ln n}\ \sqrt{R} = \begin{cases} \sqrt{2\ln n}\ \sqrt{m_0} & \text{for amplitude} \\ 2\sqrt{2\ln n}\ \sqrt{m_0} & \text{for height} \end{cases} \tag{6.7}$$

In the above equation, the extreme wave height is obtained as a function of the number of waves, n. For practical purposes, however, it may be more meaningful to express the extreme wave height in terms of time rather than as a function of number of waves. This expression can be made by using the formulation of the average number of zero-crossing per unit time derived in Eq. (4.42). That is,

$$n = \frac{1}{2\pi}\sqrt{m_2/m_0} \tag{6.8}$$

where m_j is the jth moment of the wave spectrum.

Then, the probable extreme wave amplitude, $\bar{y}_n$, is expressed as a function of time T:

$$\bar{y}_n = \sqrt{2\ln\left\{\frac{(60)^2 T}{2\pi}\sqrt{\frac{m_2}{m_0}}\right\}}\sqrt{m_0} \tag{6.9}$$

where T is time in hours.

An example of probable extreme wave height as a function of time will be shown later along with design extreme wave height.

For the non-narrow-band case, the probability density function and cumulative distribution function derived in Eqs. (3.34) and (3.37), respectively, should be used in Eq. (6.4). By writing the dimensionless extreme value as $\zeta_n = y_n/\sqrt{m_0}$, the solution of the equation is found approximately for bandwidth parameters less than 0.9. That is, for $\epsilon < 0.9$, it is possible to use the following approximation for a large number of observations, n (Ochi 1973):

$$\Phi\left(\frac{\sqrt{1-\epsilon^2}}{\epsilon}\zeta_n\right) \approx 1$$

$$\Phi\left(\frac{\zeta_n}{\epsilon}\right) \approx 1 \tag{6.10}$$

By neglecting terms of small order of magnitude, Eq. (6.4) becomes,

$$-\frac{\sqrt{1-\epsilon^2}\,(1+\sqrt{1-\epsilon^2})}{2}\exp\{-\zeta_n^2/2\} + n(1-\epsilon^2)\exp\{\zeta_n^2\} + o(\zeta_n) = 0 \tag{6.11}$$

Then, the solution of Eq. (6.11) can be obtained as the probable extreme wave amplitude for non-narrow-band waves. In dimensional form, we have

$$\bar{y}_n = \sqrt{2 \ln\left\{\frac{2\sqrt{1-\epsilon^2}}{1+\sqrt{1-\epsilon^2}} n\right\}} \sqrt{m_0} \qquad (6.12)$$

As is shown in the above equation, the probable extreme amplitude is a function of bandwidth parameter ϵ. Figure 6.3 shows the probable extreme wave amplitude in dimensionless form as a function of the number of observations for various bandwidth parameters. The figure indicates that there is no significant difference in the probable extreme amplitude up to $\epsilon = 0.6$, irrespective of the number of observations.

Since the range of ϵ-values of ocean waves spans from 0.45 to 0.80, it may safely be concluded that the effect of the bandwidth parameter can be ignored in the estimation of extreme wave amplitude (or height).

Next, let us express the probable extreme wave amplitude, given Eq. (6.12), which is applicable for a non-narrow-band spectrum in terms of time instead of number of observations. The expected number of positive maxima per unit time is given by Eq. (3.33), i.e.

$$n = \frac{1}{4\pi}\left(\frac{1+\sqrt{1-\epsilon^2}}{\sqrt{1-\epsilon^2}}\right)\sqrt{m_2/m_0} \qquad (6.13)$$

Then, from Eqs. (6.12) and (6.13), the probable extreme wave amplitude to occur in time T hours becomes

$$\bar{y}_n = \sqrt{2 \ln\left\{\frac{(60)^2 T}{2\pi}\sqrt{m_2/m_0}\right\}} \sqrt{m_0} \qquad (6.14)$$

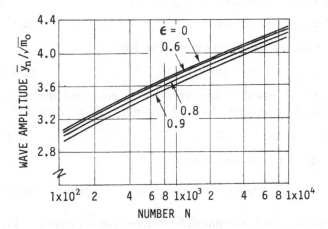

Fig. 6.3. Probable extreme wave amplitude (dimensionless) for various bandwidth parameters ϵ as a function of number of observations (Ochi 1981).

Note that the above equation is exactly the same as that for narrow-band spectra given in Eq. (6.9). This leads to an important conclusion: the magnitudes of extreme waves in a specified period of time are the same irrespective of the bandwidth parameter. In other words, the narrow-band assumption is acceptable for the estimation of extreme wave amplitude in terms of time.

The probable extreme wave amplitude (or height), $\overline{Y}_n$, is interpreted as being that most likely to occur, since it is the modal value of the probability density function. It is then important to examine the possibility that the extreme wave amplitude exceeds the probable extreme amplitude. Results of computations show that the probability is high for a large sample size n. That is, from Eq. (6.1) and by obtaining the cumulative distribution function from Eq. (6.5) along with Eq. (6.7), we have approximately,

$$\lim_{n\to\infty}\Pr\{\text{extreme wave}>\overline{y}_n\}=\lim_{n\to\infty}\{1-G(\overline{y}_n)\}=\lim_{n\to\infty}\{1-(F(\overline{y}_n))^n\}$$

$$=\lim_{n\to\infty}\left\{1-\left(1-\frac{1}{n}\right)^n\right\}=1-e^{-1}=0.632 \qquad (6.15)$$

The above equation implies that there is a 63.2 percent chance that the largest wave will exceed $\overline{y}_n$. This probability is extremely high; hence, it is highly desirable in the design of marine systems to consider a sufficiently large wave amplitude (or height) for which the probability of being exceeded is very small. In other words, a very small number α should be chosen which may be called the *risk parameter*, and the extreme wave $\hat{y}_n$ evaluated for which the following relationship holds (Ochi 1973):

$$\lim_{n\to\infty}\Pr\{\text{extreme wave}>\hat{y}_n\}=\lim_{n\to\infty}\{1-G(\hat{y}_n)\}$$

$$=\lim_{n\to\infty}\{1-(F(\hat{y}_n))^n\}=\alpha \qquad (6.16)$$

Considering that α is small and n is large, we have

$$F(\hat{y}_n)=(1-\alpha)^{1/n}\sim 1-(\alpha/n)+o(\alpha^2) \qquad (6.17)$$

On the other hand, based on the narrow-band assumption, the cumulative distribution function of the Rayleigh distribution $F(y_n)$ becomes

$$F(\hat{y}_n)=1-\exp\{-\hat{y}_n^2/R\} \qquad (6.18)$$

Hence, from Eqs. (6.17) and (6.18), we have

$$\hat{y}_n=\sqrt{\ln(n/\alpha)}\ \sqrt{R}=\sqrt{2\ln(n/\alpha)}\ \sqrt{m_0} \quad \text{for amplitude} \quad (6.19)$$

From comparison with the probable extreme value derived in Eq. (6.7), it is clear that the extreme value with the risk parameter α, where $\alpha\ll 1$, can be evaluated through a modification of the number of observations by multiplying by $1/\alpha$. Thus, for non-narrow-band waves, the design extreme value from Eq. (6.12) becomes

$$\hat{y}_n = \sqrt{2 \ln\left\{\frac{2\sqrt{1-\epsilon^2}}{1+\sqrt{1-\epsilon^2}}\frac{n}{\alpha}\right\}} \sqrt{m_0} \quad \text{for amplitude} \qquad (6.20)$$

The value of the risk parameter is at the designer's discretion, although results of many computations indicate that $\alpha=0.01$ appears to be appropriate in practice. For supplementing this statement, Figure 6.4 is prepared. The figure shows the design extreme wave amplitude (in dimensionless form, $\hat{y}_n/\sqrt{m_0}$) for various α-values along with the probable extreme wave amplitude under the narrow-band assumption.

As an example, let us consider the case $n=300$ which is approximately the number of waves observed in one hour in severe seas. The probable extreme value (dimensionless) is 3.40, compared with 4.52 and 4.70 for design extreme values with $\alpha=0.01$ and 0.005, respectively. The design extreme value with $\alpha=0.01$ is approximately 33 percent greater than the probable extreme wave, but the design extreme value with $\alpha=0.005$ is only 4.0 percent greater than that with $\alpha=0.01$.

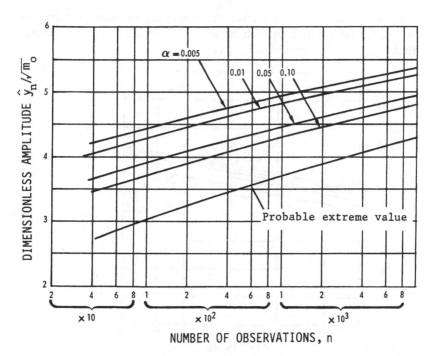

Fig. 6.4. Design extreme wave amplitude (dimensionless) for various α-values as a function of number of waves (Ochi 1973).

In estimating extreme waves for the design of marine systems, the question arises as to the number of encounters by a marine system with a particular sea severity in its lifetime. The design extreme waves with the risk parameter discussed so far are for a short-term sea state. Suppose a marine system is designed to withstand a severe wave height $\hat{y}_n$ with $\alpha=0.01$ in a specified sea, this implies that the design extreme wave provides a 99 percent assurance of safety when the marine system encounters this sea state once in its lifetime. Therefore, if the marine system is expected to encounter seas of this severity five times, for example, in its lifetime, it is necessary to divide the risk parameter by 5 in order to maintain 99 percent safety assurance throughout its lifetime.

Next, let us express design extreme waves in terms of time rather than number of waves. As stated earlier, the narrow-band assumption is acceptable when the extreme values are estimated as a function of time. Hence, from Eq. (6.14), the extreme wave amplitude to be used in the design considerations of marine systems becomes

$$\hat{y}_n = \sqrt{2 \ln\left\{ \frac{(60)^2 T}{2\pi\alpha} \sqrt{m_2/m_0} \right\} } \sqrt{m_0} \qquad (6.21)$$

Figure 6.5 shows design extreme amplitudes for various values of $(m_2/m_0)^{1/2}$ with $\alpha=0.01$. As can be seen, for a given time, the extreme amplitude increases with increasing $(m_2/m_0)^{1/2}$.

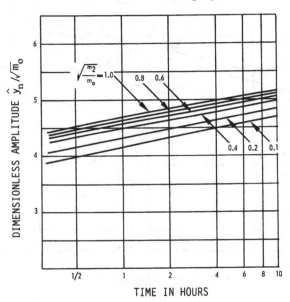

Fig. 6.5. Design extreme wave amplitude (dimensionless) for various $(m_2/m_0)^{1/2}$-values as a function of time (Ochi 1973).

In order to see the effect of time duration on the magnitude of extreme waves in a given sea, Figure 6.6 shows an example of probable extreme wave amplitude and design extreme wave amplitude with $\alpha=0.01$ as a function of time. The significant wave height of this sea state is 7.54 m. As can be seen, the magnitude of extreme wave amplitudes increases substantially during the first half to one hour, and thereafter increases slowly with time. This is the general trend of the extreme value irrespective of sea severity.

In the foregoing evaluation of extreme wave height, it is assumed that the wave peaks (positive maxima) are statistically independent. One way to evaluate the effect of statistical dependence of wave peaks on the magnitude of extreme values is to assume that waves are a stationary random process and positive maxima are subject to the Markov chain condition (Ochi 1979b). For a random process subject to the Markov chain condition, Epstein (1949) shows that the cumulative distribution function of the extreme value in n observations can be presented as

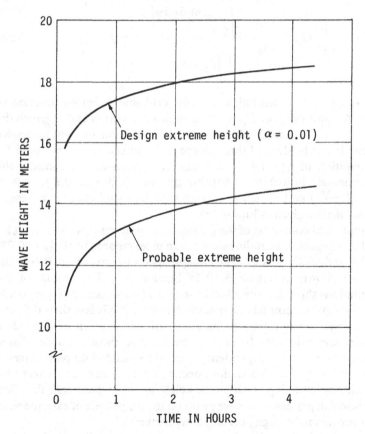

Fig. 6.6. Probable and design extreme wave amplitudes as a function of time (significant wave height 7.54 m).

$$G_n(\zeta) = \Pr\{\text{extreme value in } n \text{ observations} < \zeta\}$$

$$= G_{n-1}(\zeta) \frac{F_{\xi,\eta}(\zeta,\zeta)}{F_\xi(\zeta)} \tag{6.22}$$

where $F_{\xi,\eta}(\ )$ = cumulative distribution function of two successive maxima ξ and η

$F_\xi(\ )$ = cumulative distribution function of maxima ξ.

By applying Eq. (6.22) successively down to $n=1$, we can write $G_n(\zeta)$ as

$$G_n(\zeta) = \frac{\{F_{\xi,\eta}(\zeta,\zeta)\}^{n-1}}{\{F_\xi(\zeta)\}^{n-2}} \tag{6.23}$$

and from this the probable extreme value can be obtained as the value which satisfies the following equation:

$$\frac{d^2}{d\zeta^2} G_n(\zeta) = \frac{d^2}{d\zeta^2} \frac{\left(\int_0^\zeta \int_0^\zeta f(\xi,\eta) \, d\xi \, d\eta\right)^{n-1}}{\left(\int_0^\zeta f(\xi) \, d\xi\right)^{n-2}} = 0 \tag{6.24}$$

Here, $f(\xi)$ in the denominator is the probability density function of positive maxima given in Eq. (3.34) in which ξ is written as ζ. $f(\xi,\eta)$ in the numerator is the joint probability density function of two successive maxima. It can be derived through the same procedure as employed for the derivation of $f(\xi)$, but in this case it is necessary to consider the joint normal probability distribution of $(x_1, 0, \ddot{x}_1, x_2, 0, \ddot{x}_2)$, where $x_2(t) = x_1(t + \overline{T}_m)$ and $\overline{T}_m$ is the average time interval between successive positive maxima given in Eq. (4.33).

A numerical evaluation of the probable extreme value is carried out for a total of 14 spectra, including nine ocean wave records ($\epsilon = 0.42$ to 0.78) and five artificially made square-shape spectra ($\epsilon = 0.11$ to 0.40), the latter are shown in Figure 3.10 in Section 3.3. The results of the computations show that the effect of statistical dependence of wave peaks exists for wave spectra whose spectral bandwidth ϵ is less than 0.5, and that extreme values estimated based on the assumption of dependent peaks are approximately 10 percent greater than those estimated based on the assumption of independent peaks. Since bandwidth parameters of ocean waves less than 0.5 seldom occur in deep water areas except for very mild sea states (significant wave height less than 2 meters), the effect of statistical dependence of wave peaks on the magnitude of extreme wave height appears to be negligibly small in practice.

The principle for estimating extreme wave height discussed so far is based on the cumulative distribution function constructed from the time history of wave data. We may recall that the envelope process is introduced in Section 3.2.2, and that the statistical properties of the time history of the wave envelope may be approximately identical to those of wave amplitude if waves are assumed to be a narrow-band random process. Regarding estimation of extreme wave height based on the envelope process concept, Pierce (1985) finds from analysis of measured wave data that the envelope maxima are generally 5 to 9 percent larger than the time history maxima.

On the other hand, Naess (1982) develops a probability distribution applicable to extreme values of the envelope process and finds that the envelope method can be applied for estimating extreme values of random processes whose spectral bandwidth ϵ is less than 0.4. However, wave spectra with ϵ less than 0.4 cannot be observed in the ocean, in practice. Further, Yim et al. (1992) develop three practical methods for evaluating the expected maxima of a Gaussian time series for ocean system analysis, and conclude that the employment of the envelope concept does not improve the accuracy in estimating the extreme values. Judging from the results of these studies, it appears that the envelope concept does not contribute appreciably to estimating extreme wave height.

6.3 ESTIMATION OF EXTREME WAVE HEIGHT AND SEA STATE FROM DATA

In the previous section, estimation of extreme wave height is based on the Gaussian, with either the narrow-band or non-narrow-band, random process concept. We often encounter the situation wherein we have measured (or hindcast) data from which we want to estimate the extreme wave height without knowledge of the probability distribution. For example, we want to estimate the extreme wave height from measured (or hindcast) data obtained in an area where water depth is finite. As another example, we want to estimate the extreme sea state (significant wave height) expected to occur in 50 years from accumulated significant wave height data. In these examples, the initial probability distributions are not known.

In order to estimate the extreme values under this situation, let us consider again Eq. (6.4) which should be satisfied by the probable extreme value, $\bar{y}_n$. That is

$$f'(\bar{y}_n)F(\bar{y}_n)+(n-1)\{f(\bar{y}_n)\}^2=0 \tag{6.25}$$

We may write the above equation as

$$1=-\frac{(n-1)\ \{f(\bar{y}_n)\}^2}{f'(\bar{y}_n)\ F(\bar{y}_n)} \tag{6.26}$$

and by dividing by $1-F(\bar{y}_n)$, we have

$$\frac{1}{1-F(\bar{y}_n)} = -\frac{n-1}{F(\bar{y}_n)} \frac{f(\bar{y}_n)}{1-F(\bar{y}_n)} \frac{f(\bar{y}_n)}{f'(\bar{y}_n)} \tag{6.27}$$

It is assumed that the initial distribution satisfies the L'Hôpital rule in the form of

$$\frac{f(\bar{y}_n)}{1-F(\bar{y}_n)} = -\frac{f'(\bar{y}_n)}{f(\bar{y}_n)} \qquad \text{for large } \bar{y}_n \tag{6.28}$$

then, Eqs. (6.27) and (6.28) yield

$$\frac{1}{1-F(\bar{y}_n)} = \frac{n-1}{F(\bar{y}_n)} \tag{6.29}$$

Since n is large and $\bar{y}_n$ is also large in the sample space, the right side of the above equation may be written approximately as n. Hence, we have

$$\frac{1}{1-F(\bar{y}_n)} = n \tag{6.30}$$

The left side of Eq. (6.30) is defined as the return period. Thus, Eq. (6.30) implies that, in practice, the probable extreme value expected to occur in n observations can be evaluated from the initial cumulative distribution function $F(x)$ (x is replaced by y_n) for which the return period is equal to n. In practice, the $\bar{y}_n$ is evaluated by taking the logarithm of Eq. (6.30) for convenience. The extreme wave height for design consideration, $\hat{y}_n(\alpha)$, can also be evaluated from Eq. (6.30) by replacing n by n/α.

It is a common practice to determine the initial cumulative distribution function by representing measured data by some known distribution function such as the log-normal or Weibull distribution, and the extreme value is estimated based on this presumed distribution. Often, however, the extreme value is determined by simply extending the cumulative distribution function obtained by plotting the measured data. In this case, loss of accuracy in estimating the extreme value is inevitable to some extent due to the visual extension of the plotted data line. Examples of the application of Eq. (6.30) for estimating extreme wave height and sea state are presented below.

The first example is wave height data obtained during hurricane CAMILLE: the histogram of wave height (dimensionless) is shown in Figure 5.17. The wave data are measured for 30.4 minutes during which 180 wave heights are observed, and the significant wave height is 12.3 m. As discussed with reference to Figure 5.17, the data may be represented by the Rayleigh probability distribution; hence, the extreme values are estimated based on the Rayleigh distribution; the parameter of which is obtained from spectral analysis. Figure 6.7 shows the logarithm of the return period computed based on the Rayleigh distribution. The probable extreme wave height expected to occur in 30.4 minutes is estimated from $\bar{y}_n$ ($n=180$) as 19.8 m. The observed extreme wave height is 22.6 m which exceeds the probable extreme wave height by 14 percent; however, the extreme wave height for design considerations with the risk parameter 0.01 is 27.1 m, far greater than the observed value.

Another example is estimation of the extreme sea state expected to occur in 50 years from the significant wave height data given in Table 5.1. Figure 6.8 shows the logarithm of the return period of the data. Included in the figure is the logarithm of the return period of the generalized gamma probability distribution shown in Figure 5.3. We first estimate the extreme significant wave height by visually extending the data points shown in Figure 6.8. Since 5412 observations of data are taken over a period of 3 years, the expected number in 50 years is 90 200 ($\ln n=11.41$). As shown in the figure, the probable extreme significant wave height $\bar{y}_n$ for this value is 8.05 m. On the other hand, the extreme value for design consideration of a marine system $\hat{y}_n(\alpha)$ with $\alpha=0.01$ is 9.20 m. It is recognized from Eq. (6.30) that the extreme significant wave height for design consideration in 50 years with the risk parameter $\alpha=0.01$ is equivalent to the probable extreme significant wave height in $50/\alpha$ years, namely 5000 years.

Next, let us estimate the extreme significant wave height by representing the cumulative distribution function by the generalized gamma distribution shown in Figure 5.3. By applying the theoretical cumulative distribution function of the generalized gamma distribution given in

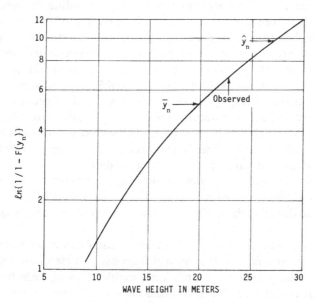

Fig. 6.7. Example of estimation of extreme wave height from hurricane CAMILLE data.

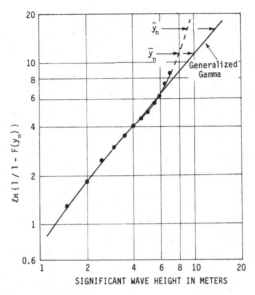

Fig. 6.8. Example of estimation of extreme sea state (significant wave height)
(data from Bouws 1978).

Eq. (5.5), the probable and design extreme significant wave heights for
$\alpha=0.01$ become 10.1 m and 13.6 m, respectively. These values are substantially greater than those estimated by visually extending the return
period computed from data. It is noted that estimation of the extreme
values by extending the return period is simple, however, the data in the
sample space of large cumulative distributions are very sparse; usually on
the order of less than 0.2 percent of the data. Therefore, some uncertain-
ties in estimating extreme values are unavoidable in this approach.
Furthermore, in this example, extreme values expected to occur in 50
years are estimated from data accumulated in only 3 years. This may be
another reason why such a large discrepancy is observed between extreme
values estimated by extending the data points and those estimated by the
probability distribution function representing the data.

In order to improve estimation of long-term extreme sea state,
Kerstens et al. (1988) analyze accumulated significant wave height data
by applying the Bayesian theorem. The basic concept of this approach is
as follows.

The parameters involved in the probability distribution are usually
considered as constants to be determined from data. In Bayesian statistics, however, these parameters are random variables obeying some probability law. For convenience, let us consider the single parameter θ which
is a continuous-type random variable having the probability density function $h(\theta)$. This probability density function is called the *prior probability*

density function. Then, following Bayes' formula (see Appendix A), we may write

$$f(\theta|x) = \frac{h(\theta) f(x|\theta)}{f(x)} \tag{6.31}$$

Here, $f(\theta|x)$ is called the *posterior probability density function* of θ. There exists the concept that the θ-value of $f(x|\theta)$ in Eq. (6.31) may be determined taking into consideration successive sets of data rather than a fixed value. If so regarded, $f(x|\theta)$ may be considered the likelihood function of θ for a given x, and it is written as $L(\theta|x)$. This subject is beyond the scope of this text, but readers who are interested may refer to Fisher's treatise (Fisher 1922). In any event, we may modify Bayes' theorem as

$$f(\theta|x) = Ch(\theta)L(\theta|x) \tag{6.32}$$

where C is a normalization constant.

The above equation provides the foundation for processing new data such that the parameter may be successively revised as more data become available. To elaborate on this statement, if we initially have a set of sample data, denoted by x_I, we may write

$$f(\theta|x_I) = C_I h(\theta)L(\theta|x_I) \tag{6.33}$$

Upon obtaining a second set of sample data, x_{II}, which is assumed to be statistically independent of the first set, we have

$$f(\theta|x_I,x_{II}) = C_I h(\theta)L(\theta|x_I)L(\theta|x_{II})$$
$$= C_{II} f(\theta|x_I)L(\theta|x_{II}) \tag{6.34}$$

The above procedure can be repeated, and at the Jth data sample, we have the following relationship:

$$f(\theta|x_I, x_{II}, \ldots, x_J) = C_J f(\theta|x_I, x_{II}, \ldots, x_{J-1})L(\theta|x_J) \tag{6.35}$$

This procedure shows how θ can be continuously revised as new data become available, Kerstens *et al.* apply this approach for estimating the extreme significant wave height expected in 100 years by using data accumulated over five years but revising the value of the parameter involved in the probability distribution function by using data obtained every year. They consider two probability distributions; the Weibull and the Type I asymptotic distribution which will be discussed in Section 6.5.

Application of the Bayesian approach undoubtedly contributes significantly in estimating extreme sea states for the design consideration of marine systems. It should be noted, however, that this method provides more accurate information only on the parameter(s) of the assigned probability distribution. Therefore, for the most desirable approach in estimating extreme sea states, care must be taken first in selecting the probability distribution function which most accurately represents the significant wave height data, and then the Bayesian approach can be applied to the selected distribution.

6.4 EXTREME WAVE HEIGHT IN A NON-STATIONARY SEA STATE

Estimation of extreme wave height presented in the previous sections deals with waves in short-term steady-state (stationary) seas; i.e. the wave spectrum as well as the significant wave height are constant in a given sea. It is of interest, however, to estimate the largest wave height in the long term when sea severity is changing. A typical example of this situation is the estimation of extreme wave height during a hurricane including the growing and decaying stages of the sea condition. Estimation of this can be achieved through preparation of the probability density function of long-term wave heights following the method presented in Section 3.8, and then estimating the extreme value by applying the method discussed in the previous section. Borgman (1973), on the other hand, develops a concise approach for estimating extreme wave height during a hurricane as outlined below.

Let $[0,I]$ be the overall time interval of interest during a hurricane, and divide this interval into n segments in which the sea state is steady-state. We may write the time duration of the jth segment as ΔI_j, and the sea state as $S(\tau_j)$, where τ_j is an index representing the jth segment. Furthermore, by letting $\overline{T}_0(\tau_j)$ be the average zero-crossing period in the jth segment, the number of waves in the jth segment is given by $\Delta I_j/\overline{T}_0(\tau_j)$. Next, following the definition of the cumulative distribution function of the extreme value given in Eq.(6.2), we may write the probability of extreme wave height in the jth segment as

$$\Pr\{H_{\max}<h_e\}=G(h_e)=\{F[h_e,S(\tau_j)]\}^{\Delta I_j/\overline{T}_0(\tau_j)} \qquad (6.36)$$

By applying this concept to all n-segment seas, the probability of extreme wave height in the entire time interval becomes

$$\Pr\{H_{\max}<h_e|[0,I]\}=\prod_{j=1}^{n}\{F[h_e,S(\tau_j)]\}^{\Delta I_j/\overline{T}_0(\tau_j)} \qquad (6.37)$$

Next, we consider the limit of this probability by letting the segment time interval ΔI_j be as small as possible, and thereby $S(\tau_j)$, $\overline{T}_0(\tau_j)$ becomes a continuous function of τ. Hence, we may write

$$\Pr\{H_{\max}<h_e|[0,I]\}=G\{h_e|[0,I]\}$$

$$=\exp\left\{\int_0^I \frac{1}{\overline{T}_0(\tau)}\ln F[h_e,S(\tau)]d\tau\right\} \qquad (6.38)$$

The above equation is the cumulative distribution function of the largest wave height h_e in time interval $[0,I]$. $F[h_e,S(\tau)]$ is the cumulative distribution function of the wave height in a given sea state. Borgman assumes the initial distribution $F[h_e,S(\tau)]$ to be the Rayleigh probability distribution, and numerically evaluates $G(h_e)$ for eighteen hurricanes. The results of computations show that $\ln\ln\{1/G(h_e)\}$ versus h_e is found to be a straight line, i.e.

$$\ln\ln\{1/G(h_e)\} = -Ah_e^2 + B \qquad (6.39)$$

This being the case, we may write

$$G(h_e) = \exp\left\{-e^{-A(h_e^2 - B/A)}\right\} \qquad \text{for } h_e > 0 \qquad (6.40)$$

The form of this cumulative distribution function resembles that of the Type I asymptotic extreme value distribution which will be discussed in the next section, but the distribution function is a function of the square of the random variable h_e, and the sample space is $(0,\infty)$ instead of $(-\infty,\infty)$ for the Type I asymptotic distribution.

Krogstad (1985) presents Eq. (6.38) in terms of sea state (significant wave height H_s) instead of time history. That is, let $\pi(H_s)$ be the probability density function of the sea state during a hurricane. Then, the cumulative distribution function applicable to the extreme wave in time interval $[0,I]$ is given by

$$\Pr\{H_{\max} < h_e\} = G(h_e)$$

$$= \exp\left\{I\int_0^\infty \pi(H_s)\frac{1}{\overline{T}_0(H_s)}\ln F[h_e,H_s]\mathrm{d}H_s\right\} \qquad (6.41)$$

where I is the time interval considered.

Krogstad extends the method by including the probability distribution of wave period associated with the extreme wave height.

6.5 ASYMPTOTIC DISTRIBUTIONS OF LARGEST WAVES AND SEA STATES

It is sometime necessary to estimate extreme waves and sea states from data consisting of the largest values taken during a fixed time interval; namely, the largest values in regularly sampled data such as hourly, daily, monthly maxima. As an example, Table 6.1 shows statistical data of the daily largest significant wave heights obtained by the Coastal Engineering Research Center Field Station in North Carolina. The data are at a location 450 m off the shoreline where the water depth is 8.5 m on average. Observation of significant wave height was carried out four times a day

Table 6.1. *Daily maximum significant wave height data obtained by the Coastal Engineering Research Center at Duck, North Carolina.*

Significant wave height (m)	Number of observations
0.2–0.4	11
0.4–0.6	151
0.6–0.8	158
0.8–1.0	175
1.0–1.2	109
1.2–1.4	116
1.4–1.6	91
1.6–1.8	63
1.8–2.0	47
2.0–2.2	40
2.2–2.4	27
2.4–2.6	19
2.6–2.8	21
2.8–3.0	12
3.0–3.2	6
3.2–3.4	7
3.4–3.6	6
3.6–3.8	2
	Total 1061
	(in 42 months)

over 42 months and the table shows the accumulation of the largest value in a day.

It is stated in Section 5.1.1 that, in general, there is no theoretical basis for selecting any particular probability distribution to characterize significant wave height data. However, if the data consists of the largest values every day (or week or month), and if the number of observations is sufficiently large, then the extreme value may be estimated through the asymptotic distribution developed for extreme values.

The asymptotic distribution of extreme values have been developed by many statisticians, Fréchet, Fisher, Tippett, van Mises, Gnedenko, etc., and their work is summarized by Gumbel (1958). The asymptotic formula is developed as the limiting distribution of the largest (or the smallest) value in a sample of size N, where N is very large. By writing $N = kn$, where k is a fixed constant, we may interpret this limiting distribution to be equivalent to the limiting distribution of the largest value in a sample of size n (where $n \rightarrow \infty$) with fixed k. Based on this interpretation, the statistical properties of the largest value in a fixed period of time may be estimated by applying the asymptotic distribution of the extreme value.

The asymptotic distribution of extreme values was first developed by Fréchet (1927), and the same distribution and two other distributions were derived by Fisher and Tippett (1928). Gnedenko (1943) showed that these are the only three distributions which satisfy the condition required for the asymptotic distribution. These distributions are called Type I (Fisher–Tippett), Type II (Fréchet) and Type III (Fisher–Tippett) asymptotic extreme value distributions.

For predicting extreme values of waves and sea states, Type I and Type III asymptotic distributions may be applicable, but application of the Type II distribution is not appropriate for the analysis of waves and sea states because of the condition imposed on the distribution function.

Although asymptotic extreme value distributions have been frequently applied for analysis of wave data, it is noted that the probability distribution of the daily maxima wave height or significant wave height is rarely accurately represented by asymptotic distributions. This may be partially attributed to the fact that the data are not taken from a sample space representing the statistical properties requisite for the asymptotic extreme value distribution. The asymptotic distribution deals with maxima data in a large number of observations (theoretically infinite) from a consistent sample space. However, this is not the case in practice.

An extensive review of extreme significant wave height for the design considerations of offshore structures is made by Muir and El-Shaarawl (1986). They also state that data will seldom be accurate enough for very complicated statistical methods to be utilized to the fullest extent.

6.5.1 Type I asymptotic extreme value distribution

The Type I asymptotic distribution is applicable to maxima data whose initial probability distribution is of an exponential type, defined by von Mises as a distribution which satisfies the following condition:

$$\lim_{x \to \infty} \frac{d}{dx} \left[\frac{1 - F(x)}{f(x)} \right] = 0 \tag{6.42}$$

Since almost all initial probability distributions considered for analysis of waves and sea states belong to this category, the Type I asymptotic distribution can be applied for analysis of wave and sea state data. Let us assume that the initial cumulative distribution function is given in the form of

$$F(x) = 1 - \exp\{-q(x)\} \tag{6.43}$$

where $q(x)$ is a positive real-valued function satisfying the condition required for $f(x)$ to be a cumulative distribution function.

By using the relationship given in Eq. (6.30), we can write the initial cumulative distribution function as

$$F(x)=1-e^{-q(x)}\left\{\frac{1}{n}e^{q(\bar{y}_n)}\right\}=1-\frac{1}{n}e^{-\{q(x)-q(\bar{y}_n)\}} \qquad (6.44)$$

Then, from the definition given in Eq. (6.1), the cumulative distribution function of the extreme value for large n becomes

$$G(y_n)=\lim_{n\to\infty}\left(1-\frac{1}{n}e^{-\{q(y_n)-q(\bar{y}_n)\}}\right)^n$$

$$=\exp\{-e^{-\{q(y_n)-q(\bar{y}_n)\}}\} \qquad (6.45)$$

Since the probability distribution function of extreme values is much more concentrated around its modal value, $\bar{y}_n$, than in the case with the initial probability density function, the term $q(y_n)-q(\bar{y}_n)$ may be expanded by the Taylor series. Then, by neglecting higher-order terms, Eq. (6.45) becomes

$$G(y_n)=\exp\{-e^{-\{q'(\bar{y}_n)(y_n-\bar{y}_n)\}}\} \qquad (6.46)$$

Here, neither $q'(y_n)$ nor $\bar{y}_n$ are known in reality; hence let us express Eq. (6.46) in the following form,

$$G(z)=\exp\{-e^{-z}\} \qquad (6.47)$$

where

$$z=q'(\bar{y}_n)(y_n-\bar{y}_n) \qquad (6.48)$$

The mean and variance of the random variable z can be obtained with the aid of its characteristic function as

$$E[z]=\gamma \text{ (Euler's constant, 0.577)}$$
$$\text{Var}[z]=\pi^2/6 \qquad (6.49)$$

Then, by taking the mean and variance of z given in Eq. (6.48), we can derive the following relationship:

$$q'(\bar{y}_n)=\frac{\pi/\sqrt{6}}{\sqrt{\text{Var}[y_n]}}$$

$$\bar{y}_n=E[y_n]-\frac{\gamma}{q'(\bar{y}_n)} \qquad (6.50)$$

Since the mean and variance of y_n can be evaluated from the data, the cumulative distribution function of the extreme value $G(y_n)$ given in Eq. (6.46) can be evaluated with the aid of Eqs. (6.50).

In summary, the cumulative distribution function of the Type I asymptotic extreme value distribution is given in the following form:

$$G(y_n)=\exp\{-e^{-\alpha(y_n-u)}\}, \quad -\infty<y_n<\infty \qquad (6.51)$$

where

$$\alpha = \frac{\pi/\sqrt{6}}{\sqrt{\mathrm{Var}[y_n]}}$$

$$u = E[y_n] - \frac{\sqrt{6}}{\pi}\gamma\sqrt{\mathrm{Var}[y_n]} \qquad (6.52)$$

The question arises as to the sample space $(-\infty,\infty)$ of $G(y_n)$ given in Eq. (6.51). The sample space of waves and sea states are non-negative and thereby the cumulative distribution function must be truncated at $y_n=0$. The truncated $G(y_n)$ applicable to the sample space $(0,\infty)$ is given by

$$G_*(y_n) = \frac{1}{1-\exp\{-e^{\alpha u}\}}[\exp\{-e^{-\alpha(y_n-u)}\} - \exp\{-e^{\alpha u}\}]$$

$$0 \leqslant y_n < \infty \qquad (6.53)$$

and the probability density function $g(y_n)$ for the sample space $(0,\infty)$ becomes

$$g_*(y_n) = \frac{1}{1-\exp\{-e^{\alpha u}\}}[\alpha\exp\{-\alpha(y_n-u)\}\exp\{-e^{-\alpha(y_n-u)}\}]$$

$$0 \leqslant y_n < \infty \qquad (6.54)$$

Results of computations have shown, however, that as far as the significant wave height is concerned, the truncated negative portion is very small. For example, for the significant wave height data shown in Table 6.1, the probability of the negative portion, $G(y_n=0)$, is less than 1 percent.

As seen in Eq. (6.51), the asymptotic Type I extreme value distribution has two parameters, α and u, which have a functional relationship with the mean and variance of y_n. If the sample size n is large, the parameters may best be evaluated by using the sample mean and variance in Eq. (6.52). For a small sample size, the following equations, derived through the maximum likelihood method, may be used:

$$(1/n)\sum_{i=1}^{n}\exp\{-\alpha(y_i-u)\} = 1$$

$$(1/\alpha) + \frac{\sum_{i=1}^{n}y_i\exp\{-\alpha y_i\}}{\sum_{i=1}^{n}\exp\{-\alpha y_i\}} = (1/n)\sum_{i=1}^{n}y_i \qquad (6.55)$$

By using an iteration procedure, a solution which satisfies the second equation yields the maximum likelihood estimator $\hat{\alpha}$, and therefrom the estimator $\hat{u}$ can be obtained from the first equation. For estimating the

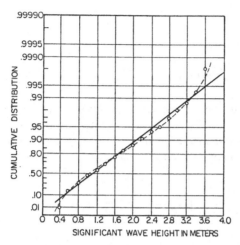

Fig. 6.9. Comparison of daily maximum significant wave height data and Type I asymptotic extreme value distribution plotted on extreme value probability paper.

parameters involved in the Type I asymptotic extreme value distribution, the results of several studies are available. These include Lettenmair and Burges (1982), Isaacson and Mackensie (1981), Carter and Challenor (1983), among others. In particular, estimation of extreme significant wave heights is discussed in the latter two references.

Figure 6.9 shows a comparison between the cumulative distribution function of the daily maximum significant wave height data given in Table 6.1 and the Type I distribution (straight line) whose parameters are evaluated by Eq. (6.52) from a sample size of 1061. The distribution functions are plotted on extreme value probability paper. As seen in the figure, except for small and large values of significant wave height, the major portion of the distribution function agrees reasonably well with the Type I asymptotic distribution. However, because of the discrepancy between the cumulative distribution functions at large significant wave heights, the estimation of extreme values by extending the Type I distribution results in a substantial overestimation. The discrepancy of the cumulative distribution functions observed in Figure 6.9 is not an exception; many examples of maxima data (not necessarily limited to wave height data) exhibit a trend similar to that presented in Figure 6.9.

6.5.2 Type III asymptotic extreme value distribution

The Type III asymptotic extreme value distribution is associated with initial distributions which are bounded toward the extreme value. Thus, the distribution is used for estimating the extreme maxima (or minima) where there exists an upper (or lower) bound in the sample space. The direct derivation of the Type III distribution is extremely complicated; however, it can be derived from the Type I distribution through the change of random variable technique. For convenience, we may write the random variables of the Type I distribution as X_n and that of the Type III distribution as Y_n, and consider the following transformation from X_n to Y_n:

$$x_n - u = -\ln\left(\frac{w - y_n}{w - v}\right) \qquad -\infty < y_n < \infty \qquad (6.56)$$

where w is the upper limit of y_n and v is a parameter to be determined from the data.

We may write the parameter α in Eq. (6.51) as k in the Type III distribution to avoid possible confusion. The transformation yields

$$G(y_n) = \exp\left\{-\left(\frac{w - y_n}{w - v}\right)^k\right\} \qquad -\infty < y_n < w$$

$$0 < v < w \qquad 0 < k < \infty \qquad (6.57)$$

This is the cumulative distribution function of the Type III asymptotic extreme value distribution having an upper limit w. Note that the sample space of the distribution given in Eq. (6.57) is $(-\infty, w)$. For the sample space $(0, w)$, the probability distribution should be truncated at $y_n = 0$, and this results in the following modified Type III asymptotic distribution:

$$G_*(y_n) = \frac{1}{1 - \exp\left\{\left(\frac{w}{w - v}\right)^k\right\}}\left[\exp\left\{-\left(\frac{w - y_n}{w - v}\right)^k\right\} - \exp\left\{-\left(\frac{w}{w - v}\right)^k\right\}\right]$$

$$0 < y_n < w \qquad (6.58)$$

Accordingly, the probability density function becomes,

$$g_*(y_n) = \frac{1}{1 - \exp\left\{\left(\frac{w}{w - v}\right)^k\right\}} \frac{k(w - y_n)^{k-1}}{(w - v)^k} \exp\left\{-\left(\frac{w - y_n}{w - v}\right)^k\right\}$$

$$0 < y < w \qquad (6.59)$$

In the case where w is known, the parameters may be evaluated through the Type I asymptotic distribution; i.e. $k = \alpha$ and v can be evaluated by the following procedure. By writing

$$z = -\ln(w - y_n) \qquad (6.60)$$

the cumulative distribution given in Eq. (6.57) can be written as

$$G(z) = \exp\left\{-e^{-k\{z + \ln(w - v)\}}\right\} \qquad (6.61)$$

Then, from comparison between Eqs. (6.51) and (6.61),

$$u = -\ln(w - v) \qquad (6.62)$$

Since w is known, the parameters k and v can be evaluated following the same procedure as shown for estimating the parameters of the Type I asymptotic distribution.

In the case where the value of the upper bound w is unknown, the three parameters in Eq. (6.57) may be determined through the maximum likelihood method, the skewness method or by the nonlinear regression method. The maximum likelihood method, however, is too complicated to apply to this distribution in practice. Estimation of the parametric values through skewness may be considered only for a sample of large size. Skewness of the Type III distribution can be expressed solely as a function of the parameter k as follows:

$$\text{skewness} = -\frac{\Gamma\left(1+\frac{3}{k}\right)-3\Gamma\left(1+\frac{2}{k}\right)\Gamma\left(1+\frac{1}{k}\right)+2\Gamma^2\left(1+\frac{1}{k}\right)}{\left\{\Gamma\left(1+\frac{2}{k}\right)-\Gamma^2\left(1+\frac{1}{k}\right)\right\}^{3/2}} \quad (6.63)$$

By equating Eq. (6.63) to the following sample skewness computed from data, the parameter k can be evaluated.

$$\text{sample skewness} = \frac{\frac{1}{n}\sum_{i=1}^{n}(y_{ni}-\bar{y}_n)^3}{\left\{\frac{1}{n}\sum_{i=1}^{n}(y_{ni}-\bar{y}_n)^2\right\}^{3/2}} \quad (6.64)$$

where

$$\bar{y}_n = \frac{1}{n}\sum_{i=1}^{n}y_{ni}$$

The parameters w and v are then subsequently determined by equating the theoretical mean and variance of the distribution to the sample mean and variance, respectively. That is,

$$E[y_n] = w - (w-v)\Gamma\left(1+\frac{1}{k}\right)$$

$$\text{Var}[y_n] = (w-v)^2\left\{\Gamma\left(1+\frac{1}{k}\right)-\Gamma^2\left(1+\frac{1}{k}\right)\right\} \quad (6.65)$$

This method of estimating the parameters may sometimes be difficult in practice in that the sample skewness asymptotically increases very slowly with increase in k-values, and thereby the k-value cannot be determined precisely for a specified skewness (Isaacson and Mackenzie 1981). The example data given in Table 6.1 is not an exception.

Another method to estimate the parameters is to apply nonlinear multiple regression analysis through an iteration procedure (Ochi *et al.* 1986, Mesa 1985). Taking the logarithm of Eq. (6.57) twice results in

$$\ln\{-\ln G(y_n)\} = k\ln\left(\frac{w-y_n}{w-v}\right) \qquad (6.66)$$

The left side of Eq. (6.66) can be evaluated from data, and the right side is linear in the parameter k, and monotonic in the parameters w and v and is determined through regression analysis. Figure 6.10 shows a comparison between the cumulative distribution function obtained from data given in Table 6.1 and that computed by applying Eq. (6.66). The computed cumulative distribution function displays a reasonable fit to the majority of the data, but some discrepancy does exist at the lower values.

In order to improve the agreement between the Type III distribution and the observed data, Eq. (6.66) is modified as follows:

$$\ln\{-\ln G(y_n)\} = k\ln\left(\frac{w-y_n}{w-v}\right) + \Delta(y_n) \qquad (6.67)$$

Here $\Delta(y_n)$ is the difference between the theoretical distribution and the observed data, and is expressed in the form of polynomials given by

$$\Delta(y_n) = a + b(y_n - y_0) + c(y_n - y_0)^2 + d(y_n - y_0)^3 \qquad (6.68)$$

The values of a, b, c, d and y_0 are determined again by employing the nonlinear regression procedure. Thus, from Eq. (6.67), the modified Type III asymptotic distribution can be written as

$$G(y_n) = \exp\left\{-\left(\frac{w-y_n}{w-v}\right)^k e^{\Delta(y_n)}\right\} \qquad (6.69)$$

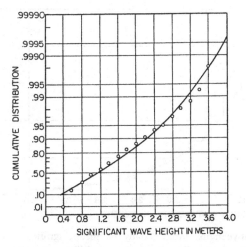

Fig.6.10. Comparison between daily maximum significant wave height data and Type III asymptotic extreme value distribution based on a non-linear regression method.

Note that $G(y_n)$ given in Eq. (6.69) satisfies the conditions required of the cumulative distribution function. Hence, the addition of $\Delta(y_n)$ does not affect the basic characteristics of the original Type III asymptotic distribution. A comparison between the cumulative distribution function given in Eq. (6.69) and that obtained from data is shown in Figure 6.11. A comparison between the probability density function of the modified Type III and the histogram is shown in Figure 6.12 in which good agreement can be seen over the entire variate range.

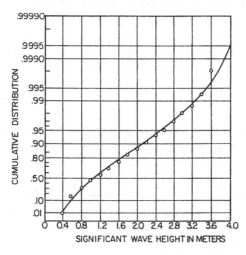

Fig. 6.11. Daily maximum significant wave height data and modified Type III asymptotic cumulative distributions (Ochi *et al.* 1986).

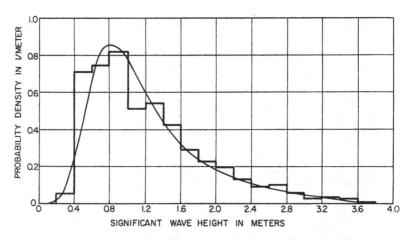

Fig. 6.12. Comparison between histogram of daily maximum significant wave height and modified Type III asymptotic probability density function (Ochi *et al.* 1986).

7 DIRECTIONAL CHARACTERISTICS OF RANDOM SEAS

7.1 INTRODUCTION

Wave spectra discussed in earlier chapters represent wave energy at a certain location in the ocean where the wave energy is an accumulation of the energy of all waves coming from various directions. The spectrum may therefore be called the *point spectrum*. The spectral analyses and prediction methods of wave heights and periods presented in the preceding chapters assume that wave energy is traveling in a specific direction, commonly considered the same direction as the wind. In this respect, the wave spectrum may be considered as a *uni-directional spectrum*.

In reality, however, wind-generated wave energy does not necessarily propagate in the same direction as the wind; instead, the energy usually spreads over various directions, though the major part of the energy may propagate in the wind direction. Thus, for an accurate description of random seas, it is necessary to clarify the spreading status of energy. The wave spectrum representing energy in a specified direction is called the *directional spectrum*, denoted by $S(\omega, \theta)$.

Information on wave directionality is extremely significant for the design of marine systems such as ships and ocean structures. This is because the responses of a system in a seaway computed using a uni-directional wave spectrum are not only overestimated but the associated coupled responses induced by waves from other directions are also disregarded.

In order to obtain information on wave directionality, extensive data acquisition techniques have been developed through application of different types of instruments, and numerous papers have been published on analysis techniques. Each of these acquisition techniques has both merits and limitations.

A comprehensive evaluation of reliability, performance and data quality of various directional wave measurement systems was carried out by Allender *et al.* (1989) by applying data obtained in severe sea conditions in the North Sea under an international project called Wave Directional Measurement Calibration (WADIC) Project. A comprehensive evaluation was also carried out on three measurement systems and various techniques from data obtained in laboratory experiments (Benoit and Teisson 1994).

For stochastic analysis of wave directionality, it is extremely important to clarify the basis supporting the estimation technique from analysis of data obtained by a particular type of instrument. This subject will be discussed in detail in Section 7.2 for the three most commonly employed instruments for estimating wave directionality; namely, the wave probe array, floating buoys and pressure and current meters.

A study of the angular energy spreading function of ocean waves was carried out in detail by Longuet-Higgins *et al.* (1961) and Cartwright (1961). From the analysis of measured data, they proposed an angular spreading function which is, by and large, the basis of various formulations developed later. The analysis of the spreading function developed by Longuet-Higgins *et al.* will be discussed in Section 7.3.

Measurements of wave directionality are extremely laborious, time-consuming and costly. Hence, it is highly desirable to estimate the directional spreading characteristics as accurately as possible from a limited amount of data. To do this, several approaches have been developed, primarily applying techniques developed in statistical inference theory. Three approaches commonly considered in this area of estimation are presented in Section 7.4.

In Section 7.5, several formulations of the energy spreading function currently being considered for the design of marine systems are summarized.

7.2 PRINCIPLE OF EVALUATION OF DIRECTIONAL WAVE SPECTRA

7.2.1 Wave probe array

Directional wave spectra can be evaluated from analysis of records obtained from several wave probes deployed over a given area of the sea. A variety of probe arrangements has been proposed depending on wave conditions at the measurement site. For example, fewer probes may be acceptable at a geographical location where the principal wind direction is known than at a location where the wind direction is variable.

As an example of probe arrangement, Figure 7.1 shows a typical five-probe arrangement called the CERC five-gage array. Four probes are arranged in a circle of radius 25.6 m with one probe at the center. The background theory supporting the evaluation of directional wave spectra by employing the wave probe array is given by Borgman and Panicker (1970) and Panicker and Borgman (1970) as follows.

Let wave probes P, Q, R, . . . be deployed arbitrarily in the ocean as shown in Figure 7.2. We first consider two wave probes P and Q with distance ℓ between them. The wave profiles measured at P and Q may be expressed in the following form given in Eq. (1.15):

$$\eta_p(\mathbf{r},t) = \mathrm{Re} \int\int e^{i(\mathbf{k}\cdot\mathbf{r}_p - \omega t + \epsilon)}\,\mathrm{d}A(\omega,\theta)$$

$$\eta_q(\mathbf{r},t) = \mathrm{Re} \int\int e^{i(\mathbf{k}\cdot\mathbf{r}_q - \omega t + \epsilon)}\,\mathrm{d}A(\omega,\theta)$$

$$= \mathrm{Re} \int\int e^{i(\mathbf{k}\cdot\mathbf{r}_p + k\ell\cos(\beta - \theta) - \omega t + \epsilon)}\,\mathrm{d}A(\omega,\theta) \qquad (7.1)$$

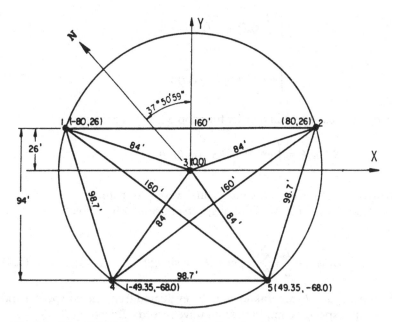

Fig. 7.1. The CERC five-gage array for directional wave measurement (Panicker and Borgman 1970).

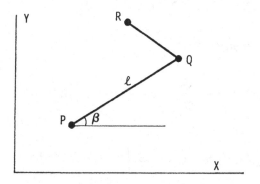

Fig. 7.2. Sketch indicating the array of wave gages.

where

$$\mathbf{k \cdot r}_p = k(x_p \cos\theta + y_p \sin\theta)$$

$$\mathbf{k \cdot r}_q = k(x_q \cos\theta + y_q \sin\theta)$$

$$dA(\omega,\theta) = \sqrt{2S(\omega,\theta)d\omega\,d\theta}$$

The cross-correlation function between P and Q can be evaluated from Eq. (7.1) as

$$R_{pq}(\theta,\tau) = \int_{-\infty}^{\infty} \frac{1}{2T} \eta_p(\mathbf{r},t)\,\eta_q(\mathbf{r},t+\tau)dt$$

$$= \int_{-\pi}^{\pi} e^{ik\ell\cos(\beta-\theta)} R_{pp}(\theta,\tau)d\theta \qquad (7.2)$$

Then, the cross-spectral density function $S_{pq}(\omega,\theta)$ becomes

$$S_{pq}(\omega,\theta) = \int_{-\pi}^{\pi} e^{ik\ell\cos(\beta-\theta)} S_{pp}(\omega,\theta)d\theta \qquad (7.3)$$

where $S_{pp}(\omega,\theta) = S(\omega,\theta)$, the directional wave spectrum.

Let us express the directional spectrum $S(\omega,\theta)$ in a Fourier series in terms of θ:

$$S(\omega,\theta) = \frac{a_0}{2} + \sum_{n=1}^{N} (a_n \cos n\theta + b_n \sin n\theta) \qquad (7.4)$$

where a_0, a_n and b_n are unknown to be evaluated from the co-spectra and quadrature spectra computed from wave records. By letting $\beta - \theta = \phi$ and integrating with respect to ϕ, the cross-spectrum $S_{pq}(\omega)$ can be written from Eqs. (7.3) and (7.4) as

$$S_{pq}(\omega) = \int_{-\pi}^{\pi} \frac{a_0}{2} e^{ik\ell\cos\phi}d\phi + \int_{-\pi}^{\pi}\sum_{n=1}^{\infty} \{a_n \cos n(\beta-\phi)$$

$$+ b_n \sin n(\beta-\phi)\} e^{ik\ell\cos\phi}d\phi$$

$$= \int_{-\pi}^{\pi} \frac{a_0}{2} e^{ik\ell\cos\phi}d\phi + \sum_{n=1}^{\infty} (a_n \cos n\beta + b_n \sin n\beta)\int_{-\pi}^{\pi}\cos n\phi\, e^{ik\ell\cos\phi}d\phi$$

$$+ \sum_{n=1}^{\infty} (a_n \sin n\beta - b_n \cos n\beta)\int_{-\pi}^{\pi}\sin n\phi\, e^{ik\ell\cos\phi}d\phi \qquad (7.5)$$

Note that

$$\int_{-\pi}^{\pi} \cos n\phi \, e^{ik\ell \cos \phi} d\phi = 2\pi i^n \mathcal{J}_n(k\ell)$$

$$\int_{-\pi}^{\pi} \sin n\phi \, e^{ik\ell \cos \phi} d\phi = 0 \tag{7.6}$$

where $\mathcal{J}_n(k\ell)$ is the Bessel function of order n.

Thus, the cross-spectrum $S_{pq}(\omega)$ can be expressed in a Fourier series as follows:

$$S_{pq}(\omega) = \pi \left\{ a_0 \mathcal{J}_0(k\ell) + 2\sum_n i^n (a_n \cos n\theta + b_n \sin n\theta) \mathcal{J}_n(k\ell) \right\} \tag{7.7}$$

We may write the above equation as

$$S_{pq}(\omega) = \pi \left[\frac{a_0}{2} A_0 + \sum_n i^n (a_n A_n + b_n B_n) \right] \tag{7.8}$$

where A_0, A_n and B_n are known quantities given by

$$A_0 = \mathcal{J}_0(k\ell)$$
$$A_n = 2 \cos n\beta \, \mathcal{J}_n(k\ell)$$
$$B_n = 2 \sin n\beta \, \mathcal{J}_n(k\ell) \tag{7.9}$$

Thus, the real and imaginary parts of the cross-spectrum $S_{pq}(\omega)$ can be written as follows:

$$C_{pq}(\omega) = \pi \left[\frac{a_0}{2} A_0 - (a_2 A_2 + b_2 B_2) + (a_4 A_4 + b_4 B_4) + \dots \right]$$

$$Q_{pq}(\omega) = \pi [(a_1 A_1 + b_1 B_1) - (a_3 A_3 + b_3 B_3) + \dots] \tag{7.10}$$

where a_0, a_1, b_1, b_2, etc., are unknown.

Equations (7.10) shows the result of cross-spectral analysis of records obtained by two wave gages P and Q. We may carry out a similar analysis by employing several wave probes. Then, we have two equations from cross-spectral analysis for any pair of gages, and independently one equation from auto-spectral analysis. Hence, by arranging n ($n \geq 2$) gages, we have $n(n-1)+1$ known quantities. On the other hand, we have $(2m+1)$ unknowns for m terms of the Fourier series. This implies that spectral analysis of three wave gages yields seven known quantities, which agrees with the number of unknowns for a Fourier series with $m=3$. However, for four gages, we have 13 known quantities which is a greater number than the 11 unknowns for a Fourier series expansion to $m=5$. One way to evaluate the unknown coefficients in this case is to apply the least squares analysis method.

As an example of application of this approach, Panicker and Borgman present contour curves of the directional spectrum (shown in Figure 7.3). The figure shows a typical bimodal spectrum off the Paciffic coast with directional angle between 50 and 90 degrees.

7.2.2 Floating buoys

The directional wave spectrum $S(\omega,\theta)$ may be expressed as a product of the point spectrum $S(\omega)$ and a function representing the spreading characteristics, $D(\omega,\theta)$, called the *energy spreading function*. Let us first express the directional spectrum $S(\omega,\theta)$ in a Fourier series in terms of θ as

$$S(\omega,\theta)=\frac{a_0}{2}+\sum_{n=1}^{\infty}(a_n\cos n\theta+b_n\sin n\theta) \tag{7.11}$$

where

$$a_n=\frac{1}{\pi}\int_{-\pi}^{\pi}S(\omega,\theta)\cos n\theta\,d\theta$$

$$b_n=\frac{1}{\pi}\int_{-\pi}^{\pi}S(\omega,\theta)\sin n\theta\,d\theta$$

Next, we write

$$S(\omega,\theta)=S(\omega)D(\omega,\theta) \tag{7.12}$$

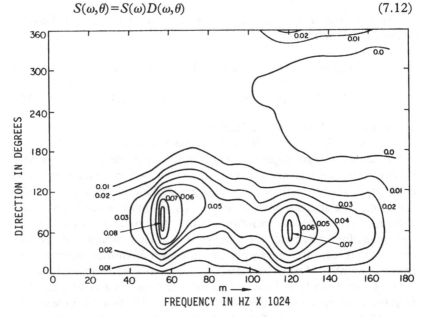

Fig. 7.3. Contour plot of directional spectra obtained from analysis of data measured by the CERC five-gage array (Panicker and Borgman 1970).

where

$$\int_{-\pi}^{\pi} D(\omega,\theta)\, d\theta = 1$$

From Eqs. (7.11) and (7.12), the energy spreading function $D(\omega,\theta)$ may be expressed in the following Fourier series:

$$D(\omega,\theta) = \frac{1}{\pi}\left\{\frac{1}{2} + \sum_{n=1}^{\infty}(A_n \cos n\theta + B_n \sin n\theta)\right\} \qquad (7.13)$$

where

$$A_n = \int_{-\pi}^{\pi} D(\omega,\theta) \cos n\theta\, d\theta = \frac{1}{S(\omega)}\int_{-\pi}^{\pi} S(\omega,\theta) \cos n\theta\, d\theta$$

$$B_n = \int_{-\pi}^{\pi} D(\omega,\theta) \sin n\theta\, d\theta = \frac{1}{S(\omega)}\int_{-\pi}^{\pi} S(\omega,\theta) \sin n\theta\, d\theta$$

The above equations are the basis for the assessment of wave directional characteristics employing a floating buoy; the coefficients A_n and B_n are evaluated through auto- and cross-spectral analysis of wave displacement (or acceleration), slopes and curvatures.

(a) Pitch–roll buoy

One of the simplest but well-designed devices is the pitch–roll buoy developed by Longuet-Higgins et al. (1961). The buoy is a circular disk 1.71 m in diameter inside which an accelerometer and gyroscopes are accommodated for measurement of vertical acceleration (or converted to displacement) and slopes of incident waves in two rectangular directions. The theoretical background for obtaining the wave directionality through this device is as follows.

Let us first consider that we have three components; wave vertical displacement η which is converted from acceleration, and slopes $\partial\eta/\partial x$ and $\partial\eta/\partial y$. For convenience, these three records are denoted 1, 2 and 3, respectively. These components can be theoretically written as

$$\eta = \mathrm{Re}\int\int e^{i(\mathbf{k}\cdot\mathbf{r}-\omega t+\epsilon)}dA(\omega,\theta)$$

$$\frac{\partial\eta}{\partial x} = \mathrm{Re}\int\int ik \cos\theta\, e^{i(\mathbf{k}\cdot\mathbf{r}-\omega t+\epsilon)}dA(\omega,\theta)$$

$$\frac{\partial\eta}{\partial y} = \mathrm{Re}\int\int ik \sin\theta\, e^{i(\mathbf{k}\cdot\mathbf{r}-\omega t+\epsilon)}dA(\omega,\theta) \qquad (7.14)$$

As an example, let us evaluate the cross-correlation function between η and $\partial\eta/\partial x$:

$$R_{12}(\tau)=\lim_{T\to\infty}\frac{1}{2T}\int_{-T}^{T}\eta(t)\frac{\partial}{\partial x}\eta(t+\tau)dt$$

$$=\lim_{T\to\infty}\frac{1}{2T}\int_{-T}^{T}\left(\mathrm{Re}\int_{-\pi}^{\pi}\int_{0}^{\infty}e^{i(\mathbf{k}\cdot\mathbf{r}-\omega t+\epsilon)}dA(\omega,\theta)\right)$$

$$\times\left(\mathrm{Re}\int_{-\pi}^{\pi}\int_{0}^{\infty}ik\cos\theta(e^{i\{\mathbf{k}\cdot\mathbf{r}-\omega(t+\tau)+\epsilon\}})dA(\omega,\theta)\right)dt$$

$$=\int_{-\pi}^{\pi}ik\cos\theta(R(\tau,\theta))d\theta \qquad (7.15)$$

where $R(\tau,\theta)$ is the directional auto-correlation function.

By taking the Fourier transform of $R_{12}(\tau)$, the cross-spectral density $S_{12}(\omega)$ becomes

$$S_{12}(\omega)=C_{12}(\omega)+iQ_{12}(\omega)=\int_{-\pi}^{\pi}ik\cos\theta(S(\omega,\theta))d\theta \qquad (7.16)$$

Thus, we can write

$$C_{12}(\omega)=0$$

$$Q_{12}(\omega)=\int_{-\pi}^{\pi}k\cos\theta(S(\omega,\theta))d\theta \qquad (7.17)$$

By carrying out auto- as well as cross-spectral analysis similar to that given in Eq. (7.16) on all three components, we have

$$C_{11}(\omega)=\int_{-\pi}^{\pi}S(\omega,\theta)d\theta=S(\omega) \qquad\qquad Q_{11}(\omega)=0$$

$$C_{22}(\omega)=k^2\int_{-\pi}^{\pi}\cos^2\theta(S(\omega,\theta))d\theta \qquad Q_{22}(\omega)=0$$

$$C_{33}(\omega)=k^2\int_{-\pi}^{\pi}\sin^2\theta(S(\omega,\theta))d\theta \qquad Q_{33}(\omega)=0$$

$$C_{12}(\omega)=0 \qquad\qquad Q_{12}(\omega)=k\int_{-\pi}^{\pi}\cos\theta(S(\omega,\theta))d\theta$$

$$C_{13}(\omega)=0 \qquad\qquad Q_{13}(\omega)=k\int_{-\pi}^{\pi}\sin\theta(S(\omega,\theta))d\theta$$

$$C_{23}(\omega) = k^2 \int_{-\pi}^{\pi} \sin\theta\cos\theta(S(\omega,\theta))\mathrm{d}\theta \quad Q_{23}(\omega) = 0 \qquad (7.18)$$

where

k = wave number = ω^2/g
$C_{ii}(\omega) = S_i(\omega)$ = auto-spectrum of the component i
$C_{ij}(\omega)$ = co-spectrum of components i and j
$Q_{ij}(\omega)$ = quadrature spectrum of components i and j.

From comparison of Eqs. (7.13) and (7.18), A_n and B_n can be obtained as follows:

$$A_1 = \frac{1}{k}\frac{Q_{12}(\omega)}{C_{11}(\omega)}$$

$$B_1 = \frac{1}{k}\frac{Q_{13}(\omega)}{C_{11}(\omega)}$$

$$A_2 = \frac{1}{k^2}\frac{C_{22}(\omega) - C_{33}(\omega)}{C_{11}(\omega)}$$

$$B_2 = \frac{2}{k^2}\frac{C_{23}(\omega)}{C_{11}(\omega)} \qquad (7.19)$$

Thus, it is clear that the coefficients of the Fourier series representing the energy spreading function can be obtained by carrying out auto- and cross-spectral analysis of the records of displacement and two slopes. Since the number of measured components is limited to three, the Fourier expansion can be made only up to $n=2$. This sometimes causes the directional spreading function to become negative (a minor amount) at the tail portion of the spreading function. In order to let the partial sum of the Fourier series expansion be equivalent to the infinite sum, Longuet-Higgins *et al.* (1961) write Eq. (7.12) as

$$D(\omega,\theta) = \frac{1}{\pi}\left\{\frac{1}{2} + \sum_{n=1}^{N} w_n(A_n\cos n\theta + B_n\sin n\theta)\right\} \qquad (7.20)$$

and they set $w_1 = 2/3$ and $w_2 = 1/6$ for $N=2$.

In the case that the vertical acceleration obtained by the accelerometer is used in the analysis, the formula for evaluating A_n and B_n must be modified by writing the vertical acceleration (in g-units) as

$$\alpha = \mathrm{Re}\int\int e^{i(\mathbf{k}\cdot\mathbf{r} - \omega t + \epsilon)}\mathrm{d}A_\alpha(\omega,\theta) \qquad (7.21)$$

where $\mathrm{d}A_\alpha(\omega,\theta) = \sqrt{2S_\alpha(\omega,\theta)}\ \mathrm{d}\omega\,\mathrm{d}\theta$

$S_\alpha(\omega,\theta)$ = directional spectrum of vertical acceleration.

The vertical displacement can then be written as

$$\eta = -\mathrm{Re}\int\int \frac{g}{\omega^2} e^{i(\mathbf{k}\cdot\mathbf{r}-\omega t+\epsilon)}\mathrm{d}A_\alpha(\omega,\theta) \tag{7.22}$$

and thereby the two slopes become

$$\frac{\partial\eta}{\partial x} = -\mathrm{Re}\int\int (i\cos\theta)\, e^{i(\mathbf{k}\cdot\mathbf{r}-\omega t+\epsilon)}\mathrm{d}A_\alpha(\omega,\theta)$$

$$\frac{\partial\eta}{\partial y} = -\mathrm{Re}\int\int (i\sin\theta)\, e^{i(\mathbf{k}\cdot\mathbf{r}-\omega t+\epsilon)}\mathrm{d}A_\alpha(\omega,\theta) \tag{7.23}$$

We may denote α, $\partial\eta/\partial x$ and $\partial\eta/\partial y$ as 1, 2 and 3, respectively, and the auto-spectra and cross-spectra carry the subindex α to indicate that the analysis is associated with acceleration record. For instance, $S_\alpha(\omega)$ represents the acceleration spectrum. The coefficients A_n and B_n in Eq. (7.13) become, in this case,

$$A_1 = -Q_{12\alpha}(\omega)/C_{11\alpha}(\omega)$$
$$B_1 = -Q_{13\alpha}(\omega)/C_{11\alpha}(\omega)$$
$$A_2 = \{-C_{22}(\omega)+C_{33}(\omega)\}/C_{11\alpha}(\omega)$$
$$B_2 = -2C_{23}(\omega)/C_{11\alpha}(\omega) \tag{7.24}$$

where

$$C_{11\alpha}(\omega) = S_\alpha(\omega) = \text{acceleration auto-spectrum}$$

$$C_{22}(\omega) = -\int_{-\pi}^{\pi} \cos^2\theta(S_\alpha(\omega,\theta))\mathrm{d}\theta$$

$$C_{33}(\omega) = -\int_{-\pi}^{\pi} \sin^2\theta(S_\alpha(\omega,\theta))\mathrm{d}\theta$$

$$C_{23}(\omega) = -\int_{-\pi}^{\pi} \sin\theta\cos\theta(S_\alpha(\omega,\theta))\mathrm{d}\theta$$

$$Q_{12\alpha}(\omega) = -\int_{-\pi}^{\pi} \cos\theta(S_\alpha(\omega,\theta))\mathrm{d}\theta$$

$$Q_{13\alpha}(\omega) = -\int_{-\pi}^{\pi} \sin\theta(S_\alpha(\omega,\theta))\mathrm{d}\theta$$

Because of the definition given in Eqs. (7.21) and (7.23), the sign of the coefficients in Eq. (7.24) differs from that given in Cartwright and Smith (1964). However, the mathematical formulation of the co-spectra and quadrature spectra carry the negative sign except for $C_{11\alpha}(\omega)$ as shown above. Therefore, in practice, the coefficients A_1, B_1, A_2 and B_2 are the same as given in the reference.

Ocean directional wave data have been routinely obtained by the National Data Buoy Center by employing pitch–roll buoys. Measurements and analyses of these data are discussed in detail by Steele *et al.* (1992).

(b) Cloverleaf buoy

As shown in the previous section, we can evaluate the energy spreading function as a Fourier series with two terms by employing the pitch–roll buoy. In order to improve accuracy in evaluating wave directionality, it is essential to increase the number of terms in the Fourier series. The cloverleaf buoy shown in Figure 7.4, developed by Cartwright and Smith (1964), is unique and the best available for this purpose. In the buoy, three rotatable circular floats are attached at each apex of an equilateral triangular frame. The distance between floats is 2 m. The measurement outputs of this buoy are vertical acceleration by a gyro-stabilized accelerometer, slopes in two rectangular directions at each float, roll and pitch angles at the center of the buoy, and compass direction of the entire system.

Let $\eta_x(A)$, $\eta_x(B)$, $\eta_x(C)$, $\eta_y(A)$, $\eta_y(B)$, and $\eta_y(C)$ be the time history of the slopes in the X- and Y-directions at floats A, B and C, respectively. Then, the time history of the curvature at the center of the buoy can be obtained taking into consideration the distance between floats as follows:

$$\frac{\partial^2 \eta}{\partial x^2} = \{2\eta_x(A) - \eta_x(B) - \eta_x(C)\}/2\sqrt{3}$$

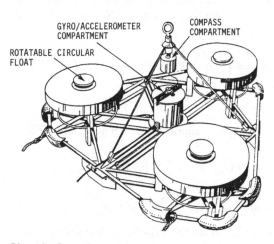

Fig. 7.4. Cloverleaf wavebuoy.

$$\frac{\partial^2 \eta}{\partial y^2} = \{\eta_y(C) - \eta_y(B)\}/2$$

$$\frac{\partial^2 \eta}{\partial x \partial y} = \{\eta_x(C) - \eta_x(B)\}/2 \tag{7.25}$$

On the other hand, we have the mathematical presentation of the vertical acceleration of the buoy in Eq. (7.21) and two slopes $\partial \eta/\partial x$ and $\partial \eta/\partial y$ in Eq. (7.23). These are denoted 1, 2 and 3, respectively. The curvatures can be obtained by differentiating the slopes with respect to x and/or y as follows:

$$\frac{\partial^2 \eta}{\partial x^2} = \mathrm{Re} \int \int (k \cos \theta)^2 e^{i(\mathbf{k \cdot r} - \omega t + \epsilon)} \mathrm{d}A_\alpha(\omega, \theta)$$

$$\frac{\partial^2 \eta}{\partial y^2} = \mathrm{Re} \int \int (k \sin \theta)^2 e^{i(\mathbf{k \cdot r} - \omega t + \epsilon)} \mathrm{d}A_\alpha(\omega, \theta)$$

$$\frac{\partial^2 \eta}{\partial x \partial y} = \mathrm{Re} \int \int (k^2 \sin \theta \cos \theta \; e^{i(\mathbf{k \cdot r} - \omega t + \epsilon)} \mathrm{d}A_\alpha(\omega, \theta) \tag{7.26}$$

These equations are denoted 4, 5 and 6, respectively. Then, auto- and cross-spectral analysis is carried out on six components (1 to 6) following the same procedure as the example shown in Eqs. (7.15) through (7.17), and the results are presented in the form

$$S_{ij}(\omega) = \int_{-\pi}^{\pi} h(k, \theta) S_\alpha(\omega, \theta) \mathrm{d}\theta \tag{7.27}$$

where $S_\alpha(\omega, \theta)$ is the directional acceleration spectrum, and $h(k, \theta)$ is tabulated in Table 7.1. For example, the cross-spectrum between acceleration α (1) and curvature $\partial^2 \eta/\partial y^2$ (5) is denoted $S_{15\alpha}(\omega)$ and its real and imaginary parts are

$$C_{15\alpha}(\omega) = \int_{-\pi}^{\pi} k \sin^2 \theta S_\alpha(\omega, \theta) \mathrm{d}\theta$$

$$Q_{15\alpha}(\omega) = 0 \tag{7.28}$$

Thus, the following equations can be derived using Table 7.1.

$$C_{14\alpha}(\omega) + C_{15\alpha}(\omega) = k \int_{-\pi}^{\pi} S_\alpha(\omega, \theta) \mathrm{d}\theta = k S_\alpha(\omega)$$

$$C_{44}(\omega) + 2C_{45}(\omega) + C_{55}(\omega) = k^2 S_\alpha(\omega) \tag{7.29}$$

where $C_{44}(\omega)$ and $C_{55}(\omega)$ are the auto-spectra of the curvatures of $\partial^2 \eta/\partial x^2$ and $\partial^2 \eta/\partial y^2$, respectively. We can now obtain the coefficients A_3, B_3, etc.

Table 7.1. *Function $h(k, \theta)$ for evaluating auto and cross-spectra applicable to analysis of data obtained by cloverleaf buoy measurements.*

	1 Acceleration α		2 Slope $\dfrac{\partial\eta}{\partial x}$		3 $\dfrac{\partial\eta}{\partial y}$		4 Curvature $\dfrac{\partial^2\eta}{\partial x^2}$		5 $\dfrac{\partial^2\eta}{\partial y^2}$		6 $\dfrac{\partial^2\eta}{\partial x\partial y}$	
	$C(\omega)$	$Q(\omega)$	$C(\omega)$	$Q(\omega)$	$C(\omega)$	$Q(\omega)$	$C(\omega)$	$Q(\omega)$	$C(\omega)$	$Q(\omega)$	$C(\omega)$	$Q(\omega)$
1 α	1	0	0	$-\cos\theta$	0	$-\sin\theta$	$k\cos^2\theta$	0	$k\sin^2\theta$	0	$k\sin\theta\cos\theta$	0
2 $\dfrac{\partial\eta}{\partial x}$			$-\cos^2\theta/x$	0	$-\sin\theta\cos\theta$	0	0	$-k\cos^3\theta$	0	$-k\sin^2\theta\cos\theta$	0	$-k\sin\theta\cos^2\theta$
3 $\dfrac{\partial\eta}{\partial y}$					$-\sin^2\theta$	0	0	$-k\sin\theta\cos^2\theta$	0	$-k\sin^3\theta$	0	$-k\sin^2\theta\cos\theta$
4 $\dfrac{\partial^2\eta}{\partial x^2}$							$k^2\cos^4\theta$	0	$k^2\sin^2\theta\cos^2\theta$	0	$k^2\sin\theta\cos^3\theta$	0
5 $\dfrac{\partial^2\eta}{\partial y^2}$									$k^2\sin^4\theta$	0	$k^2\sin^3\theta\cos\theta$	0
6 $\dfrac{\partial^2\eta}{\partial x\partial y}$											$k^2\sin^2\theta\cos^2\theta$	0

of the Fourier series with the aid of Eqs. (7.28) and (7.29) along with Table 7.1. For instance, from Eq. (7.13)

$$A_3 = \frac{1}{S_\alpha(\omega)} \int_{-\pi}^{\pi} S_\alpha(\omega,\theta) \cos 3\theta \, d\theta$$

$$= \frac{1}{S_\alpha(\omega)} \int_{-\pi}^{\pi} S_\alpha(\omega,\theta) \, (\cos^3\theta - 3\sin^2\theta\cos\theta) d\theta$$

$$= \{3Q_{25}(\omega) - Q_{24}(\omega)\} / \{C_{14\alpha}(\omega) + C_{15\alpha}(\omega)\} \tag{7.30}$$

Similarly, we obtain B_3, A_4 and B_4 as follows:

$$B_3 = \{Q_{35}(\omega) - 3Q_{34}(\omega)\} / \{C_{14}(\omega) + C_{15}(\omega)\}$$
$$A_4 = \{C_{44}(\omega) - 6C_{45}(\omega) + C_{55}(\omega)\} / \{C_{44}(\omega) + 2C_{45}(\omega) + C_{55}(\omega)\}$$
$$B_4 = 4\{C_{46}(\omega) - C_{56}(\omega)\} / \{C_{44}(\omega) + 2C_{45}(\omega) + C_{55}(\omega)\} \tag{7.31}$$

Thus, we have the coefficients of the Fourier series given in Eq. (7.13) up to $n=4$. As in the case of the pitch–roll buoy, the partial sum of the Fourier series representing the energy spreading function obtained by employing the cloverleaf buoy should be weighted. The weighting factors for $N=4$ are given by Longuet-Higgins et al. as $w_1=8/9$, $w_2=28/45$, $w_3=56/165$ and $w_4=14/99$.

7.2.3 Pressure and current meters

Another technique to evaluate the directional spreading function $D(\omega,\theta)$ is to measure the pressure and two rectangular components of the wave particle velocities U and V. The instrument used for this is often called the PUV-meter, and is suitable for measurement of wave directionality in areas of finite water depth. The basic principle for evaluating the directionality by the PUV-meter is similar to that for the pitch–roll buoy. That is, we may write the pressure and the two rectangular velocity components as

$$p = \text{Re} \int_{-\pi}^{\pi} \int_{0}^{\infty} K_p(\omega) \, e^{i(\mathbf{k}\cdot\mathbf{r} - \omega t + \epsilon)} dA(\omega,\theta)$$

$$u = \text{Re} \int_{-\pi}^{\pi} \int_{0}^{\infty} K_c(\omega) \cos\theta \, e^{i(\mathbf{k}\cdot\mathbf{r} - \omega t + \epsilon)} dA(\omega,\theta)$$

$$v = \text{Re} \int_{-\pi}^{\pi} \int_{0}^{\infty} K_c(\omega) \sin\theta \, e^{i(\mathbf{k}\cdot\mathbf{r} - \omega t + \epsilon)} dA(\omega,\theta) \tag{7.32}$$

where

$$K_p(\omega) = \rho g \frac{\cosh k(z+h)}{\cosh kh}$$

$$K_c(\omega) = \omega \frac{\cosh k(z+h)}{\sinh kh} \tag{7.33}$$

and $z =$ depth where measurement is made
(negative downward)
  $h =$ water depth
From Eq. (7.32) the auto- and cross-spectral density function of these three components can be written as follows:

$$S_{pp}(\omega) = K_p^2(\omega) \int_{-\pi}^{\pi} S(\omega,\theta)d\theta = K_p^2(\omega)S(\omega)$$

$$S_{uu}(\omega) = K_c^2(\omega) \int_{-\pi}^{\pi} S(\omega,\theta)\cos^2\theta\, d\theta$$

$$S_{vv}(\omega) = K_c^2(\omega) \int_{-\pi}^{\pi} S(\omega,\theta)\sin^2\theta\, d\theta$$

$$C_{pu}(\omega) = K_p(\omega)K_c(\omega) \int_{-\pi}^{\pi} S(\omega,\theta)\cos\theta\, d\theta$$

$$C_{pv}(\omega) = K_p(\omega)K_c(\omega) \int_{-\pi}^{\pi} S(\omega,\theta)\sin\theta\, d\theta$$

$$C_{uv}(\omega) = K_c^2(\omega) \int_{-\pi}^{\pi} S(\omega,\theta)\sin\theta\cos\theta\, d\theta \tag{7.34}$$

By using these spectral density functions, the coefficients of the Fourier series given in Eq. (7.13) can be evaluated as

$$A_1 = \frac{1}{S(\omega)} \int_{-\pi}^{\pi} S(\omega,\theta)\cos\theta\, d\theta = \frac{C_{pu}(\omega)}{K_p(\omega)\,K_c(\omega)} \frac{K_p^2(\omega)}{S_{pp}(\omega)}$$

$$= \frac{K_p(\omega)\,C_{pu}(\omega)}{K_c(\omega)\,S_{pp}(\omega)}$$

$$B_1 = \frac{1}{S(\omega)} \int_{-\pi}^{\pi} S(\omega,\theta)\sin\theta\, d\theta = \frac{1}{K_p(\omega)\,K_c(\omega)} \frac{C_{pv}(\omega)}{S(\omega)}$$

$$= \frac{K_p(\omega)\,C_{pv}(\omega)}{K_c(\omega)\,S_{pp}(\omega)}$$

$$A_2 = \frac{1}{S(\omega)} \int_{-\pi}^{\pi} S(\omega,\theta) \cos 2\theta \, d\theta = \frac{1}{K_c^2(\omega)} \frac{S_{uu}(\omega) - S_{vv}(\omega)}{S(\omega)}$$

$$= \frac{K_p^2(\omega)}{K_c^2(\omega)} \frac{S_{uu}(\omega) - S_{vv}(\omega)}{S_{pp}(\omega)}$$

$$B_2 = \frac{1}{S(\omega)} \int_{-\pi}^{\pi} S(\omega,\theta) \sin 2\theta \, d\theta = \frac{1}{K_c^2(\omega)} \frac{2C_{uv}(\omega)}{S(\omega)}$$

$$= \frac{2K_p^2(\omega)}{K_c^2(\omega)} \frac{C_{uv}(\omega)}{S_{pp}(\omega)} \tag{7.35}$$

Thus, the energy spreading function can be obtained from Eq. (7.20) in the form of a Fourier series with terms up to $n=2$.

7.3 ANALYSIS OF DIRECTIONAL ENERGY SPREADING FUNCTION

The directional energy spreading function presented in the form of a Fourier series is shown in Eq. (7.13). Longuet-Higgins et al. (1961) carry out an analysis of the energy spreading function and propose that the spreading function is proportional to an even-numbered power of one-half of the cosine of the directional angle. That is,

$$D(\omega,\theta) = \left| \cos \frac{1}{2} (\theta - \bar{\theta}) \right|^{2s} G(s) \tag{7.36}$$

where

$$\bar{\theta} = \text{average direction of energy spreading} = \tan^{-1}(B_1/A_1)$$

$$= \tan^{-1} \frac{Q_{13}(\omega)}{Q_{12}(\omega)} = \tan^{-1} \frac{\int_{-\pi}^{\pi} S(\omega) \sin\theta \, d\theta}{\int_{-\pi}^{\pi} S(\omega) \cos\theta \, d\theta} \tag{7.37}$$

The function $G(s)$ is a normalizing factor determined from the condition that the integration of $D(\omega,\theta)$ from $-\pi$ to π with respect to θ is unity. In integrating Eq. (7.36) with respect to θ, it is convenient to apply the following formula:

$$(\cos x)^{2s} = \frac{1}{2^{2s-1}} \left[\sum_{r=0}^{s-1} \binom{2s}{r} \cos 2(s-r)x + \frac{1}{2} \binom{2s}{s} \right] \tag{7.38}$$

Integration of the first term (the summation term) of the above equation from $-\pi$ to π with respect to x becomes zero, and integration of the second term is a constant which yields

$$G(s) = \frac{2^{2s-1}}{\pi} \frac{\{\Gamma(s+1)\}^2}{\Gamma(2s+1)} \tag{7.39}$$

Thus, Longuet-Higgins *et al.* show that the directional spreading function can be presented as follows:

$$D(\omega,\theta) = \frac{1}{\pi} 2^{2s-1} \frac{\{\Gamma(s+1)\}^2}{\Gamma(2s+1)} |\cos \frac{1}{2}(\theta-\bar{\theta})|^{2s} \tag{7.40}$$

where the parameter s is a positive real number which controls the degree of concentration of the spreading energy around the mean value $\bar{\theta}$. In reality, however, s depends on frequency and wind speed as well as fetch length, as will be shown later. Figure 7.5 shows $D(\omega,\theta)$ multiplied by π for various s-values with $\bar{\theta}=0$.

It is of interest to examine how well the spreading function given in Eq. (7.40) agrees with the formula in a Fourier series as shown in Eq. (7.13). For this, Mitsuyasu *et al.* (1975) carry out the following analysis.

First, the spreading function given in Eq. (7.13) is rewritten by letting $A_n = C_n \cos \epsilon$ and $B_n = C_n \sin \epsilon$. That is,

$$D(\omega,\theta) = \frac{1}{\pi}\left\{\frac{1}{2} + \sum_{n=1}^{\infty} C_n \cos(n\theta - \epsilon)\right\} \tag{7.41}$$

where

$$C_n = \sqrt{A_n^2 + B_n^2}$$

$$\epsilon = \tan^{-1}(B_n/A_n)$$

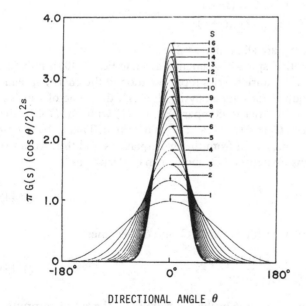

Fig. 7.5. $\pi D(\omega,\theta)$ for various s-values with $\bar{\theta}=0$.

Here, $\tan^{-1}(B_n/A_n)$ is approximated by $\bar{\theta}$, and furthermore for convenience $\bar{\theta}=0$. Then, Eq. (7.41) may be simply written as

$$D(\omega,\theta)=\frac{1}{\pi}\left\{\frac{1}{2}+\sum_{n=1}^{\infty}C_n\cos n\theta\right\} \tag{7.42}$$

where C_n are known quantities, since A_n and B_n are evaluated from analysis of the data.

Next, Eq. (7.40) is rewritten by letting $\bar{\theta}=0$. By applying Eq. (7.38), $D(\omega,\theta)$ becomes

$$D(\omega,\theta)=\frac{1}{\pi}\,2^{2s-1}\frac{\{\Gamma(s+1)\}^2}{\Gamma(2s+1)}\left[\frac{1}{2^{2s-1}}\sum_{r=0}^{s-1}\binom{2s}{r}\cos(s-r)\theta+\frac{1}{2}\binom{2s}{r}\right]$$

$$=\frac{1}{\pi}\left[\frac{1}{2}+\frac{s}{s+1}\cos\theta+\frac{s(s-1)}{(s+1)(s+2)}\cos 2\theta\right.$$

$$\left.+\frac{s(s-1)(s-2)}{(s+1)(s+2)(s+3)}\cos 3\theta+...\right]$$

$$=\frac{1}{\pi}\left[\frac{1}{2}+\sum_{n=1}^{\infty}C_n\cos n\theta\right] \tag{7.43}$$

where

$$C_n=\frac{s(s-1)(s-2)...(s-n+1)}{(s+1)(s+2)...(s+n)}\qquad n=1,2,3,...$$

and $(s-1)$, $(s-2)$, etc. are all absolute values.

Thus, the theoretical spreading function given in Eq. (7.40) is reduced to the same form as a Fourier series representation of the energy spreading function developed for data analysis, although the value of s in Eq. (7.43) is unknown. In order to compare Eq. (7.41) with Eq. (7.43), the relationship between C_1 and C_2, C_1 and C_3, is obtained. That is, by letting $n=1$ and $n=2$, s is eliminated from the two equations and thus from the theoretical relationship between C_1 and C_2 can be derived as

$$C_2=\frac{C_1(2C_1-1)}{2-C_1} \tag{7.44}$$

Similarly, the relationship between C_1 and C_3 becomes

$$C_3=\frac{C_1(2C_1-1)(3C_1-2)}{(2-C_1)(3-2C_1)} \tag{7.45}$$

Since C_1, C_2 and C_3 can be evaluated from data, the result computed from data and theory can be compared. Figure 7.6 shows an example of the relationship between theory and data obtained by Mitsuyasu *et al.* (1975) by using the cloverleaf buoy. As seen in the figure, agreement between the theoretical and measured relationship between C_1 and C_2 is excellent, but the measured, C_3 value for a specified C_1 is greater than the theoretical value. It may be concluded, therefore, that the theoretically derived energy spreading function given in Eq. (7.41) is a fairly good first approximation for the wave energy distribution function.

Mitsuyasu *et al.* (1975) carry out further analysis for evaluating the value of s in Eq. (7.40). Suppose the energy spreading function given in Eq. (7.41) is a perfect fit to the measured data, then all s-values evaluated from C_n in Eq. (7.43) must be the same. However, this is not the case in practice. Let s_n be the s-value evaluated from C_n. Mitsuyasu *et al.* find that only s_1 and s_2 are nearly equal, but s_3 and s_4 are different from s_1, perhaps due to an inaccuracy involved in the measurement of wave curvatures. Therefore, the averaged value of s_1 and s_2, denoted $\bar{s}$, is used for finding a functional relationship with frequency.

Figure 7.7 shows the functional relationship between the parameter $\bar{s}$ and the dimensionless frequency $\bar{f}$ obtained from five sets of measured data. Here, the dimensionless frequency is defined as $\bar{f}=2\pi fU/g=U/c$, where U is the wind speed and c is the wave celerity. The arrows in the figure indicate the dimensionless modal frequency $\bar{f}_m$ of each point spectrum, where the parameter $\bar{s}$ shows the maximum value. From the results shown in Figure 7.7, Mitsuyasu *et al.* give the following formula for $\bar{s}$:

$$\bar{s}=\begin{cases}11.5\,(\bar{f}^5/\bar{f}_m^{7.5}) & \text{for } \bar{f}\leqslant\bar{f}_m \\ 11.5\,\bar{f}^{-2.5} & \text{for } \bar{f}>\bar{f}_m\end{cases} \qquad (7.46)$$

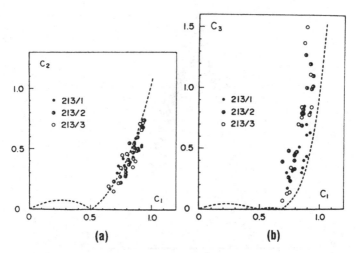

Fig. 7.6. Comparison between theoretical and measured ratios of angular harmonic amplitudes C_1/C_2 and C_1/C_3 (Mitsuyasu *et al.* 1975).

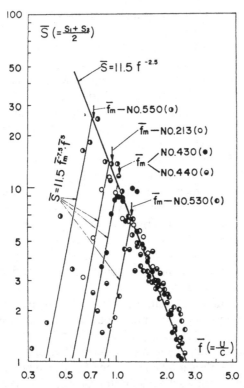

Fig. 7.7. Functional relationship between parameter $\bar{s}$ and dimensionless frequency $\bar{f}$ (Mitsuyasu *et al.* 1975).

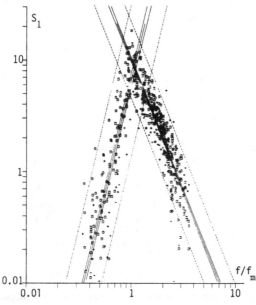

Fig. 7.8. s_1-value as a function of f/f_m (Hasselmann *et al.* 1980).

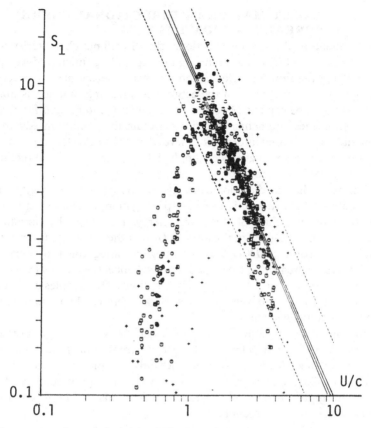

Fig. 7.9. s_1-value as a function of U/c (Hasselmann *et al.* 1980).

Hasselmann *et al.* (1980) carry out an extensive analysis of directional wave data obtained in the North Sea (JONSWAP site); in particular, the s-value associated with C_1 in Eq. (7.43), denoted as s_1. They find that s_1 is not only a function of the frequency but also of U/c_m. Here, c_m is the wave celerity for the modal frequency f_m. Figure 7.8 shows the s_1-value plotted against f/f_m, and Figure 7.9 shows the s_1-value plotted against U/c, where c is the celerity in deep water. Included in these figures are regression lines with 95 percent limits for the mean values and the standard deviation around the regression. From the results of the analysis, Hasselmann *et al.* propose the s_1-value to be

$$
s_1 = \begin{cases} (9.77\pm0.43)(f/f_m)^{-(2.33\pm0.06)-(1.45\pm0.45)(U/c_m-1.17)} \\ \qquad\qquad\qquad\qquad \text{for } U\geqslant c_m \text{ and } f>f_m \\ (6.97\pm0.83)(f/f_m)^{4.06\pm0.22} \\ \qquad\qquad\qquad\qquad \text{for } U\geqslant c_m \text{ and } f<f_m \end{cases} \qquad (7.47)
$$

7.4 ESTIMATION OF DIRECTIONAL ENERGY SPREADING FROM DATA

As discussed in the previous sections, the directional characteristics of ocean waves can be evaluated through the spreading function developed by applying the Fourier series. Although this approach has a theoretical background, the accuracy of the estimated spreading function is not as good as might be expected because of the limited amount of information acquired in measurements. In order to obtain the maximum advantage from data with limited information, several methods developed in statistical inference theory have been applied for estimating the directional energy spreading function.

It is noted that the directional energy spreading function, $D(\omega,\theta)$ is not a probability density function of wave propagation; instead, it represents the state of spreading of the wave energy. However, the directional spreading function given in the form of $D(\theta)$ or the spreading function for a specified frequency, $D(\theta|\omega)$, are non-negative integrable functions, and the integrated area with respect to θ is unity; hence, they satisfy the condition required to be probability density functions. This enables us to make use of the techniques developed in statistical inference theory for estimating the directional spreading function.

Although many methods for estimating the wave energy spreading function have been developed based on different approaches, the commonly acknowledged appropriate methods for practical use are the maximum likelihood method, the maximum entropy method and the Bayesian approach method. These three methods will be discussed in detail in the following sections.

7.4.1 Maximum likelihood method

Application of the maximum likelihood method (ML-method) developed in statistical inference theory for estimating stochastic wave properties from measured data is presented in Section 3.9.2. The same approach may be applied for estimating the parameters of the Fourier series representing the directional energy spreading function. Data obtained through measurements using the pitch–roll buoy yield six spectra from which the parameters A_1, B_1, A_2 and B_2 can be evaluated as shown in Eq. (7.19). It is inevitable that some errors may be involved in estimating these parameters because of possible errors in acquiring data during measurements as well as in evaluating spectra from a finite length of data.

Long and Hasselmann (1979) and Long (1980) consider the problem of estimating the directional spreading function $D(\omega,\theta)$ from a set of estimates which are obtained by integrating the directional spectral density function $S(\omega,\theta)$ when the density function is subject to statistical variability. In order to solve this problem, they develop a method to accept or reject a hypothesized true (expected) spreading function through the statistical test applicable for data obtained by a pitch–roll buoy. The basic principle is as follows.

By suppressing the dependency of frequency, the sample spectral density functions given in Eq. (7.18), denoted by C_{11}, C_{22}, C_{33}, C_{12}, Q_{13} and Q_{23}, are assumed to follow a joint normal probability distribution with zero mean and a covariance matrix Σ evaluated from the sample. Then, the estimates of the Fourier coefficient $D(\theta)$ are defined as

$$\hat{\mathbf{A}} = (\hat{A}_1, \hat{B}_1, \hat{A}_2, \hat{B}_2)^T \tag{7.48}$$

where elements of the vector $\mathbf{A}$ are given in Eq. (7.19). By letting

$$k = \{(C_{22} + C_{33})/C_{11}\}^{1/2} \tag{7.49}$$

we may write the elements of the estimate as

$$\hat{A}_1 = Q_{12}/\{C_{11}(C_{22} + C_{33})\}^{1/2}$$
$$\hat{B}_1 = Q_{13}/\{C_{11}(C_{22} + C_{33})\}^{1/2}$$
$$\hat{A}_2 = (C_{22} - C_{33})/(C_{22} + C_{33})$$
$$\hat{B}_2 = 2C_{23}/(C_{22} + C_{33}) \tag{7.50}$$

Here, the estimate $\hat{\mathbf{A}}$ is also normally distributed with zero mean and a covariance matrix, the latter can be evaluated from the data.

If we have a hypothesized directional spectrum which is considered to be true, we can evaluate the error ϵ_A involved in estimating $\hat{\mathbf{A}}$ by subtracting the vector associated with the hypothesized directional spectrum. By assuming the error vector ϵ_A to be normally distributed, the random variable $\epsilon_A \Sigma^{-1} \epsilon_A$ obeys the χ^2 distribution with four degrees of freedom for the present case, and thereby we may accept or reject the hypothesized directional spectrum with a specified level of confidence following the routine procedure in the goodness-of-fit technique. Hasselmann et al. (1980) apply the method to a large body of measured data for analysis of directional wave spectra, and claim that unimodal and bimodal angular distributions can be fairly well estimated from information obtained by the pitch–roll buoy.

Another application of the statistical inference concept for reliable estimation of directional wave spectra from data obtained by the pitch–roll buoy is to develop the statistical estimator for the six spectra given in Eq. (7.18). In estimating the value of each spectra C_{11}, C_{22}, Q_{12}, etc. for a given frequency, the average value of sample data is most commonly considered. It is of interest, however, to determine if C_{11}, C_{22}, etc. when estimated based on the ML-method yield a better estimation of the spreading function.

Glad and Krogstad (1992) carry out an analytical study on this subject and find that the commonly used average sample spectra are indeed the

ML-estimators of the spectra if the mean wave number is considered as an independent variable and is evaluated by Eq. (7.49).

Glad and Krogstad furthermore modify the ML-estimators for which the wave number is evaluated by the linear wave theory dispersion relation. The results of application of those modified ML-estimators for estimating the directional spreading function, however, show that the improvement over the average sample spectra is marginal.

Skarsoulis and Athanassoulis (1993) also carry out a study on the ML-estimators applicable to the analysis of pitch–roll buoy data similar to that done by Glad and Krogstad. The results of their study also show that the average sample spectra are ML-estimators if the mean wave number is evaluated by Eq. (7.49). In their analysis, the ML-estimators are evaluated numerically from the joint probability density function of the average sample spectra which is the Wishart probability distribution function. The likelihood function is then obtained based on the Wishart distribution and from this the ML-estimators are obtained by the routine procedure discussed in Section 3.9.2.

Besides the ML-method commonly considered in statistical inference theory, several techniques have been developed for estimating the directional energy spreading function from measured data. These comprise the data adaptive technique; however, they are still known as the ML-method.

This category of the ML-method was first introduced by Capon et al. (1967) and Capon (1969) for estimating the directional properties of a propagating seismic wave. They consider a signal consisting of a single plane wave in the presence of noise and assume that the noise components have a multi-dimensional normal distribution and apply the ML-method through the likelihood function of the multi-dimensional normal distribution. This method yields a high-resolution estimation using a wave-number window which is not fixed, but is variable at each wave number considered. Furthermore, this method has the feature that it does not make an *a priori* assumption about the shape of the directional spectrum.

Davis and Regier (1977) apply Capon's concept for estimating directional wave spectra from measured data obtained by multi-element arrays. The results of application of the ML-method developed indeed show a very high resolution for estimating the directional spectrum. Unfortunately, their method is applicable only for wave gage array measurements.

Isobe et al. (1984) extend Davis and Regier's approach for estimating directional wave spectra by developing a method applicable for data obtained by a combination of different instruments; namely, data that is a combination of pressure, water particle velocity, wave profile, acceleration, slope, etc. This method is referred to by the authors as the *extended maximum likelihood method* (the EML-method). The method is applicable not only for analysis of data from an array but also for data obtained at one point. The principle and practical application of the EML-method are given below.

Throughout the following derivation of directional wave spectra, wave-number frequency spectra are employed. The wave profile at a location $\mathbf{x}$ can be presented as

$$\eta(\mathbf{x},t) = \int_\omega \int_\mathbf{k} e^{i(\mathbf{k}\cdot\mathbf{x}-\omega t)} dA(\mathbf{k},\omega) \tag{7.51}$$

where

$$E[dA(\mathbf{k},\omega)dA^\star(\mathbf{k}',\omega')] = \begin{cases} S(\mathbf{k},\omega)d\mathbf{k}\,d\omega & \text{if } \mathbf{k}=\mathbf{k}' \text{ and } \omega=\omega' \\ 0 & \text{otherwise} \end{cases}$$

On the other hand, the cross-spectral density function between wave profiles measured at two locations separated by a distance $\mathbf{r}$ can be written as

$$S(\mathbf{r},\omega) = \int_\mathbf{k} S(\mathbf{k},\omega)e^{-i\mathbf{k}\cdot\mathbf{r}}d\mathbf{k} \tag{7.52}$$

The wave number vector $\mathbf{k}$ can be expressed in terms of wave energy propagation direction, θ, and frequency ω, through the dispersion equation; hence, $S(\mathbf{k},\omega)$ represents a directional wave spectrum. Equation (7.52) implies that if cross-spectra are available for an infinite number of $\mathbf{r}$, the wave directional spectra $S(\mathbf{k},\omega)$ can be evaluated by the inverse Fourier transform of $S(\mathbf{r},\omega)$. In order to obtain the relationship similar to that given in Eq. (7.52) but applicable for various wave kinematic quantities such as pressure, velocity, acceleration, etc., Isobe et al. (1984) define the *transfer function*, denoted by $H(\mathbf{k},\omega)$, and express it in the form

$$H(\mathbf{k},\omega) = G(\mathbf{k},\omega)(\cos\theta)^\alpha(\sin\theta)^\beta \tag{7.53}$$

Here, $G(\mathbf{k},\omega)$ represents a function showing a linear relationship between the kinematic quantity and surface wave profile. For example, the velocity in the x-direction is given by letting $G(\mathbf{k},\omega) = \omega\cosh kz/\sinh kh$, $\alpha=1$ and $\beta=0$, where h is the water depth.

By using the transfer function, the wave kinematic quantity $\xi(\mathbf{x},t)$ may be presented from Eq. (7.51) as

$$\xi(\mathbf{x},t) = \int_\omega \int_\mathbf{k} H(\mathbf{k},\omega)e^{i(\mathbf{k}\cdot\mathbf{x}-\omega t)}dA(\mathbf{k},\omega) \tag{7.54}$$

Table 7.2. *Transfer function H(* **k***, ω) for several kinematic quantities (Isobe et al. 1984).*

	$H(\mathbf{k}, \omega)$	$G(\mathbf{k}, \omega)$	α	β
Surface				
Elevation	1	1	0	0
Vertical velocity	$-i\omega$	$-i\omega$	0	0
Vertical acceleration	$-\omega^2$	$-\omega^2$	0	0
Particle velocity				
x–direction	$\omega\cos\theta\dfrac{\cosh kz}{\sinh kh}$	$\omega\dfrac{\cosh kz}{\sinh kh}$	1	0
y–direction	$\omega\sin\theta\dfrac{\cosh kz}{\sinh kh}$	$\omega\dfrac{\cosh kz}{\sinh kh}$	0	1
z–direction	$-i\omega\dfrac{\sinh kz}{\sinh kh}$	$-i\omega\dfrac{\sinh kz}{\sinh kh}$	0	0
Particle acceleration				
x–direction	$-i\omega^2\cos\theta\dfrac{\cosh kz}{\sinh kh}$	$-i\omega^2\dfrac{\cosh kz}{\sinh kh}$	1	0
y–direction	$-i\omega^2\sin\theta\dfrac{\cosh kz}{\sinh kh}$	$-i\omega^2\dfrac{\cosh kz}{\sinh kh}$	0	1
z–direction	$-\omega^2\dfrac{\sinh kz}{\sinh kh}$	$-\omega^2\dfrac{\sinh kz}{\sinh kh}$	0	0

A list of transfer functions for several interesting kinematic quantities are taken from Isobe's publication and shown in Table 7.2.

By applying the expression in Eq. (7.54), a cross-spectra between two quantities ξ_m and ξ_n, denoted by $S_{mn}(\omega)$, at different locations $\mathbf{x}_m$ and $\mathbf{x}_n$, respectively, can be obtained in a similar fashion as shown in Eq. (7.52). That is,

$$S_{mn}(\omega)=\int_{\mathbf{k}} H_m(\mathbf{k},\omega)H_n^*(\mathbf{k},\omega)\,e^{-i\mathbf{k}(\mathbf{x}_n-\mathbf{x}_m)}S(\mathbf{k},\omega)\mathrm{d}\mathbf{k} \qquad (7.55)$$

Next, an estimated directional wave spectrum will be derived based on Eq. (7.55) by applying the maximum likelihood technique. For this, the desired directional wave-number spectrum is presented as a linear combination of cross-spectra $S_{mn}(\omega)$. We may write

$$\hat{S}(\mathbf{k},\omega)=\sum_m\sum_n\alpha_{mn}(\mathbf{k})S_{mn}(\omega) \qquad (7.56)$$

Here, $\alpha_{mn}(\mathbf{k})$ and $\alpha_{nm}(\mathbf{k})$ are complex conjugates so that $\hat{S}(\mathbf{k},\omega)$ is a real-valued function. Further, it is assumed that $\alpha_{mn}(\mathbf{k})$ may be presented as

$$\alpha_{mn}(\mathbf{k})=\gamma_m(\mathbf{k})\,\gamma_n^*(\mathbf{k}) \qquad (7.57)$$

and thereby Eq. (7.56) can be written as

$$\hat{S}(\mathbf{k},\omega) = \sum_m \sum_n \gamma_m(\mathbf{k}) S_{mn}(\omega) \gamma_n^*(\mathbf{k}) \tag{7.58}$$

By applying Eq. (7.55) to the above equation, the estimated energy spreading wave-number spectrum becomes

$$\hat{S}(\mathbf{k},\omega) = \int_{\mathbf{k}'} w(\mathbf{k},\mathbf{k}') S(\mathbf{k}',\omega) d\mathbf{k}' \tag{7.59}$$

where

$$w(\mathbf{k},\mathbf{k}') = \sum_m \sum_n \gamma_m(\mathbf{k}) H_m(\mathbf{k}',\omega)\, e^{i\mathbf{k}'\cdot\mathbf{x}_m} H_n^*(\mathbf{k}',\omega) e^{-i\mathbf{k}'\cdot\mathbf{x}_n} \gamma_n^*(\mathbf{k})$$

Furthermore, for convenience, $w(\mathbf{k},\mathbf{k}')$ may be written as

$$w(\mathbf{k},\mathbf{k}') = \sum_m \sum_n \gamma_m(\mathbf{k}) T_{mn}(\mathbf{k}') \gamma_n^*(\mathbf{k}) \tag{7.60}$$

where

$$T_{mn}(\mathbf{k}') = H_m(\mathbf{k}',\omega)\, e^{i\mathbf{k}'\cdot\mathbf{x}_m} H_n^*(\mathbf{k}',\omega)\, e^{-i\mathbf{k}'\cdot\mathbf{x}_n}$$

In Eq. (7.59), $S(\mathbf{k}',\omega)$ is the unknown true energy spreading spectrum, and $w(\mathbf{k},\mathbf{k}')$ is called the *wave-number window function*. The smaller the window function, the higher the resolution of $S(\mathbf{k},\omega)$ that can be obtained; the best estimation may be achieved when $w(\mathbf{k},\mathbf{k}')$ is approximated by the delta-function centered at $\mathbf{k}=\mathbf{k}'$. Hence, let $w(\mathbf{k},\mathbf{k}')$ have the following constraint under $\mathbf{k}=\mathbf{k}'$:

$$w(\mathbf{k},\mathbf{k}) = \sum_m \sum_n \gamma_m(\mathbf{k}) T_{mn}(\mathbf{k}) \gamma_n^*(\mathbf{k}) = 1 \tag{7.61}$$

Further, let us minimize $\hat{S}(\mathbf{k},\omega)$ given in Eq. (7.59) under the condition given in Eq. (7.61). This is equivalent to maximizing the following quantity:

$$\frac{w(\mathbf{k},\mathbf{k})}{\hat{S}(\mathbf{k},\omega)} = \frac{\displaystyle\sum_m \sum_n \gamma_{\mathrm{m}}(\mathbf{k}) T_{mn}(\mathbf{k}) \gamma_n^*(\mathbf{k})}{\displaystyle\sum_m \sum_n \gamma_{\mathrm{m}}(\mathbf{k}) S_{mn}(\omega) \gamma_n^*(\mathbf{k})} \tag{7.62}$$

This, in turn, may be likened to the problem of finding the maximum eigenvalue λ which satisfies the following relationship for the given matrices $T_{mn}(\mathbf{k})$ and $S_{mn}(\omega)$:

$$\sum_n T_{mn}(\mathbf{k})\, \gamma_n^* = \lambda \sum_n S_{mn}(\omega) \gamma_n^* \tag{7.63}$$

and hence, we have

$$\sum_m \sum_n S_{\ell m}^{-1}(\omega) T_{mn}(\mathbf{k}) \gamma_n^* = \lambda \gamma_\ell^* \qquad (7.64)$$

where $S_{\ell m}^{-1}(\omega)$ is the inverse matrix of $S_{\ell m}(\omega)$.

Thus, from Eq. (7.62), the estimated spectra is inversely proportional to the maximum eigenvalue λ_{max}. That is,

$$\hat{S}(\mathbf{k},\omega) \propto 1/\lambda_{max} \qquad (7.65)$$

where λ_{max} can be obtained as

$$\lambda_{max} = \sum_m \sum_n H_m^*(\mathbf{k},\omega) \, e^{-i\mathbf{k}\cdot\mathbf{x}_m} S_{mn}^{-1}(\omega) H_n(\mathbf{k},\omega) \, e^{i\mathbf{k}\cdot\mathbf{x}_n} \qquad (7.66)$$

By applying this relationship for all frequencies, the estimated energy distribution as a function of wave number and frequency ω is

$$\hat{S}(\mathbf{k},\omega) = C \Big/ \left[\sum_m \sum_n S_{mn}^{-1}(\omega) H_m^*(\mathbf{k},\omega) H_n(\mathbf{k},\omega) \, e^{i\mathbf{k}\cdot(\mathbf{x}_n - \mathbf{x}_m)} \right] \qquad (7.67)$$

where C is a constant determined such that the integration of $\hat{S}(\mathbf{k},\omega)$ with respect to $\mathbf{k}$ yields a point spectrum. That is

$$S(\omega) = \int_{\mathbf{k}} S(\mathbf{k},\omega) d\mathbf{k} \qquad (7.68)$$

By applying the transfer function given in Eq. (7.53) and by using the dispersion relationship for a specified frequency, the estimated wave-number frequency spectrum $\hat{S}(\mathbf{k},\omega)$ can be converted to a spectrum presented as a function of angle θ and frequency ω as follows:

$$\hat{S}(\theta,\omega) = C' \Big/ \left[\sum \sum S_{mn}^{-1}(\omega) G_m^*(k,\omega) G_n(k,\omega) \right.$$

$$\left. \times (\cos\theta)^{\alpha_m + \alpha_n} (\sin\theta)^{\beta_m + \beta_n} e^{i\mathbf{k}\cdot(\mathbf{x}_n - \mathbf{x}_m)} \right] \qquad (7.69)$$

where the constant C' is determined such that

$$S(\omega) = \int_0^{2\pi} \hat{S}(\theta,\omega) d\theta \qquad (7.70)$$

The function $G(k,\omega)$ in Eq. (7.69) is based on linear wave theory as given in Table 7.2. Isobe et al. (1984) however, recommend that it be evaluated from measured data in practice. For example, for data obtained by the pitch–roll buoy, by denoting $\eta = 1$, $\partial\eta/\partial x = 2$ and $\partial\eta/\partial y = 3$, $G(k,\omega)$ should be evaluated from the auto-spectra by

$$G(k,\omega) = [\{S_{22}(\omega) + S_{33}(\omega)\}/S_{11}(\omega)]^{1/2} \qquad (7.71)$$

and for data obtained by the cloverleaf buoy, by denoting $\partial^2\eta/\partial x^2=4$, $\partial^2\eta/\partial y^2=5$ and $\partial^2\eta/\partial x\,\partial y=6$, we have

$$G(k,\omega)=[\{S_{44}(\omega)+S_{55}(\omega)+S_{66}(\omega)\}/S_{11}(\omega)]^{1/2} \tag{7.72}$$

Furthermore, by defining a function $\phi_{mn}(\omega)$ as

$$\phi_{mn}(\omega)=S_{mn}(\omega)/\{G_m(k,\omega)G_n^*(k,\omega)\} \tag{7.73}$$

the inverse of $S_{mn}(\omega)$ in Eq. (7.69) can be written as

$$S_{mn}^{-1}(\omega)=\phi_{mn}^{-1}(\omega)/\{G_m^*(k,\omega)G_n(k,\omega)\} \tag{7.74}$$

Since $\phi_{mn}(\omega)$ is a Hermitian matrix, $\phi_{mn}^{-1}(\omega)$ is also Hermitian and can be expressed in the following form,

$$\phi_{mn}^{-1}(\omega)=a_{mn}(\omega)-ib_{mn}(\omega) \tag{7.75}$$

where a_{mn} and b_{mn} are real-valued matrices, and

$$a_{nm}(\omega)=a_{mn}(\omega) \quad \text{and} \quad b_{nm}(\omega)=-b_{mn}(\omega) \tag{7.76}$$

From Eqs. (7.74) and (7.76), Eq. (7.69) can be written as

$$\hat{S}(\theta,\omega)=C'/\left[\sum_m\sum_n(\cos\theta)^{\alpha_m+\alpha_n}(\sin\theta)^{\beta_m+\beta_n}\Big(a_{mn}(\omega)\cos\{\mathbf{k}\cdot(\mathbf{x}_n-\mathbf{x}_m)\}\right.$$
$$\left.+b_{mn}(\omega)\sin\{\mathbf{k}\cdot(\mathbf{x}_n-\mathbf{x}_m)\}\Big)\right] \tag{7.77}$$

Thus, the procedure for practical application of the EML-method is summarized by Isobe *et al.* (1984) as follows:

(1) For a given set of data, compute cross-spectra $S_{mn}(\omega)$ for all available combinations of the measured elements.

(2) Determine $G_m(k,\omega)$ for a given ω; examples of which are shown in Eqs. (7.71) and (7.72).

(3) Evaluate $a_{mn}(\omega)$ and $b_{mn}(\omega)$ by applying Eqs. (7.73) through (7.75).

(4) Compute $\hat{S}(\theta,\omega)$ given in Eq. (7.77). The constant C' can be evaluated through the relationship given in Eq. (7.70).

Figures 7.10 and 7.11 show examples of comparisons between the preliminary known (true) directional spreading and those computed by applying the EML-method to data obtained through the numerical simulations technique. Figure 7.10 shows examples when the directional spreading is sharply concentrated. Case (a) is an example of the computation carried out for a pitch–roll buoy, while case (b) is that for a cloverleaf buoy. The ordinate $\bar{s}$ is a dimensionless spectrum $S/S_{\max}$, where $S_{\max}$ is the magnitude of the maximum true directional spectrum. Note that for comparison the frequency ω may be omitted without loss in generality. Included also in these comparisons are the energy spreading functions computed by the Fourier series expansion method presented in Section 7.2.2 (a) and (b), respectively. It can be seen in these figures that the results computed by applying the EML-method show excellent agreement with the true directional spreading.

Figure 7.11 shows comparisons similar to those in Figure 7.10 but the directional spreading is of the moderately spreading category. In this example, the EML-method yields excellent results, but the Fourier series expansion method also shows good agreement between the computed and target spreadings, particularly for the cloverleaf buoy data.

In regard to the reliability of the estimated directional spectrum $\hat{S}(\omega,\theta)$ by applying the ML-method, several interesting studies have been carried out. This will be discussed next. If the cross-spectral matrix is inversely reconstructed from $\hat{S}(\omega,\theta)$ and this matrix in turn is used as an input to the ML-method for estimating the directional spectrum of the second generation, the estimated spectrum is different from $\hat{S}(\omega,\theta)$ in most cases. In order to improve this inconsistent trend, a method called the *iterative maximum likelihood method* (IML-method) is developed by Pawka (1983) for estimating directional wave spectra through a linear array, and by Oltman-Shay and Guza (1984) for a pitch–roll buoy. Krogstad *et al.* (1988) use the following iterative scheme for estimating the directional spreading function:

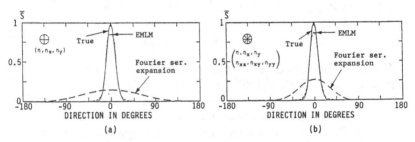

Fig. 7.10. Comparison between directional wave spectra obtained by applying EML method and simulation data (direction is sharply concentrated case) (Isobe *et al.* 1984).

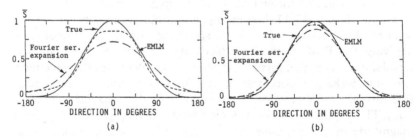

Fig. 7.11. Comparison between directional wave spectra obtained by applying EML method and simulation data (direction is moderately spreading case) (Isobe *et al.* 1984).

$$D_n = D_{n-1} + \gamma(\hat{D} - D'_{n-1}) \tag{7.78}$$

where $\hat{D}$ = directional spreading function estimated by the
       ML-method

D'_{n-1} = directional spreading function obtained by using
       cross-spectra computed based on $\hat{D}_{n-1}$. For the first
       iteration ($n=1$), D'_0 is that computed based on $\hat{D}$

D_{n-1} = directional spreading function for the $(n-1)$th iteration.
       For the first iteration ($n=1$), $D_0 = \hat{D}$

D_n = directional spreading function obtained through the
       nth iteration

γ = relaxation parameter. $\gamma = 1.0$ to 1.2 is proposed by
       Krogstad *et al.*

The IML-method certainly improves the estimated directional spectrum as shown in the example in Figure 7.12(a). Krogstad *et al.* (1988)
show that the rapidity with which the iteration converges to a stable spectrum depends strongly on the relaxation parameter γ and the number of
required iterations as shown in Figure 7.12(b). As seen, divergence
occurs abruptly for γ greater than 1.5.

Another interesting approach is the development of a modified version
of the ML-method in which the data are partitioned into signal and noise.
Marsden and Juszko (1987) present a methodology called the *eigenvector
method* (EV-method) for estimating the directional spectra from data
obtained by a pitch–roll buoy. This concept considers the partitioning
of the signal and noise components through the diagonalization of the
cross-spectral matrix. There exist three possible eigenvectors for the
pitch–roll buoy data. The eigenvectors corresponding to the two smallest

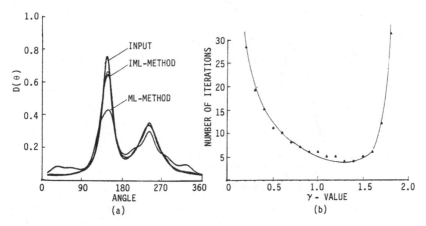

Fig. 7.12. (a) Comparison between directional wave spectra obtained by
            applying ML and EML methods, and (b) relaxation parameter γ
            and number of necessary iterations (Krogstad *et al.* 1988).

eigenvalues are assumed to span the noise of the measurement, while the eigenvector corresponding to the largest eigenvalue is assumed to span the signal. The directional spectrum is then estimated by minimizing the noise component. This method is identical to the ML-method if the three eigenvectors are considered to span the signal.

An iteration scheme is also applied to the EV-method. An example application of the method to field data analysis is shown in Figure 7.13. Figure 7.13(a) shows a point spectrum constructed from data obtained off Newfoundland which indicates double peaks; the large wave energy at the frequency 0.1 Hz may be attributed to that of swell. Figure 7.13(b) shows the directional wave spectra for frequency 0.1 Hz obtained by applying the ML, IML, EV and the iterative EV-methods. The figure clearly shows propagation of wave energy in two directions; one toward 60 degrees, the other toward 170 degrees. Since analysis at the frequency

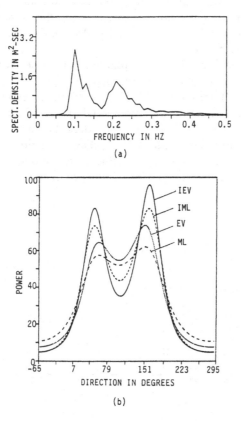

Fig. 7.13. (a) Wave point spectrum, and (b) comparison of directional wave spectra estimated by ML, IML, EV and IEV methods for frequency 0.1 Hz (Marsden and Juszko 1987).

of 0.2 Hz shows energy is propagated toward 60 degrees only, this appears to be the local wind-generated sea; hence, the energy toward 170 degrees is considered to be swell. It can be seen in the figure that the iterative EV-method seems to be superior to the other methods in this example.

7.4.2 Maximum entropy method

The maximum entropy method (the ME-method) was originally developed as a measure of uncertainty in statistical information theory, and the method has been applied for determining the probability density function of a random variable from data in which information about the distribution is inadequate. The basic concept of this method is that the probability distribution which maximizes the entropy (defined below) has the least bias on the density function under some constraints; the moment constraints, for example (Jaynes 1957). Here, the entropy for a continuous-type random variable X with the probability density function $f(x)$ is defined as follows:

$$H = -\int f(x) \ln f(x)\, dx = E[-\ln f(x)] \tag{7.79}$$

where the integration is made over the sample space of X. As an example of moment constraints, it is commonly considered that the theoretical moments of the distribution are equal to those evaluated from sample data.

In order to maximize the entropy H, it is common practice to apply the Lagrangian multipliers method. For this, the entropy is written as

$$H = -\int f(x) \left\{ \ln f(x) + \sum_{j=0}^{n} \lambda_j x^j \right\} dx \tag{7.80}$$

and H is maximized with respect to $f(x)$ by letting $\partial H/\partial f = 0$. The maximization yields an estimated probability density function as

$$\hat{f}(x) = \exp\left\{ -1 - \sum \lambda_j x^j \right\} \tag{7.81}$$

where λ_j is determined so that the theoretical moments evaluated based on the estimated probability density function are equal to the sample moments; i.e.

$$\int x^k \exp\left\{ -1 - \sum \lambda_j x^j \right\} dx = m_k \qquad k = 1, 2, \dots \tag{7.82}$$

where $m_0 = 1$, m_k ($k \neq 0$) are sample moments from data.

Kobune and Hashimoto (1986) apply the ME-method for estimating wave directional energy spreading for a pitch–roll buoy. For the entropy given in Eq. (7.79), they substitute the directional spreading function

$D(\theta|\omega)$ for $f(x)$, and derive the following estimated spreading function through maximizing the entropy:

$$\hat{D}(\theta|\omega)=\exp\left\{-\lambda_0-\sum\lambda_j\alpha_j(\theta)\right\} \qquad (7.83)$$

The Lagrangian multipliers λ_j are determined from a Fourier series expansion of the energy spreading function. We write the coefficients of the Fourier series given in Eq. (7.13) in the following form:

$$\beta_i=\int_0^{2\pi}\hat{D}(\theta|\omega)\alpha_j(\theta)d\theta \qquad (7.84)$$

where $\hat{D}(\theta|\omega)$ = spreading function for a given ω
   $\alpha_i = \cos\theta, \sin\theta, \cos 2\theta, \sin 2\theta$ for $i = 1, 2, 3$ and 4, respectively
   $\beta_i = A_1, B_1, A_2$ and B_2.

Then, from Eqs. (7.83) and (7.84), we have the following nonliear equations with respect to λ_j:

$$\int_0^{2\pi}\{\beta_i-\alpha_i(\theta)\}\exp\left\{-\sum_{j=1}^{4}\lambda_j\alpha_j(\theta)\right\}d\theta=0$$

$$i=1, 2, 3 \text{ and } 4 \qquad (7.85)$$

The unknown Lagrangian multiplier λ_j in Eq. (7.83) can be obtained as a solution of Eq. (7.85). For this, Kobune and Hashimoto (1986) apply the Newton–Raphson method developed for multiple variables. Upon evaluating λ_j, the multiplier λ_0 in Eq. (7.83) can be obtained from

$$\lambda_0=\ln\left(\int_0^{2\pi}\exp\left\{-\sum_{j=1}^{4}\lambda_j\alpha_j(\theta)\right\}d\theta\right) \qquad (7.86)$$

Note that if we consider only $i=1$ and 2, the estimated spreading function becomes the formula proposed by Borgman (1969). That is, for $i=1$ and 2, we may write Eq. (7.83) as

$$\hat{D}(\theta|\omega)=\exp\{-\lambda_0-\lambda_1\cos\theta-\lambda_2\sin\theta)\} \qquad (7.87)$$

By letting $-\lambda_1=a\cos\theta$ and $\lambda_2=a\sin\theta$, λ_0 can be calculated from Eq. (7.86) as

$$\lambda_0=\ln\int_0^{2\pi}\exp\{a\cos(\theta-\theta_0)\}d\theta=\ln\{2\pi I_0(a)\} \qquad (7.88)$$

where $I_0(a)$ is a modified Bessel function.
   Thus, Eq. (7.87) becomes

$$D(\theta|\omega) = \frac{1}{2\pi I_0(a)} \exp\{a\cos(\theta-\theta_0)\} \tag{7.89}$$

This formula is called the circular normal probability distribution and is proposed by Borgman (1969) as a directional spreading function.

An example of a comparision of the directional energy spreading functions in dimensionless form computed by using numerical simulation models is shown in Figure 7.14. Computations are carried out by applying the maximum entropy (ME) method, the extended maximum likelihood (EML) method and the Fourier series expansion (FS) method.

Hashimoto *et al.* (1994) further extend the ME-method taking into consideration errors involved in cross-spectral density functions.

The ME-method is applied by Nwogu (1989) for estimating the directional spectra from the results obtained by an array of wave probes. Using his approach the cross-spectral density function of waves computed from data obtained at two locations is given by

$$S_{mn}(\omega) = \int_{-\pi}^{\pi} \exp\{ik(x_n - x_m)\} S(\omega,\theta)d\theta \tag{7.90}$$

where $S(\omega,\theta) = S(\omega)D(\omega,\theta)$.

Equation (7.90) may be expressed as

$$\phi_j(\omega) = \frac{S_{mn}(\omega)}{S(\omega)} = \int_{-\pi}^{\pi} q_j(\theta)D(\omega,\theta)d\theta \qquad j=1, 2, ..., (M+1) \tag{7.91}$$

where $M = N(N-1)$; N is the number of wave probes, and

$$\Phi_j(\omega) = \begin{cases} \text{Re}\, S_{mn}(\omega)/\{S_m(\omega)S_n(\omega)\}^{1/2} & \text{for } j=1, 2, ..., (M/2) \\ \text{Im}\, S_{mn}(\omega)/\{S_m(\omega)S_n(\omega)\}^{1/2} & \text{for } j=(M/2)+1, ...M \\ 1 & \text{for } j=M+1, ... \end{cases}$$

$$q_j(\theta) = \begin{cases} \cos\{kr_j\cos(\beta_j - \theta)\} & \text{for } j=1, 2, ..., (M/2) \\ \sin\{kr_j\cos(\beta_j - \theta)\} & \text{for } j=(M/2)+1, ..., M \\ 1 & \text{for } j=M+1 \end{cases}$$

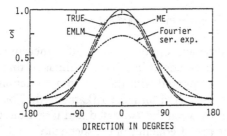

Fig. 7.14. Example of directional wave spectrum estimated by applying ME method (Kobune and Hashimoto 1986).

$$r_j = \{(x_n - x_m)^2 + (y_n - y_m)^2\}^{1/2}$$
$$\beta_j = \tan^{-1}\{(y_n - y_m)/(x_n - x_m)\}$$

The following is not given in Nwogu's paper, but is included here to supplement Eq. (7.91).

(i) As stated in connection with Eq. (7.10), the number of cross-spectra for N gages is $N(N-1)+1$ which is given as $j = M+1$ in Eq. (7.91). For example, for $N=4$, we have 13 known quantities in which the real parts are $j = 1, 2, ..., 7$, and the imaginary parts are $j = 8, 9, ..., 13$.

(ii) $q_j(\theta)$ can be expressed in the form of a Bessel function of the first kind; i.e.

$$\cos\{kr_j\cos(\beta_j - \theta)\} = \mathcal{J}_0(kr_j) + 2\sum_{n=1}^{\infty}(-1)^n\mathcal{J}_{2n}(kr_j)\cos 2n(\beta_j - \theta)$$

$$\sin\{kr_j\cos(\beta_j - \theta)\} = 2\sum_{n=0}^{\infty}(-1)^n\mathcal{J}_{2n+1}(kr_j)\cos(2n+1)(\beta_j - \theta) \tag{7.92}$$

Following the procedure of the maximum entropy method, the estimated directional spreading function becomes

$$\hat{D}(\theta,\omega) = \exp\left\{-1 + \sum_{j=1}^{M+1}\lambda_j q_j(\theta)\right\} \tag{7.93}$$

The Lagrangian multipliers λ_j are determined from the constraint, which is given for the present problem as follows:

$$\Phi_k(\omega) = \int_{-\pi}^{\pi}\exp\left\{-1 + \sum_{j=1}^{M+1}\lambda_j q_j(\theta)\right\}q_k(\theta)\mathrm{d}\theta \tag{7.94}$$

where $k = 1, 2, ..., (M+1)$.

Figure 7.15 shows a comparison between measured and estimated directional wave spreading functions. The directional waves are generated in the experimental tank in an attempt to simulate a JONSWAP wave spectrum with the following angular spreading function for $s=5$:

$$D(\theta) = \frac{1}{\pi}\frac{\Gamma^2(s+1)}{\Gamma(s+\frac{1}{2})}\cos^{2s}(\theta) \qquad |\theta| < \pi/2 \tag{7.95}$$

The wave spreading function evaluated from the measured data obtained by five wave probes arranged in a layout similar to that of the CERC probe array is then compared with the spreading functions estimated by applying the ME-method and the ML-method. The figure shows excellent agreement between the estimated and target directional spreadings obtained for the peak frequency of the spectrum.

There is another maximum entropy method developed by Burg (1967) in spectral analysis, and this method may also be applied for estimating directional wave spectra. Although this method is known as an ME-method, the definition of maximum entropy and the basic principle are different from those of the ME-method developed in statistical inference theory. The entropy in this case is defined based on a steady-state Gaussian random process in the following form:

$$H = \frac{1}{\omega_*} \int\limits_0^\infty \ln S(\omega) d\omega = \frac{1}{\omega_*} \int\limits_0^\infty \ln \left[\sum_{\tau=-\infty}^\infty \phi(\tau) \, e^{-i\omega\tau\Delta t} \right] d\omega \qquad (7.96)$$

where

$S(\omega)$ = spectral density function
$\phi(\tau)$ = auto-correlation function
ω_* = constant
Δt = sampling rate.

The desired spectrum can be obtained by maximizing the entropy H with respect to the unknown auto-correlation function $\phi(\tau)$, with a constraint on the covariance matrix. Note that the maximizing procedure is carried out through the auto-correlation function.

Lygre and Krogstad (1986) apply this approach for estimating the directional spreading function of data obtained by a pitch–roll buoy. In their approach, the spreading function is written in the form of a Fourier series as

$$D(\theta|\omega) = \frac{1}{2\pi} \sum_{n=-\infty}^\infty C \, e^{in\theta} \qquad (7.97)$$

where $C_0 = 1$ and $C_{-n} = C_n^*$ (conjugate of C_n), and the maximum entropy is defined in this case as

$$H = \int\limits_{-\pi}^\pi \ln\{D(\theta|\omega)\} d\theta \qquad (7.98)$$

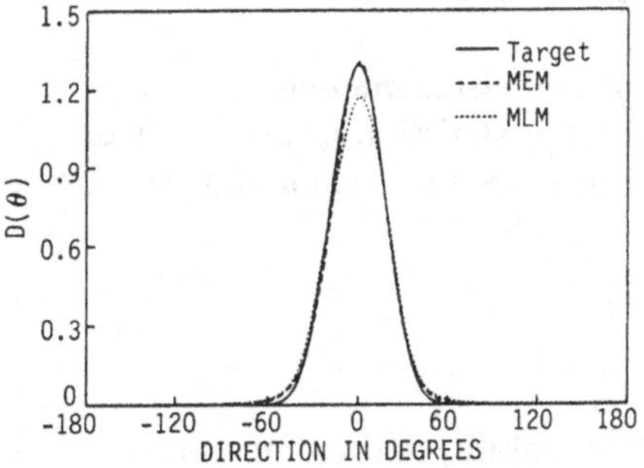

Fig. 7.15. Comparison between directional wave spectra estimated by applying ME and ML methods (Nwogu 1989).

Although $D(\theta|\omega)$ is not a spectrum, the same concept as that developed by Burg is applied here with the following constraints:

$$\int_{-\pi}^{\pi} D(\theta|\omega)\, e^{-ik\theta}\, d\theta = c_k \qquad (7.99)$$

For the analysis of pitch–roll buoy data, the Fourier series expansion given in Eq. (7.97) is limited to C_1 and C_2. In this case, the estimated directional spreading function is given by

$$\hat{D}(\theta|\omega) = \frac{1}{2\pi} \frac{1 - \phi_1 \hat{C}_1^* - \phi_2 \hat{C}_2^*}{|1 - \phi_1 e^{-i\theta} - \phi_2 e^{-i2\theta}|^2} \qquad (7.100)$$

The parameters ϕ_1 and ϕ_2 are evaluated from

$$\phi_1 = \frac{\hat{C}_1 - \hat{C}_1^* \hat{C}_2}{1 - |\hat{C}_1|^2}$$

$$\phi_2 = \hat{C}_2 - \hat{C}_1 \phi_1 \qquad (7.101)$$

where $\hat{C}_1$ and $\hat{C}_2$ are estimates of the Fourier coefficients, and $\hat{C}_1^*$ and $\hat{C}_2^*$ are the conjugates of $\hat{C}_1$ and $\hat{C}_2$, respectively. From the cross- and auto-spectra shown in Eq. (7.18), the estimates $\hat{C}_1$ and $\hat{C}_2$ are given by Long (1980) as follows:

$$\hat{C}_1 = \frac{Q_{12} + iQ_{13}}{(C_{11}\, C_{22} + C_{33})^{1/2}}$$

$$\hat{C}_2 = \frac{(C_{22} - C_{33}) - i2C_{23}}{C_{22} + C_{33}} \qquad (7.102)$$

Figure 7.16 shows an example of the estimated directional spreading function computed by Eq. (7.100) using wave data obtained in the Norwegian Sea (Lygre and Krogstad 1986). Also included in the figure is the estimate obtained by applying the ML-method. The vertical line in the figure indicates the wind direction at the time of measurements.

7.4.3 Application of a Bayesian method
The Bayesian inference method in statistics, in particular the following modified Bayes' theorem, is introduced in Section 6.3 for estimating the long-term extreme sea state:

$$f(\theta|x) = C\, h(\theta)\, L(\theta|x) \qquad (7.103)$$

where

$f(\theta|x)$ = posterior probability density function
$h(\theta)$ = prior probability density function
$L(\theta|x)$ = likelihood function of θ for a given x
C = normalization constant.

The above equation states that the posterior distribution is proportional to the product of the likelihood function and the prior probability distribution $h(\theta)$. The modified Bayesian method given in Eq. (7.103) is applied by Hashimoto and Kobune (1988) for estimating the energy spreading function, although in their approach the prior distribution $h(\theta)$ is not given as a function of θ per se, but is derived through analysis, as presented below.

It is assumed that the directional spreading function $D(\theta|\omega)$ consists of K equally spaced piecewise constant functions over the directional range from 0 to 2π. By letting

$$\ln D(\theta_k|\omega) = x_k(\omega) \qquad k = 1, 2, \ldots, K \tag{7.104}$$

the directional spreading function is presented as

$$D(\theta|\omega) = \sum_{k=1}^{K} e^{x_k(\omega)} I_k(\theta|\omega) \tag{7.105}$$

where

$$I_k(\theta|\omega) = \begin{cases} 1 & (k-1)\Delta\theta < \theta < k\Delta\theta \\ 0 & \text{otherwise} \end{cases}$$

Next, by using the dispersion relationship $\omega^2 = kg \tanh kh$ (h=water depth), the cross-spectrum given in Eq. (7.55) is presented in the following form as a function of frequency ω and angle θ.

$$S_{mn}(\omega) = \int H_j(\omega, \theta) S(\omega, \theta) d\theta \tag{7.106}$$

Then, by applying Eq. (7.105), $S_{mn}(\omega)$ may be written as

$$S_{mn}(\omega) = \int H_j(\omega, \theta) S(\omega) D(\theta|\omega) d\theta$$

$$= S(\omega) \sum e^{x_k(\omega)} \int H_j(\omega, \theta) I_k(\theta|\omega) d\theta$$

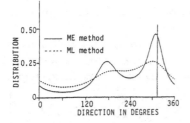

Fig. 7.16. Comparison of directional wave spectra estimated by applying ML and ME methods (Lygre and Krogstad 1986).

$$= S(\omega)\left(\sum_{k=1}^{K} \alpha_{j,k}(\omega)\, e^{x_k(\omega)}\right) \tag{7.107}$$

where

$$\alpha_{j,k}(\omega) = \int H_j(\omega,\theta) I_k(\theta|\omega)\, d\theta$$

Note that the term in the brackets is equivalent to the directional spreading function for a specified frequency.

Assuming the cross-spectra obtained from measured data carry an error $\epsilon(\omega)$, the directional spreading function is written as

$$\phi_j(\omega) = \sum_{k=1}^{K} \alpha_{j,k}(\omega)\, e^{x_k(\omega)} + \epsilon_j(\omega) \qquad j=1,2,\ldots,2N \tag{7.108}$$

In applying Eq. (7.108) to data analysis, the errors ($j=1, 2, \ldots, N$) are assumed to be independent and obey a normal distribution with zero mean and unknown variance σ^2. Estimation of the directional spreading function can be done through maximizing the following likelihood function with respect to x_k and σ^2.

$$L(x_1, x_2, \ldots, x_k; \sigma^2) = \frac{1}{(2\pi\sigma^2)^N}$$

$$\times \exp\left[-\frac{1}{2\sigma^2}\sum_{j=1}^{2N}\left\{\phi_j(\omega) - \sum_{k=1}^{K}\alpha_{j,k}(\omega)\, e^{x_k(\omega)}\right\}^2\right] \tag{7.109}$$

In maximizing the above likelihood function one additional condition should be considered, since the likelihood function is developed assuming that the directional spreading function $D(\theta|\omega)$ is a piecewise constant function. In order to have a continuous smooth spreading function, the second derivative of $\ln D(\theta|\omega) = x_k$ should be close to zero. This results in the maximization of Eq. (7.109) being equivalent to maximizing the following quantity:

$$L(x_1, x_2, \ldots, x_k; \sigma^2)\exp\left\{-\frac{u^2}{2\sigma^2}\sum_{k=1}^{K}(x_k - 2x_{k-1} + x_{k-2})^2\right\} \tag{7.110}$$

where u is a parameter which will be determined later along with the unknown variance σ^2.

From a comparison of the above formula with the right-hand side of Eq. (7.103), the above formula can be considered to be a posterior distribution if the prior probabiity distribution of $\mathbf{x} = (x_1, x_2, \ldots, x_k)$ is given by

$$h(\mathbf{x}|u^2, \sigma^2) = \left(\frac{u}{\sqrt{2\pi}\,\sigma}\right)^K \exp\left\{-\frac{u^2}{2\sigma^2}\sum_{k=1}^{K}(x_k - 2x_{k-1} + x_{k-2})^2\right\} \tag{7.111}$$

For the present case, the Bayes formula given in Eq. (7.103) may be written in the following form:

$$f(\mathbf{x}|u^2,\sigma^2)=C\,h(\mathbf{x}|u^2,\sigma^2)L(\mathbf{x},\sigma^2) \tag{7.112}$$

The estimate of $\mathbf{x}$ can be achieved by maximizing the above equation, which in turn is equivalent to evaluating the mode of $f(\mathbf{x}|u^2,\sigma^2)$. If the value of u is given, the value of $\mathbf{x}$ which maximizes Eq. (7.112) is determined by minimizing the following quantity, irrespective of σ^2.

$$\sum_{j=1}^{2N}\left\{\phi_j-\sum_{k=1}^{K}\alpha_{j,k}\,e^{x_k}\right\}^2+u^2\left\{\sum_{k=1}^{K}(x_k-2x_{k-1}+x_{k-2})^2\right\} \tag{7.113}$$

where $\alpha_{j,k}=\mathrm{Re}\,\alpha_{j,k}(\omega)$.

On the other hand, the values of u and σ^2 can be obtained by minimizing the following quantity:

$$-2\ln\int L(\mathbf{x},\sigma^2)h(\mathbf{x}|u^2,\sigma^2)\mathrm{d}x \tag{7.114}$$

The Bayesian approach for estimating the directional spectrum developed by Hashimoto and Kobune (1988) appears to yield a high resolution in the evaluation of the wave energy spreading function. The approach, however, deals with many unknown $x_k(\omega)$, and this necessitates time-consuming iterative computation in practice.

Figure 7.17 shows an example of a comparison between the originally specified (true) spreading function and that computed by following the Bayesian approach using numerical simulation data. Included in the

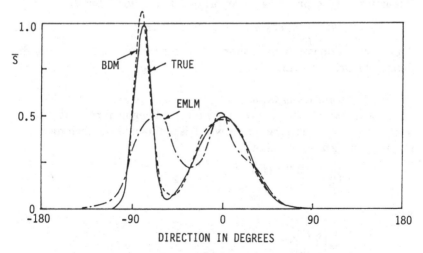

Fig. 7.17. Example of directional wave spectrum estimated by applying the Bayesian approach BDM (Hashimoto and Kobune 1988).

figure is the directional distribution function computed by applying the extended maximum likelihood (EML) method.

To date, various methods for estimating the directional wave energy spreading function from data have been developed. Each method has its own unique features and may have some shortcomings as well. Selection of a suitable method for analysis of data depends on the instruments employed for measurement and depends heavily on sea conditions at the site. A simple estimation method may be adequate for moderate spreading of wave energy due to normal wind speed without the presence of swell, while a sophisticated estimation method is necessary when swell is present or when estimating at a site where frequent change of wind direction may occur. Readers who are interested in the relative merits of the various estimation methods for analysis of data, although they are not conclusive, are referred to Benoit (1992, 1993), Benoit and Teisson (1994), Brissette and Tsanis (1994), Briggs (1984), and Young (1994).

7.5 FORMULATION OF THE WAVE ENERGY SPREADING FUNCTION

For the design and safe operation of marine systems, it is extremely important to evaluate system responses in a seaway; in particular, the evaluation of motions in six degrees of freedom is vital for floating systems. For this, it is very convenient to evaluate directional wave spectra by applying the energy spreading formulation to a point spectrum. For this, several formulations have been proposed and are summarized below.

(a) Cosine-square formula

It is assumed that spreading is proportional to $(\cos\theta)^2$. That is,

$$D(\theta) = (2/\pi)\cos^2\theta \qquad \text{for } -\pi/2 < \theta < \pi/2 \qquad (7.115)$$

This formulation is extremely simple, hence it has often been used for the design of marine system.

(b) Mitsuyasu formula

Mitsuyasu et al. (1975) propose the Longuet-Higgins formulation given in Eq. (7.40) with the parameter s, which is evaluated from their measurements by using a cloverleaf buoy:

$$D(\theta,f) = \frac{2^{2s-1}}{\pi} \frac{\{\Gamma(s+1)\}^2}{\Gamma(2s+1)} |\cos(\theta/2)|^{2s} \qquad (7.116)$$

where

$$s = \begin{cases} s_m (f/f_m)^5 & \text{for } f \leq f_m \\ s_m (f/f_m)^{-2.5} & \text{for } f > f_m \end{cases}$$

$$s_m = 11.5(2\pi f_m U/g)^{-2.5}$$

f_m = modal frequency

U = mean wind speed at 10 m height

(c) Hasselmann formula

From analysis of measured data, Hasselmann *et al.* (1980) propose the same formula as given in Eq. (7.116), but the s_1-value given in Eq. (7.47) is substituted for the s-value.

(d) Borgman's proposed formula

$$D(\theta) = \frac{1}{2\pi I_0(a)} \exp\{a\cos(\theta - \bar{\theta})\} \qquad (7.117)$$

where

a = positive constant

$\bar{\theta}$ = angle of propagation of predominant wave energy

$I_0(\ )$ = modified Bessel function of order zero.

This formula is called the *circular normal distribution* (Gumbel *et al.* 1953). The formula agrees with the first approximation of the estimated spreading function developed by Kobune and Hashimoto (1986) (see Section 7.4.2) based on the maximum entropy method.

8 SPECIAL WAVE EVENTS

8.1 BREAKING WAVES

8.1.1 Wave breaking criteria

The wave breaking phenomenon observed in the deep open ocean occurs whenever a momentarily high wave crest reaches an unstable condition. Profiles of incident waves for which breaking is imminent are shown in Figure 8.1(a) and (b); (a) is an example recorded during hurricane

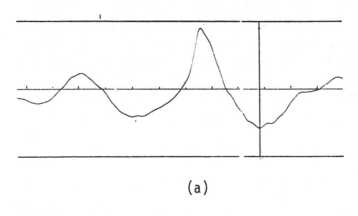

(a)

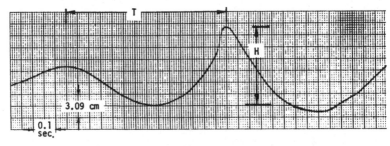

(b)

Fig. 8.1. Example of irregular waves with imminent breaking: (a) hurricane
CAMILLE; (b) laboratory test.

CAMILLE, while (b) is that obtained during experiments on breaking of irregular waves generated in a tank. Included in the figure are the local wave height, H, and period, T, associated with breaking. As seen in Figure 8.1, the period measured from peak-to-peak and that from crest-to-crest are almost equal.

Wave breaking occurs intermittently; hence the frequency of occurrence of breaking in a specified sea state is of considerable interest. As will be shown later, the severity of the sea is a major, but not the governing, factor; instead, the shape of the wave spectrum controls the frequency of occurrence to a great extent.

For estimating the frequency of occurrence of wave breaking in random seas, criteria for the breaking phenomenon must be clarified. Criteria for the breaking of ocean waves have been studied from different viewpoints by many researchers. These include Michel (1893), Dean (1968), Banner and Phillips (1974), Van Dorn and Pazan (1975), Nath and Ramsey (1976), Longuet-Higgins (1963a, 1969, 1974, 1976, 1979, 1985), Longuet-Higgins and Cokelet (1978), Longuet-Higgins and Fox (1977), Cokelet (1977), Kjeldsen and Myrhaug (1978), Ochi and Tsai (1983), Snyder and Kennedy (1983), Weissman et al. (1984), Holthuijsen and Herbers (1986), Xu et al. (1986), Ramberg and Griffin (1987), among others.

The most widely known wave breaking criterion is the limiting steepness for a Stokes wave, analytically derived by Michel (1893): breaking takes place when the wave height exceeds 14.2 percent of the Stokes limiting wave length, which is 20 percent greater than that of ordinary sinusoidal waves of the same frequency. This criterion can be written in terms of wave height, H, and period, T, as

$$H \geqslant 0.027 \, gT^2 \tag{8.1}$$

Results of laboratory experiments on breaking of regular waves generated in the experimental tank indeed demonstrate that breaking takes place when the condition given in Eq. (8.1) is satisfied. However, it is found from the results of tests on breaking of irregular waves generated in the tank that the observed number of breaking random waves is much greater than that theoretically computed based on the criterion given in Eq. (8.1), and that the following relationship provides the criterion for breaking of irregular waves (Ochi and Tsai 1983):

$$H \geqslant 0.020 \, gT^2 \tag{8.2}$$

The validity of the criterion has been confirmed by experimental studies carried out by Xu et al. (1986) in which irregular waves were generated by a blower installed in an air-sea tank, and by Ramberg and Griffin (1987) through tests in waves generated in a convergent channel. Figure 8.2 shows the breaking criteria obtained in the three different laboratory experiments. The lines given in the figure are the average of the scattered data for each experiment, over the range of wave heights tested. Included in the figure is the line indicating the breaking criterion of regular waves given in Eq. (8.1). Although the experimental conditions are quite different, the criteria obtained from these experiments are very close. As an average of the three experimental lines, Eq. (8.2) appears to be the most appropriate criterion for the breaking of irregular waves from the viewpoint of the wave height and period relationship.

A wave breaking criterion based on acceleration at the wave crest was first proposed by Phillips (1958), namely when the downward acceleration exceeded that of gravity. This criterion is used in connection with the

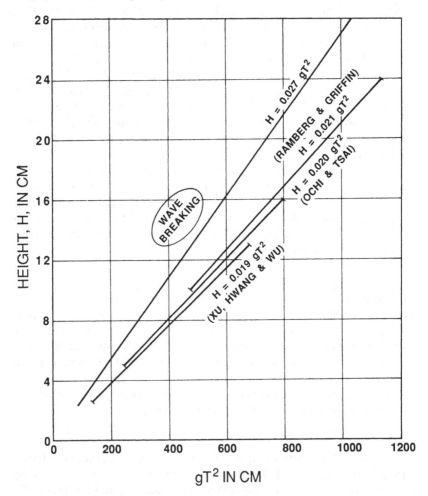

Fig. 8.2. Relationship between wave height and periods for breaking irregular waves (Ochi 1990b).

development of the equilibrium range considered for wave spectra. Snyder and Kennedy (1983) show a threshold acceleration of $-0.5g$ for white-capping waves, while Longuet-Higgins (1963a,1969) shows (theoretically) that white caps appear in a progressive wave when it reaches the limiting form at the crest in which the downward acceleration is equal to $0.5g$ in all directions within the Stokes 120-degree angle at the crest (see Figure 8.3).

Longuet-Higgins and Fox (1977) analytically evaluate the form of waves approaching their limiting steepness, and find the downward acceleration at the crest to be approximately $0.39g$. Longuet-Higgins (1985), however, points out some ambiguity in the definition of wave acceleration used as a criterion of breaking waves. He shows that the wave acceleration obtained from measurements by a fixed vertical probe is an apparent acceleration (the Eulerian acceleration), while wave particle acceleration, ideally measured by a small floating buoy, is the true acceleration (the Lagrangian acceleration). He shows that these accelerations are different for waves of finite amplitude.

Srokosz (1985), on the other hand, evaluates the probability of occurrence of wave breaking assuming that breaking occurs whenever the downward acceleration at the crests exceeds $0.4g$. He explains the relationship between this criterion and that given in Eq. (8.2); this will be discussed in detail in the next section.

Dawson et al. (1993) consider the nonlinear crest amplitudes for breaking, and the computed probability of occurrence of breaking agrees well with the experimental results if the downward crest acceleration is assumed to be $g/3$.

There is no doubt that the wave breaking phenomenon is associated with instability characteristics of waves. Studies of the stability of nonlinear gravity waves in deep water have been carried out by many researchers. These include analytical studies by Longuet-Higgins (1978a,b), Longuet-Higgins and Cokelet (1978), and experimental studies carried out by Melville (1982) and Sue et al. (1982), among others. One of the conclusions derived by Longuet-Higgins is that instability in regular waves occurs when the wave slope ak (where $a=$amplitude, $k=$wave number) approaches 0.436. this can be written in

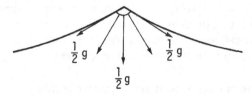

Fig. 8.3. Limiting acceleration near the crest in a Stokes 120 degree angle (Longuet-Higgins 1969).

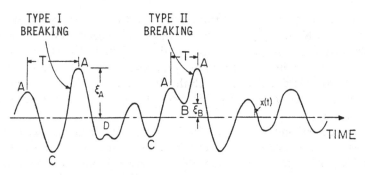

Fig. 8.4. Explanatory sketch of Type I and Type II breakings.

terms of wave height and period as $H=0.022\ gT^2$ which is very close to the relationship given in Eq. (8.2).

It is noted that wave breaking as discussed above takes place along the excursion as it crosses the zero-line. This may be called Type I breaking as illustrated in Figure 8.4. There is another type of breaking which occurs along the excursion above the zero-line (Type II) shown in Figure 8.4. This is the breaking of waves superimposed on long waves; hence, the secondary height where breaking occurs is small. Xu *et al.* (1986) observed this type of breaking in their experiments on waves generated in the air-sea tank. In their analysis for establishing the breaking criterion, the period of the long carrier waves was used, resulting in the breaking wave height being very low, $H=(0.005-0.010)\ gT^2$. Breaking wave heights of the same order are reported in field observations made in Lake Washington (Weissman *et al.* 1984). It is believed, however, that Eq. (8.2) would also hold for the Type II breaking shown in Figure 8.4 if the excursion between local crests and time interval between crests are regarded as H and T, respectively.

For a more precise description of steep wave profiles for which breaking is imminent, Kjeldsen and Myrhaug (1978) propose three parameters; crest front steepness, and vertical and horizontal asymmetry factors. On the other hand, statistical wave-by-wave analysis of records obtained from buoy observations and measurements of whitecaps in the North Sea by Holthuijsen and Herbers (1986) indicate that a separation of breaking waves (whitecaps) and non-breaking waves based on such seemingly obvious parameters as wave steepness and wave asymmetry appears to be difficult. They claim that breaking occurs at wave steepness values much less than the theoretically expected values of limiting waves. They also claim that waves in the open sea tend to break near the center of a wave group.

8.1.2 Probability of occurrence of wave breaking

The probability of occurrence of wave breaking in random seas can be evaluated by applying the breaking criterion given in Eq. (8.2) to the joint probability density function of wave height and period. The joint probability density function analytically developed for waves with a narrow-band spectrum by Longuet-Higgins (1983) is given in Section 4.2, and for waves with a non-narrow-band spectrum by Cavanié et al. (1976) in Section 4.3. Both of these joint distributions are given in closed form and thereby the breaking criterion can be incorporated into the distribution functions. In the following, the joint probability density function for a non-narrow-band spectrum will be applied for evaluating the probability of occurrence of breaking in random seas.

Let us first write the breaking criterion given in Eq. (8.2) in dimensionless form as follows:

$$\nu \geq \alpha(\overline{T}_m^2/\sqrt{m_0})\lambda^2 \tag{8.3}$$

where

ν = dimensionless excursion = $\zeta/\sqrt{m_0}$
λ = dimensionless time interval between positive maxima
  = $T/\overline{T}_m$
ζ = excursion
T = time interval between positive maxima
$\overline{T}_m$ = average time interval between successive positive maxima given in Eq. (4.33)
m_0 = area under the wave spectrum
α = 0.196 m/s².

The joint probability density function of excursion and the associated time interval applicable for Type I breaking is essentially the same as that for the positive maxima and associated time interval, which is given in Eq. (4.34). Here, the modification necessary for normalizing the probability distribution of time interval as given in Eq. (4.37) is omitted for simplicity.

By considering twice the magnitude of the positive maxima, we may write $\nu=2\xi$, where ξ is defined in Eq. (4.34) as a dimensionless positive maxima. λ is the same as τ in Eq. (4.34), and the joint probability density function may be written as a function of the bandwidth parameter ϵ as follows:

$$f(\nu,\lambda) = \frac{1}{32\sqrt{2\pi}} \frac{(1+\sqrt{1-\epsilon^2})^3}{\epsilon(1-\epsilon^2)} \frac{\nu^2}{\lambda^5}$$

$$\times \exp\left[-\frac{\nu^2}{8\epsilon^2\lambda^4}\left\{\frac{1}{16}\frac{(1+\sqrt{1-\epsilon^2})^4}{1-\epsilon^2} - \frac{1}{2}\left(1+\sqrt{1-\epsilon^2}\right)^2\lambda^2 + \lambda^4\right\}\right]$$

$$0 \leq \nu < \infty \qquad 0 \leq \lambda < \infty \tag{8.4}$$

It should be noted that Eq. (8.4) includes the positive maxima belonging to both Type I and Type II excursions. Hence, in evaluating the probability of occurrence of wave breaking associated with Type I excursions, the maxima belonging to Type II excursions have to be deleted. This can be done by taking into consideration the frequency of occurrence of each excursion.

The frequency of occurrence of Type I and Type II excursions, denoted by p_I and p_{II}, respectively, are derived in Section 3.7 with regard to the probability distribution of half-cycle excursions. These are,

$$p_I = \frac{2\sqrt{1-\epsilon^2}}{1+\sqrt{1-\epsilon^2}}$$

$$p_{II} = \frac{1-\sqrt{1-\epsilon^2}}{1+\sqrt{1-\epsilon^2}} \tag{8.5}$$

Thus, the conditional probability of wave breaking given that Type I excursion has occurred can be evaluated from Eqs. (8.4) and (8.5) taking into consideration the breaking condition given in Eq. (8.3). That is,

$$\Pr\{\text{breaking wave} \mid \text{Type I excursion}\}$$

$$= p_I \int_0^\infty \int_{\alpha\left(\frac{\overline{T}_m^2}{\sqrt{m_0}}\right)\lambda^2}^\infty f(\nu,\lambda) d\nu d\lambda \tag{8.6}$$

Then, the probability of wave breaking associated with Type I excursions can be obtained by multiplying Eq. (8.6) by the probability of occurrence of the Type I excursion. By taking account of the fact that the sum of the probabilities of Type I and Type II breaking waves is equal to unity when all waves have broken, the probability of Type I breaking is given as follows:

$$\Pr\{\text{breaking wave, Type I}\}$$

$$= \frac{p_I^2}{p_I^2 + p_{II}} \int_0^\infty \int_{\alpha\left(\frac{\overline{T}_m^2}{\sqrt{m_0}}\right)\lambda^2}^\infty f((\nu,\lambda) d\nu d\lambda \tag{8.7}$$

The joint probability density function applicable for Type II excursions and the frequency of their occurrences are not given here, but are presented in Ochi and Tsai (1983).

It is of interest to examine the effect of shape of the wave spectrum on the frequency of occurrence of wave breaking. As an example, computations are made for the two wave spectra shown in Figure 8.5. These are obtained from measurements in the North Atlantic. Spectrum (a) represents a sea of significant wave height 4.6 m, while Spectrum (b) is that for a significant wave height of 10.8 m. The joint probability density function of the Type I excursion and its associated time interval and the lines indicating breaking conditions are shown in Figure 8.6. As can be seen, the location of lines indicating the breaking condition differs substantially because of the difference in the magnitude of moments of the spectrum, and this in turn yields a difference in the probabilities of breaking; the probability being zero for Spectrum (a) and 0.13 for Spectrum (b).

In order to elaborate on the effect of moments on breaking presented

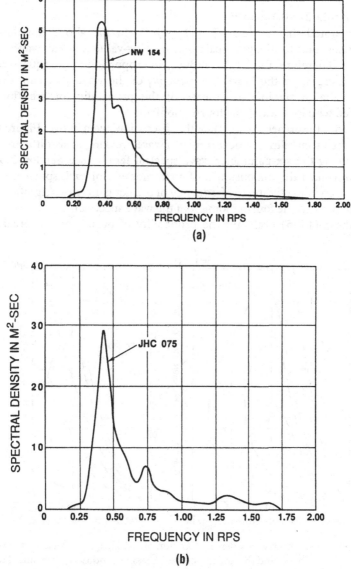

Fig. 8.5. Wave spectra used for computation of breaking waves.

in the above paragraph, let us rewrite the breaking criterion given in Eq. (8.3) as follows:

$$\nu \geqslant (4\pi)^2 \alpha \left(\frac{\sqrt{1-\epsilon^2}}{1+\sqrt{1-\epsilon^2}} \right)^2 \frac{\sqrt{m_0}}{m_2} \lambda^2$$

$$= (4\pi)^2 \alpha \frac{\sqrt{1-\epsilon^2}}{(1+\sqrt{1-\epsilon^2})^2} \frac{\lambda^2}{\sqrt{m_4}} \tag{8.8}$$

where $\alpha = 0.02g = 0.196$ m/s^2.

It is noted that $(1-\epsilon^2)^{1/2}/\{1+(1-\epsilon^2)^{1/2}\}^2$ in the above equation is almost constant for the practical range of the ϵ-value of ocean waves, say 0.4–0.8. Therefore, the ν-value reduces in proportion to the square-root of the increase in the fourth moment m_4 of the wave spectrum. This implies that a significant increase of the probability of breaking is expected with increase in the fourth moment.

In order to further substantiate the discussion given above, Figure 8.7 shows the probabilities of occurrence of wave breaking in six different sea states using the six-parameter wave spectral family (see Section 2.3.4) plotted against the dimensionless fourth moment of each spectrum. As can be seen in the figure, the frequency of breaking increases significantly with increase in the fourth moment of the wave spectrum.

Srokosz (1985) evaluates the probability of occurrence of breaking

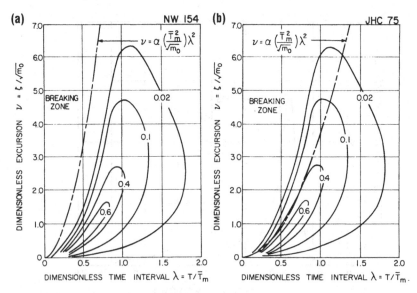

Fig. 8.6. Joint probability density function of wave height and associated time interval, and the line indicating breaking condition (Ochi and Tsai 1983).

waves taking into consideration the criterion that breaking takes place if the downward acceleration at the wave crest exceeds the limit αg where α is unknown at this stage. For evaluating the probability of occurrence of breaking, Srokosz modifies the probability density function of maxima developed by Cartright and Longuet-Higgins given in Eq. (3.28) such that the integration limit of acceleration is from $-\infty$ to $-\alpha g$ instead of $-\infty$ to 0. This results in a density function for the crest to be broken, denoted by η in dimensionless form, of

$$f(\eta) = \frac{1}{\sqrt{2\pi}} \left[\epsilon \exp\left\{ -\frac{1}{2}\gamma^2 - \frac{1}{2\epsilon^2}(\eta - \sqrt{1-\epsilon^2}\gamma)^2 \right\} \right.$$

$$\left. + (1-\epsilon^2)\eta \exp\{-\eta^2/2\} \int_\mu^\infty \exp\{-x^2/2\}\,dx \right] \quad (8.9)$$

where

$$\eta = (\text{crest to be broken})/\sqrt{m_0}$$
$$\gamma = \alpha g/\sqrt{m_4}$$
$$\mu = (\gamma - \sqrt{1-\epsilon^2}\eta)/\epsilon$$
$$\epsilon = \text{bandwidth parameter (see Eq. 3.32).}$$

By integrating η over all crest heights, Srokosz derives the probability of occurrence of wave breaking as

$$\Pr\{\text{breaking}\} = \exp\{-\alpha^2 g^2/(2m_4)\} \quad (8.10)$$

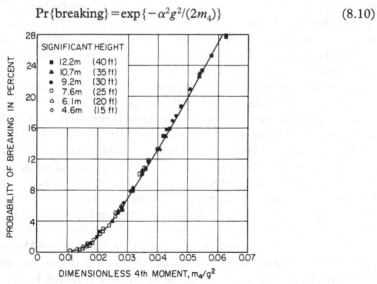

Fig. 8.7. Probability of occurrence of breaking waves plotted against the dimensionless fourth moment of the spectrum (Ochi and Tsai 1983).

which indicates that the probability of breaking depends on the fourth moment of the wave spectrum, which agrees with the trend observed in Figure 8.7.

In choosing the α-value for wave breaking, Srokosz rewrites the breaking criterion given in Eq. (8.2) by letting $H=2a$ ($a=$amplitude), and $T=2\pi/\omega$ as follows:

$$a\omega^2 \geqslant 0.4g \qquad (8.11)$$

Here, $a\omega^2$ is the magnitude of the downward acceleration at the crest of a wave in linear wave theory; hence, by choosing the unknown α constant as 0.4, the Srokosz's criterion used for evaluating the probability of occurrence of breaking waves coincides with that given in Eq. (8.2). By letting $\alpha=0.4$, the probability of wave breaking given in Eq. (8.10) becomes

$$\Pr\{\text{breaking}\} = \exp\{-0.08g^2/m_4\} \qquad (8.12)$$

The results of computations of the probability of occurrence of Type I breaking given in Eq. (8.12) agrees well with those shown in Figure 8.7.

Inasmuch as the fourth moment of the spectrum plays a significant role in evaluation of the probability of occurrence of wave breaking, care has to be taken in computing the fourth moment. Although the high frequency tail of the spectrum is generally crucial for evaluating the fourth moment, the magnitude of spectral densities considered in naval and ocean engineering for frequencies greater than $\omega=2.0$ rps is imperceptibly small in comparison with those at lower frequencies (see examples of spectra shown in Figure 2.15); hence, for spectral formulations that are used for evaluating wave properties or responses of marine systems in a seaway, the cut-off frequency $\omega=2.2$ rps for deep water and $\omega=3.5$ rps for shallow water appears to be sufficient.

Referring to Figure 8.7, it can be seen that the probability of wave breaking increases with increase in the fourth moment. This suggests that a considerable amount of wave breaking can be expected in seas represented by the Pierson–Moskowitz spectrum, since its fourth moment is theoretically infinite. However, this is not the case if the frequency range up to $\omega=2.2$ is considered in computing the frequency of occurrence of wave breaking. As an example, Figure 8.8 shows the shapes of the Pierson–Moskowitz spectrum and the two-parameter wave spectrum with modal frequency at $\omega=0.52$ for the same significant wave height of 12.2 m. The results of computations on the probability of occurrence of breaking waves show a probability of 0.001 for the Pierson–Moskowitz spectrum in contrast with a probability of 0.14 for the two-parameter spectrum. Furthermore, the probabilities of wave breaking in seas with the Pierson–Moskowitz spectrum are nearly zero irrespective of the magnitude of significant wave height if spectral densities up to $\omega=2.2$ only are considered. This result satisfies the condition required for wave spectra of

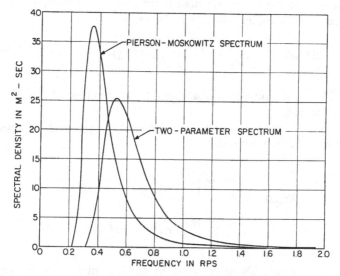

Fig. 8.8. Pierson–Moskowitz and two-parameter wave spectra used for computing the probability of wave breaking.

fully developed seas, and it may serve to substantiate the validity of the computation method for evaluating the probability of occurrence of breaking waves.

8.1.3 Energy loss resulting from wave breaking

Wind-generated waves during a storm may be completely broken so that the sea state is close to a fully developed sea, although this situation is not frequently encountered. The majority of sea conditions observed during a storm are partially developed such that waves may still have the potential to be broken.

When breaking takes place, wave energy is lost in the form of turbulence and this results in a reduction of the magnitude of spectral density at certain frequencies. Studies on the evaluation of energy loss associated with wave breaking have been carried out by Longuet-Higgins (1969) and Tung and Huang (1987) by employing different breaking criteria. In the following, estimation of energy loss and modification of the wave spectrum associated with Type I wave breaking are presented based on the criterion given in Eq. (8.2).

The average loss of energy in one wave cycle may be evaluated assuming that a breaking wave reduces its height to the limiting breaking height, H_*. The resulting energy loss is given by (Longuet-Higgins 1969)

$$\text{energy loss due to breaking} = \frac{1}{2}\rho g\left\{\left(\frac{H}{2}\right)^2 - \left(\frac{H_*}{2}\right)^2\right\} \quad (8.13)$$

Then, the loss of spectral energy density for a given wave period, denoted $\Delta S(\omega)$, can be written, ignoring the factor ρg, as,

$$\Delta S(\omega)=\int_{H_*}^{\infty}\frac{1}{2}\left\{\left(\frac{H}{2}\right)^2-\left(\frac{H_*}{2}\right)^2\right\}f(H|T)dH \qquad (8.14)$$

where $f(H|T)$=conditional probability density function of wave height for a given period.

The rate of loss of spectral energy density from the original spectral density for a given wave frequency becomes

$$\mu=\frac{\Delta S(\omega)}{S(\omega)}=\frac{\displaystyle\int_{H_*}^{\infty}\frac{1}{2}\left\{\left(\frac{H}{2}\right)^2-\left(\frac{H_*}{2}\right)^2\right\}f(H|T)\,dH}{\displaystyle\int_{0}^{\infty}\frac{1}{2}\left(\frac{H}{2}\right)^2 f(H|T)\,dH} \qquad (8.15)$$

We may write the above equation in dimensionless form by letting $H/\sqrt{m_0}=\nu$ and $T/\overline{T}_m=\lambda$ and applying the joint probability density function given in Eq. (8.4). For this, let us rewrite Eq. (8.4) as follows:

$$f(\nu,\lambda)=\frac{1}{4\sqrt{2\pi}}\frac{\alpha^3\beta^2}{\epsilon^3}\frac{\nu^2}{\lambda^5}\exp\left[-\frac{\nu^2}{8\epsilon^2\lambda^4}\{(\lambda^2-\alpha^2)^2+\alpha^4\beta^2\}\right] \quad (8.16)$$

where

$$\alpha=\tfrac{1}{2}\,(1+\sqrt{1-\epsilon^2})$$

$$\beta=\epsilon/\sqrt{1-\epsilon^2}.$$

By integrating Eq. (8.16) with respect to ν, the marginal probability density function of λ becomes

$$f(\lambda)=\int_{0}^{\infty}f(\nu,\lambda)d\nu=\frac{\alpha^2\beta^2\lambda}{\{(\lambda^2-\alpha^2)^2+\alpha^4\beta^2\}^{3/2}} \qquad (8.17)$$

Hence, the conditional probability density function $f(\nu|\lambda)$ is obtained as

$$f(\nu|\lambda)=\frac{f(\nu,\lambda)}{f(\lambda)}$$

$$=\frac{1}{4\sqrt{2\pi}}\frac{\{(\alpha^2-\lambda^2)^2+\alpha^4\beta^2\}^{3/2}}{\epsilon^3}\frac{\nu^2}{\lambda^6}\exp\left[-\frac{\nu^2}{8\epsilon^2\lambda^4}\{(\lambda^2-\alpha^2)^2+\alpha^4\beta^2\}\right]$$

$$(8.18)$$

By using the formulation given in Eq. (8.18), the ratio of loss of spectral energy due to wave breaking may be written in dimensionless form as

$$\mu=\dfrac{\displaystyle\int_{\nu_*}^{\infty}(\nu^2-\nu_*^2)\,f(\nu\mid\lambda)\,d\nu}{\displaystyle\int_{0}^{\infty}\nu^2 f(\nu\mid\lambda)\,d\nu} \tag{8.19}$$

where ν_* is the limiting breaking wave height (the minimum ν) which is given in Eq. (8.8) in dimensionless form.

Thus, the modified spectral density function, $S_*(\omega)$, taking wave breaking into account can be written, for a given frequency, as

$$S_*(\omega)=S(\omega)(1-\mu) \tag{8.20}$$

Note that modification of a spectrum due to wave breaking is an interation procedure. Modification of the original spectrum by Eq. (8.20) does not necessarily mean there will be no more breaking. This is because the limiting breaking wave height is a function of the bandwidth parameter ϵ as well as the fourth moment of the spectrum, m_4. Therefore, ϵ, m_4 and the probability of breaking must be re-computed for the modified spectrum. The procedure is repeated until the probability of wave breaking becomes sufficiently small.

An example of modification of a wave spectrum is shown in Figure 8.9. The original spectrum, with a significant wave height of 12.0 m, is

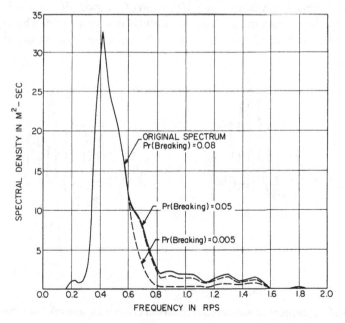

Fig. 8.9. Modification of wave spectrum resulting from wave breaking.

obtained from data taken in the North Atlantic. The computed probability of Type I wave breaking for this spectrum is 0.08. The result of the first modification yields the reduction of energy for frequencies higher than $\omega=0.8$ as shown in the figure, and the computed probability of breaking waves for the modified spectrum is 0.05. The computations for spectrum modification is repeated, and eventually a spectrum is obtained for which the probability of breaking is 0.005. For this situation, the energies at higher frequencies of the spectrum are reduced by a substantial amount, and as a result, the significant wave height is reduced to 10.9 m after all breaking takes place.

8.2 GROUP WAVES

8.2.1 Introduction

An interesting phenomenon often observed in wind-generated seas is a sequence of high waves having nearly equal periods, commonly known as *group waves*. Two examples of group waves observed in the open ocean are shown in Figure 8.10. Figure 8.10(a) is taken from Rye (1974) in which he states that the group waves were recorded in the North Sea during a storm of significant wave height 10 m, while Figure 8.10(b) shows wave groups recorded in severe seas by a weather ship at Station K in the North Atlantic.

It has been known that group waves often cause serious problems for the safety of marine systems when the period of the individual waves in the group are close to the marine system's natural motion period. This is not because the wave heights are exceptionally large, but because of motion

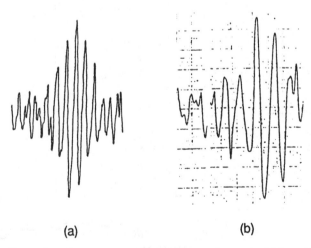

(a)   (b)

Fig. 8.10. Examples of wave groups observed at sea (example (a) from Rye 1974).

augmentation due to resonance with the wave, which may induce capsizing of the marine system. As another example, a moored marine system tends to respond to successive high waves which induce a slow drift oscillation of the system resulting in large forces on the mooring lines.

The physical explanation of the group wave phenomenon has yet to be clarified; however, Mollo-Christensen and Ramamonjiarisoa (1978) and Ramamonjiarisoa (1974) explain that the wave field does not consist of independently propagating Fourier components but instead consists wholly or in part of wave groups of a permanent type. As evidence, they present results of field and laboratory observations indicating that harmonic components of waves propagate at higher phase velocities than those predicted by linear theory, and that velocities are nearly constant for frequencies greater than the modal frequency of the spectrum (see Figure 8.11).

Many studies on stochastic analysis of group waves in random seas have been carried out, primarily concerning the frequency of occurrence of the phenomenon. These studies may be categorized into two approaches: one considers the phenomenon as a level-crossing problem associated with the envelope of a random process; the other treats a sequence of large wave heights as a Markov chain problem.

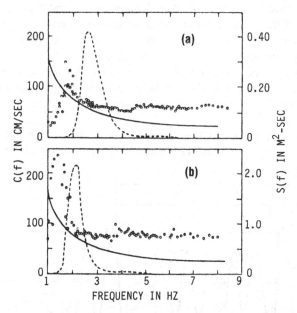

Fig. 8.11. Phase speed of wind waves measured in the laboratory. Phase speed (circles), linear theory prediction (solid line) and spectral density function (broken line): (a) wind speed 5 m/s; (b) wind speed 8 m/s (Mollo-Christensen and Ramamonjiarisoa 1978).

As stated earlier, group waves occur in a sequence of high waves having nearly equal periods; hence, we may consider the envelope of wave trains as shown in Figure 8.12, and define the exceedance of the envelope above a certain specified level as a wave group phenomenon. The principle of this approach is credited to Rice (1945, 1958) and Longuet-Higgins (1957, 1962a) although they did not apply it to wave groups *per se*; instead, they clarified the basic properties of a group phenomenon associated with a Gaussian random process.

Following this approach, Nolte and Hsu (1972), Ewing (1973), Chakrabarti *et al.* (1974), Rye (1974, 1979) and Goda (1970, 1976) have developed methods to evaluate mean values of the length of time a wave group persists as well as the mean number of wave crests in the group, etc. Longuet-Higgins (1984) gives a thorough discussion on various subjects of group waves following this approach.

In these studies, the up-crossings of the envelope above a certain level define the wave group phenomenon. It is pointed out, however, that if the time duration $\tau_{\alpha+}$ shown in Figure 8.12 is short, there may be only one wave crest or no wave crest in time $\tau_{\alpha+}$, which obviously does not constitute a wave group even though the wave envelope exceeds the specified level.

Dawson *et al.* (1991) carried out a series of excellent experiments on wave groups in random seas generated in the laboratory, and found that only 25 percent of the wave crests crossing the threshold level (taken as one-half of the significant wave height) constituted a wave group and all other wave crests were above the level but single crossings.

It is thus clear that consideration of (a) exceedance of a specified level and (b) at least two wave crests during the exceedance are required for evaluating the statistical properties of group waves. This subject will be discussed in Section 8.2.2.

Another approach for evaluating the probability of occurrence of group waves considers a sequence of high waves taking into consideration

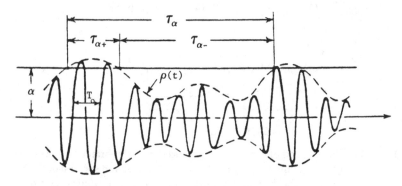

Fig. 8.12. Level crossing of the envelope of a random process.

the correlation between them based on the Markov chain concept. In general, if the magnitude of wave height depends only on the magnitude of the immediate previous waves, the waves are considered to be a Markov chain process (see Appendix B). A study to examine whether or not ocean waves follow the Markov chain characteristics was carried out by Sawhney (1962). His analysis shows that the Markov chain characteristics of waves become weak after three half cycles and are lost completely after 8 cycles.

Arhan and Ezraty (1978) show, based on the analysis of wave data obtained from a storm in the North Sea, that the correlation between successive wave heights is pronounced in waves whose heights are greater than 75 percent of significant wave height.

Rye (1974) evaluates the correlation coefficients between successive wave heights obtained off the Norwegian Coast where the water depth is about 100 m. His analysis shows that the average of the correlation coefficient is $+0.24$, where the positive sign implies that large waves tend to be succeeded by large waves. He also finds that formation of wave groups tends to be more pronounced for the growing stage of a storm than the decaying stage. These results indicate that application of the Markov chain concept is an expedient approach for evaluating the probability of occurrence of group waves.

Kimura (1980) develops a method to evaluate the probability of occurrence of group waves in a given sea based on the Markov chain concept. In order to deal with two consecutive wave heights, Kimura considers the two-dimensional Rayleigh probability distribution with the parameter k (see Eq. (3.58)) which is evaluated from the correlation coefficient obtained from the time history of wave height data. This shortcoming is improved by Longuet-Higgins (1984) and Battjes and van Vledder (1984) so his method can be incorporated with the wave spectrum. This will be discussed in Section 8.2.3.

8.2.2 Statistical properties through the envelope process approach

As stated in the preceding section, many studies address the average value of the statistical properties of group waves, such as the average number of waves in a group, the average time interval between groups, etc. Let us first evaluate the expected (average) number of crossings of a specified level α, denoted N_α, per unit time by the envelope (see Figure 8.12). This is the level up-crossing problem, and the same basic concept as considered in Eq. (4.40) can be applied except for the domain of integration. That is, by letting $f(\rho,\rho')$ be the joint probability density function of the displacement and velocity of the envelope. The average number of crossings $\overline{N}_\alpha$ can be evaluated by letting $\rho=\alpha$.

$$\overline{N}_\alpha = \int_0^\infty \dot{\rho}f(\alpha,\dot{\rho})d\dot{\rho} \qquad (8.21)$$

where ρ and $\dot{\rho}$ are statistically independent, and each obeys the Rayleigh and normal probability density function, respectively. The density functions are given by

$$f(\rho) = (\rho/m_0)\exp\{-\rho^2/2m_0\}$$
$$f(\dot{\rho}) = (1/\sqrt{2\pi\mu_2})\exp\{-\dot{\rho}^2/2\mu_2\} \qquad (8.22)$$

where

$$m_0 = \int_0^\infty S(\omega)d\omega$$

$$\mu_2 = \int_0^\infty (\omega-\overline{\omega})^2 S(\omega)d\omega$$

$\overline{\omega}$ = mean frequency of spectrum.

Note that Eq. (8.22) can be derived from the joint probability density function given in Eq. (4.9) by integrating with respect to ϕ and $\dot{\phi}$, and by letting $\mu_0 = m_0$.

From Eqs. (8.21) and (8.22), the average number of up-crossing of the level α per unit time, $\overline{N}_\alpha$, can be obtained as

$$\overline{N}_\alpha = \sqrt{\mu_2/2\pi}\,(\alpha/m_0)\,e^{-(\alpha^2/2m_0)} \qquad (8.23)$$

The inverse of $\overline{N}_\alpha$ is the average time interval between two up-crossings, denoted $\overline{\tau}_\alpha$; i.e.

$$\overline{\tau}_\alpha = \sqrt{2\pi/\mu_2}\,(m_0/\alpha)\exp\{\alpha^2/2m_0\} \qquad (8.24)$$

Note that all these formulae are for a spectrum given in terms of ω.

Next, let us evaluate the average number of waves in a group per unit time, denoted $\overline{N}_G$. For this, we evaluate the proportion of time during which the envelope exceeds the level α. This can be evaluated from the first formula given in Eq. (8.22) as

$$\Pr\{\rho > \alpha\} = \int_\alpha^\infty f(\rho)d\rho = \exp(-\alpha^2/2m_0) \qquad (8.25)$$

Then, the average time for a wave group to occur, denoted by $\overline{\tau}_{\alpha+}$, becomes

$$\overline{\tau}_{\alpha+} = \overline{\tau}_\alpha \Pr\{\rho > \alpha\} = \sqrt{2\pi/\mu_2}\,(m_0/\alpha) \qquad (8.26)$$

On the other hand, the average zero-crossing time (average wave period) for a narrow-band spectrum is given in Eq. (4.43) as

$$\overline{T}_0 = 2\pi \sqrt{m_0/m_2} \tag{8.27}$$

Hence, from Eqs. (8.26) and (8.27), we have the average number of waves in a group as

$$\overline{N}_G = \overline{\tau}_{\alpha+}/\overline{T}_0 = (1/\alpha\sqrt{2\pi})\sqrt{m_0 m_2/\mu_2} \tag{8.28}$$

For the probability distribution of the frequency of occurrence of wave groups, Nolte and Hsu (1972) and Longuet-Higgins (1984) consider the Poisson probability distribution with a parameter representing the average number $\overline{N}_\alpha$ derived in Eq. (8.23) assuming that the time interval between the occurrence of two successive wave groups is sufficiently long. That is,

$$\Pr\{n \text{ wave groups}\} = \overline{N}_\alpha \exp\{-n/\overline{N}_\alpha\} \tag{8.29}$$

Thus far, the average values of various properties associated with group waves have been evaluated through the level crossings of the envelope assuming that the envelope up-crossing of a specified level implies a wave group. Next, let us consider wave groups to consist of at least two successive high waves above a specified level, and evaluate various properties based on probability distribution functions representing these properties rather than evaluating the average values.

First, we consider the probability density function of time duration associated with wave groups. For this, we derive the probability density function of time duration $\tau_{\alpha+}$ shown in Figure 8.12, and then truncate the probability density function taking into account the condition that two or more waves exist during $\tau_{\alpha+}$. Since the derivation of the probability density function is extremely complicated, only the procedure for the analytical derivation will be outlined in the following. For details of the derivation, see Ochi and Sahinoglou (1989a).

(i) Assuming that the waves are a Gaussian random process $x(t)$ with a narrow-band spectrum, the probability that the envelope of $x(t)$ exceeds a specified level α at time t_1 with velocity $\dot{\rho}_1$, and then crosses that same level downward at time t_2 with velocity $\dot{\rho}_2$ is given approximately by the following formula:

$$\Pr\left\{\begin{array}{l} \text{up-crossing of a level } \alpha \text{ at time } t_1 \\ \text{with velocity } \dot{\rho}_1 \text{ followed by a} \\ \text{crossing at time } t_2 \text{ with velocity } \dot{\rho}_2 \end{array}\right\} = \frac{\overline{N}_{\alpha+}(\tau_{\alpha+})}{\overline{N}_{\alpha+}}$$

$$= \frac{\displaystyle\int_{-\infty}^{0}\int_{0}^{\infty} \dot{\rho}_1 \dot{\rho}_2\, f(\alpha,\dot{\rho}_1,\alpha,\dot{\rho}_2)\, \mathrm{d}\dot{\rho}_1 \mathrm{d}\dot{\rho}_2}{\displaystyle\int_{0}^{\infty} \dot{\rho}_1 f(\alpha,\dot{\rho}_1)\, \mathrm{d}\dot{\rho}_1} \tag{8.30}$$

$$0 < \dot{\rho}_1 < \infty \qquad -\infty < \dot{\rho}_2 < 0$$

Here, the numerator, $\overline{N}_{\alpha+}(\tau_{\alpha+})$, represents the expected number (per unit time) of envelope up-crossings with velocity $\dot{\rho}_1$ followed by down crossings with velocity $\dot{\rho}_2$; while the denominator, $\overline{N}_{\alpha+}$, represents the expected number (per unit time) of envelope up-crossings of the level α with velocity $\dot{\rho}_1$. $f(\alpha, \dot{\rho}_1, \alpha, \dot{\rho}_2)$ is the joint probability density function of displacement ρ and velocity $\dot{\rho}$ at time t_1 and t_2 with $\rho_1 = \rho_2 = \alpha$ in this case. By letting $t_2 - t_1 = \tau_{\alpha+}$, Eq. (8.30) is equivalent to the probability density function of $\tau_{\alpha+}$ for a specified level α.

(ii) In order to evaluate Eq. (8.30), it is necessary to first obtain the joint probability density function of the wave profile $x(t)$ and velocity $\dot{x}(t)$ at time t_1 and t_2, and then to transform this joint probability density function to that of the wave envelope and its velocity. By considering the sine and cosine components for both x and $\dot{x}$ at time t_1 and t_2, we have a set of eight random variables $(x_{c1}, \dot{x}_{c1}, x_{s1}, \dot{x}_{s1}, x_{c2}, \dot{x}_{c2}, x_{s2}, \dot{x}_{s2})$ which comprise a joint probability density function. Here, each element can be evaluated for a given wave spectrum.

(iii) The inverse operation of the 8×8 covariance matrix involved in the joint normal distribution is extremely difficult to obtain in closed form. However, this is feasible by approximating a given wave spectrum as the sum of two parts, each part being symmetric about its mean frequency as illustrated in Figure 8.13.

(iv) Next, by applying the polar coordinates given by

$$
\begin{aligned}
x_{c1} &= \rho_1 \cos \theta_1 \\
x_{s1} &= \rho_1 \sin \theta_1 \\
x_{c2} &= \rho_2 \cos \theta_2 \\
x_{c2} &= \rho_2 \sin \theta_2
\end{aligned}
\qquad (8.31)
$$

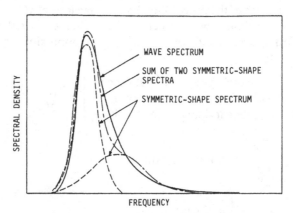

Fig. 8.13. Decomposition of wave spectrum into two symmetric-shape spectra.

the joint probability density function can be transformed to the joint density function of $\dot{\rho}_1$, $\dot{\rho}_2$, θ_1, $\dot{\theta}_1$, θ_2, and $\dot{\theta}_2$, and in turn, this joint density function can be reduced to $f(\dot{\rho}_1, \dot{\rho}_2, \dot{\theta}_1, \dot{\theta}_2)$.

(v) By integrating the joint probability density function $f(\dot{\rho}_1, \dot{\rho}_2, \dot{\theta}_1, \dot{\theta}_2)$ with respect to $\dot{\theta}_1$ and $\dot{\theta}_2$, the joint density function $f(\dot{\rho}_1, \dot{\rho}_2)$ can be obtained as a function of the specified level α, and from this the following probability density function of $\tau = t_2 - t_1$, which is written $f(\tau_{\alpha+})$, can be derived:

$$f(\tau_{\alpha+}) = C \frac{2\alpha\mu_0}{\sqrt{\pi\mu_2}} \frac{M_{22}}{(\mu_0^2 - \nu_0^2)^2}$$

$$\times \int_0^{\pi/2} \exp\left\{ -\left[\frac{\alpha^2 \{\mu_0 - \nu_0(1 - \beta^2/2)\}}{\mu_0^2 - \nu_0^2} - \frac{\alpha^2}{2\mu_0} + 4\lambda\gamma^2 \right] \right\}$$

$$\times \gamma \left[\frac{1}{\sqrt{\pi}} \exp\{-2\gamma^2\} + \sqrt{2}\gamma\,\phi(\sqrt{2}\gamma) \right] d\beta \qquad (8.32)$$

where

$$M_{22} = \mu_2(\mu_0^2 - \nu_0^2) - \nu_1^2\mu_0$$

$$\lambda = \frac{-\nu_2(\mu_0^2 - \nu_0^2) + \nu_0\nu_1^2}{\mu_2(\mu_0^2 - \nu_0^2) - \nu_1^2\mu_0} \cos\beta$$

$$\gamma = \frac{\alpha}{2(1+\lambda)\sqrt{M_{22}}} \frac{\nu_1\{\mu_0(1 - \beta^2/2) - \nu_0\}}{\sqrt{\mu_0^2 - \nu_0^2}}$$

$$\mu_0 = \int_0^\infty S(\omega)d\omega$$

$$\mu_2 = \int_0^\infty (\omega - \overline{\omega})^2 S(\omega)d\omega$$

$$\nu_0 = \int_0^\infty S(\omega)\cos(\omega - \overline{\omega})\tau d\omega$$

$$\nu_1 = \int_0^\infty (\omega - \overline{\omega})S(\omega)\sin(\omega - \overline{\omega})\tau d\omega$$

$$\nu_2 = \int_0^\infty (\omega - \overline{\omega})^2 S(\omega)\cos(\omega - \overline{\omega})\tau d\omega$$

$$\overline{\omega} = \int_0^{\infty} \omega S(\omega)\, d\omega \bigg/ \int_0^{\infty} S(\omega)\, d\omega$$

$$\tau = t_2 - t_1$$

$$\phi(z) = \frac{2}{\sqrt{\pi}} \int_0^z \exp\{-t^2\}\, dt$$

C = normalization factor to make the area under the

density function unity $= 1 \bigg/ \int_0^{\infty} f(\tau_{\alpha+})\, d\tau_{\alpha+}$

(vi) By letting $\overline{T}_0$ be the average zero-crossing period, the probability of two or more waves of time duration $\tau_{\alpha+}$ can be evaluated as follows:

(a) $\tau_{\alpha+} \leqslant \overline{T}_0$: there is either no wave crest or only one crest in $\tau_{\alpha+}$. The probability of occurrence in this situation is given by

$$p_0 = \int_0^{\overline{T}_0} f(\tau_{\alpha+})\, d\tau_{\alpha+} \tag{8.33}$$

(b) $\overline{T}_0 \leqslant \tau_{\alpha+} \leqslant 2\overline{T}_0$: there is either one or two wave crests in $\tau_{\alpha+}$. The probability of occurrence of only one crest can be evaluated by

$$p_1 = \int_{\overline{T}_0}^{2\overline{T}_0} \left\{ 1 - \left(\frac{\tau_{\alpha+}}{\overline{T}_0} - 1 \right) \right\} f(\tau_{\alpha+})\, d\tau_{\alpha+} \tag{8.34}$$

Thus, the probability of a wave group when the envelope exceeds a specified level α is given by $1 - (p_0 + p_1)$. Hence, the probability density function applicable for the time duration of the wave group, denoted by $f(\tau_G)$, can be obtained as

$$f(\tau_G) = \begin{cases} 0 & \text{for } 0 \leqslant \tau_G \leqslant \overline{T}_0 \\[2ex] \dfrac{1}{1-(p_0+p_1)} \left(\dfrac{\tau_{\alpha+}}{\overline{T}_0} - 1 \right) \left[f(\tau_{\alpha+}) \right]_{\tau_{\alpha+} = \tau_G} & \text{for } \overline{T}_0 \leqslant \tau_G \leqslant 2\overline{T}_0 \\[2ex] \dfrac{1}{1-(p_0+p_1)} \left[f(\tau_{\alpha+}) \right]_{\tau_{\alpha+} = \tau_G} & \text{for } 2\overline{T}_0 \leqslant \tau_G \end{cases} \tag{8.35}$$

The expected (average) time duration associated with wave groups denoted by $\overline{\tau}_G$, can be evaluated from

$$\overline{\tau}_G = \int_0^{\infty} \tau_G f(\tau_G)\, d\tau_G \tag{8.36}$$

As an example, computations are carried out using the wave spectrum obtained from data taken by the Norwegian Technology Research Institute in the North Sea off the Norwegian coast. The water depth at the measured site is 230 m. Figure 8.14 shows the wave spectrum for a significant wave height of 8.13 m along with the sum of two symmetric-shaped spectra used in the computation.

Figure 8.15 shows the probability density functions of time duration associated with wave groups exceeding levels of 5, 6, 7 and 8 m above the mean water level. As seen in the figure, all probability density functions originates at 10.6 s, which is the average zero-crossing period.

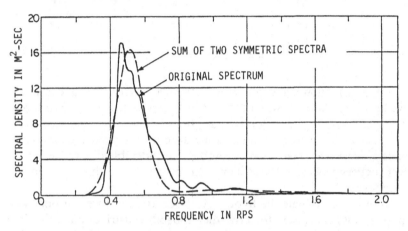

Fig. 8.14. Wave spectrum obtained from measured data and sum of two symmetric-shape spectra (significant wave height 8.13 m).

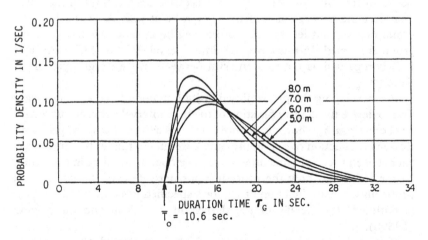

Fig. 8.15. Probability density function of time duration associated with wave groups at various levels.

Table 8.1. *Comparison between observed and computed average time duration associated with level crossing and wave group (Ochi and Sahinoglou 1989a).*

Level (m)	Sample size	Level crossing			Level crossing with at least two successive waves		
		Observed	Computed		Observed	Computed	
		Average time duration (s)	Average time duration (s)	Sample size	Average time duration (s)	Average time duration (s)	
4.0	9	15.9	3.9	5	19.0	17.9
5.0	6	10.8	3.1	1	13.7	17.0

Table 8.1 shows a comparison between observed and computed average time duration when the envelope crosses the 4 and 5 m levels. The comparison is made for two categories of envelope crossings: one is irrespective of number of waves during the crossing, while the other contains at least two crests (wave group). A considerable difference can be seen between observed and computed average time difference in the former case.

In order to evaluate the frequency of occurrence of wave groups in a given sea, it is necessary to derive the probability distribution of the time interval between successive envelope crossings of a specified level taking into consideration crossings with at least two waves. Figure 8.16 shows a sketch of the time interval between successive up-crossings of the envelope. Let the first crossing of the level α be at time A where the wave group begins, and let the second crossing be at time B where no wave group is formed. In this case, the time interval AC (instead of AB and BC) is regarded as the time interval between two wave groups, and is denoted $\tau_{\alpha G}$.

For developing the probability density function of $\tau_{\alpha G}$, it is necessary first to derive the density function of the time interval between successive level crossings, τ_α, shown in Figure 8.16. It is assumed that the probability density function of τ_α is approximately equal to that of the time interval between successive maxima of the envelope, denoted τ_m in the figure. In the following, only the procedure for derivation of the conditional probability density function of $\tau_{\alpha G}$ for a specified level, denoted $f(\tau_{\alpha G}|\alpha)$, is outlined. For details of the derivation, see Ochi and Sahinoglou (1989b).

(i) Consider a random vector x which is composed of a set of six random variables $(x_c, \dot{x}_c, \ddot{x}_c, x_s, \dot{x}_s, \ddot{x}_s)$ representing the cosine and sine components of wave displacement, velocity and acceleration. Here, the vector **x** obeys the joint normal probability distribution with zero mean and a covariance matrix whose elements can be evaluated for a given wave spectrum. The inversion of the covariance matrix can be easily obtained in this case.

(ii) By applying the polar coordinates written by $x=\rho\cos\theta$ and $x=\rho\sin\theta$, the joint probability density function $f(\rho,\dot\rho,\ddot\rho,\theta,\dot\theta,\ddot\theta)$ can be derived. Then, this joint probability density function is reduced to $f(\rho,\dot\rho,\ddot\rho,\dot\theta)$ by integrating with respect to θ and $\ddot\theta$.

(iii) The probability density function of the local positive maxima of the envelope, ξ, and phase velocity, $\dot\theta$, can be derived from

$$f(\xi,\dot\theta)=\frac{\displaystyle\int_{-\infty}^{0}\ddot\rho\,f(\xi,0,\ddot\rho,\dot\theta)\mathrm{d}\ddot\rho}{\displaystyle\int_{-\infty}^{\infty}\int_{0}^{\infty}\int_{-\infty}^{0}\ddot\rho\,f(\rho,0,\ddot\rho,\dot\theta)\,\mathrm{d}\ddot\rho\,\mathrm{d}\rho\,\mathrm{d}\dot\theta}\qquad 0<\rho<\infty\quad -\infty<\dot\theta<\infty \quad (8.37)$$

Then, from Eq. (8.37), the conditional probability density function of $\dot\theta$ given $\xi, f(\dot\theta|\xi)$ can be obtained.

(iv) Assume that $\dot\theta$ is a slowly varying function of time. Further, assume that the time interval between successive envelope peaks τ_m in Figure 8.16 is approximately equal to the time interval between two positive envelope crossings τ_α at the level α. Then, the conditional probability density function $f(\dot\theta|\xi)$ is converted to the conditional density function

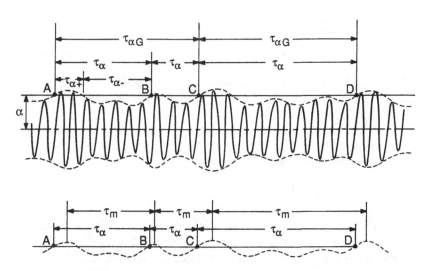

Fig. 8.16. Definition of time interval between successive wave groups, $\tau_{\alpha G}$.

$f(\tau_\alpha|\alpha)$ by letting $\xi=\alpha$ and $\dot{\theta}=2\pi k/\tau_\alpha$ where the parameter k is determined by the iteration technique. Here, the probability of the envelope crossing the level α is equal to the ratio of the average time intervals, $\bar{\tau}_{\alpha+}/\bar{\tau}_\alpha$, where $\bar{\tau}_{\alpha+}$ can be evaluated from $f(\tau_{\alpha+})$ derived earlier.

(v) The desired probability density function of the time interval between successive wave groups for a specified level α, denoted $f(\tau_{\alpha G}|\alpha)$, can be obtained by modifying $f(\tau_\alpha|\alpha)$. First consider the case that the first and third up-crossings are associated with wave groups, the situation shown in Figure 8.16. We may assume the time intervals AB and BC are statistically independent. Then, the time interval AC can be obtained as the sum of the two independent random variables each having the density function $f(\tau_\alpha|\alpha)$. Therefore, in this case, the conditional probability density function of the time interval, denoted by $f(\tau_{2\alpha}|\alpha)$, becomes

$$f(\tau_{2\alpha}|\alpha) = \int_0^{\tau_{2\alpha}} f(\tau_\alpha|\alpha) \cdot f((\tau_{2\alpha}-\tau_\alpha)|\alpha) d\tau_\alpha \qquad (8.38)$$

The same concept is applied to four up-crossings of the envelope in which the first and the fourth crossings are associated with wave groups. In this case, the conditional density function $f(\tau_{3\alpha}|\alpha)$ can be written as

$$f(\tau_{3\alpha}|\alpha) = \int_0^{\tau_{3\alpha}} f(\tau_{2\alpha}|\alpha) \cdot f((\tau_{3\alpha}-\tau_{2\alpha})|\alpha) d\tau_{2\alpha} \qquad (8.39)$$

By considering the time interval up to $\tau_{5\alpha}$, the probability density function of the time interval between successive wave groups for a specified level α, denoted $f(\tau_{\alpha G}|\alpha)$, can be written as

$$f(\tau_{\alpha G}|\alpha) = p \cdot f(\tau_\alpha|\alpha) + \frac{1-p}{4}\{f(\tau_{2\alpha}|\alpha) + f(\tau_{3\alpha}|\alpha)$$

$$+ f(\tau_{4\alpha}|\alpha) + f(\tau_{5\alpha}|\alpha)\} \qquad (8.40)$$

where p is the probability of occurrence of wave groups when the envelope exceeds a specified level $\alpha = 1 - (p_0 + p_1)$. p_0 and p_1 are given in Eqs. (8.33) and (8.34), respectively.

To evaluate the frequency of occurrence of group waves in a given sea state, the probability density function $f(\tau_{\alpha G}|\alpha)$ is approximated by the following gamma probability density function

$$f(\tau_{\alpha G}) = \frac{1}{\Gamma(m)}\lambda^m \tau_{\alpha G}^{m-1} \cdot e^{-\lambda \tau_{\alpha G}} \qquad 0 < \tau_{\alpha G} < \infty \qquad (8.41)$$

where the parameters m and λ are determined by letting the mean and variance of the distribution be identical to those of Eq. (8.40).

Based on the probability density function given in Eq. (8.41), the probability of n occurrences of wave groups in a specified time period can be evaluated as follows:

Let T_n be the waiting time up to the nth wave group. This is equal to the sum of $(\tau_{\alpha G})_1$, $(\tau_{\alpha G})_2$, ..., $(\tau_{\alpha G})_n$, where all $\tau_{\alpha G}$ are statistically independent and obey the probability density function given in Eq. (8.41). Therefore, the probability density function of T_n becomes

$$f(T_n) = \frac{1}{\Gamma(mn)} \lambda^{mn} T_n^{mn-1} e^{-\lambda T_n} \tag{8.42}$$

and from this the probability of n-occurrences of wave groups in time t can be obtained as

$$\Pr\left\{ \begin{matrix} n \text{ occurrences of wave} \\ \text{groups in time } t \end{matrix} \right\} = \Pr\{T_n \leqslant t\} - \Pr\{T_{n+1} \leqslant t\}$$

$$= \frac{\gamma(mn, \lambda t)}{\Gamma(mn)} - \frac{\gamma(m(n+1), \lambda t)}{\Gamma(m(n+1))} \quad n = 1, 2, 3, \ldots \tag{8.43}$$

where $\gamma(\ )$ is the incomplete gamma function.

As an example of practical application, Figure 8.17 shows the probability density function of the time interval of successive wave groups, $f(\tau_{\alpha G})$, for a level $\alpha = 4$ m using data taken in the North sea off Norway. The recording time of the data is 17 minutes. Included also in the figure is the approximated gamma probability density function. The probability of occurrence of wave groups in 17 minutes for the level $\alpha = 4$ m is computed for the various occurrences shown in Table 8.2. As can be seen, the probabilities of occurrence of four, five and six groups are high; the highest probability is 0.204 for five occurrences, and this result agrees with the observed number of occurrences in the record. On the other hand, if we

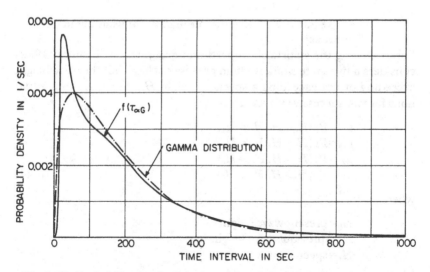

Fig. 8.17. Probability density function of time interval between two successive occurrences of wave groups (Ochi and Sahinoglou 1989b).

Table 8.2. *Comparison of predicted frequency of occurrence of wave groups and observed number of wave groups ($H_s = 8.13\,m$, $\alpha = 4.0\,m$) (Ochi and Sahinoglou 1989a).*

Prediction		Data
Number of wave groups	Probability of occurrence	Number of wave groups
0	0.001	
1	0.012	
2	0.050	
3	0.116	
4	0.180	
5	0.204	5
6	0.178	
7	0.126	
8	0.073	
9	0.036	
10	0.024	

define a wave group as the time interval between two successive envelope crossings of a specified level, then this results in the average number of occurrences of wave groups being extremely large; 37 occurrences in 17 minutes for the example shown in Table 8.2.

8.2.3 Statistical properties through the Markov chain approach

For evaluating the statistical properties of group waves, Kimura (1980) considers a two-state Markov chain process dealing with the wave height exceeding or not exceeding a specified height H_*. There are four situations for two successive waves. These are

$$p_{11} = \Pr\{H_0 < H_* | H_- < H_*\}$$
$$p_{12} = \Pr\{H_0 > H_* | H_- < H_*\}$$
$$p_{21} = \Pr\{H_0 < H_* | H_- > H_*\}$$
$$p_{22} = \Pr\{H_0 > H_* | H_- > H_*\} \tag{8.44}$$

where

H_0 = present wave height
H_- = previous wave height
H_* = specified wave height.

Then, the transition matrix can be written as

$$P = \begin{pmatrix} p_{11} & p_{12} \\ p_{21} & p_{22} \end{pmatrix} \qquad (8.45)$$

For the wave group problem, we consider successive large wave heights; hence, we are interested only in p_{22}. In addition, to evaluate the mean length of the total run (defined later) we may need p_{11}. Use of the transition matrix may not be required. This is pointed out by Longuet-Higgins (1984), who developed formulae for evaluating various statistical properties associated with wave groups in a concise form, although the results are essentially the same as those developed by Kimura. In the following, Longuet-Higgins' derivation is outlined using his notation; p_+ and p_- for the conditional probabilities p_{22} and p_{11}, respectively, defined in Eq. (8.44).

Let us first evaluate the probability of n successive waves exceeding a specified wave height H_*. The first wave exceeds H_* followed by $(n-1)$ waves higher than H_*, and then the next one must be smaller than H_*. Hence, the probability of n successive waves exceeding H_* is evaluated as

$$p(H_n) = p_+^{n-1}(1-p_+) \qquad (8.46)$$

The average number of n successive waves exceeding H_* as a group, which is called the *mean length of high runs* and denoted by $\overline{H}$, can be obtained as

$$\overline{H} = \sum_{n=1}^{\infty} np(H_n) = 1/(1-p_+) \qquad (8.47)$$

The variance of the length of high runs is given by

$$\mathrm{Var}[H_n] = \mathrm{E}[H_n^2] - (\mathrm{E}[H_n])^2 = p_+/(1-p_+) \qquad (8.48)$$

Next, let us consider a total run of length n in which the first j waves exceed H_* (high run of length j), and the remaining $(n-j)$ waves are lower than H_* (low run of length $n-j$) as illustrated in Figure 8.18. This situation is the same as $\tau_{\alpha+}$ and $\tau_{\alpha-}$ in Figure 8.12. The probability of occurrence of this situation is evaluated as

$$\mathrm{Pr}\{H_j > H_*, H_{n-j} < H_*\} = p_+^{j-1}(1-p_+)p_-^{n-j-1}(1-p_-) \qquad (8.49)$$

By accumulating the probabilities from $j=1$ to $j=n-1$, the probability of a total of n waves during two wave groups can be obtained as

$$p(G_n) = \sum_{j=1}^{n-1} p_+^{j-1}(1-p_+)p_-^{n-j-1}(1-p_-)$$
$$= (1-p_+)(1-p_-)(p_+^{n-1} - p_-^{n-1})/(p_+ - p_-) \qquad (8.50)$$

Then, the average number of high waves exceeding H_* during the time interval of two wave groups which is called the *mean length of total runs*, denoted by $\overline{G}$, is given by

$$\overline{G}=\sum_{j=2}^{\infty} n\, p(G_n)=1/(1-p_+)+1/(1-p_-) \qquad (8.51)$$

It is noted that Eq. (8.51) pertains to the time interval between two crossings of the specified level H_* which is equivalent to τ_α (not $\tau_{\alpha G}$) shown in Figure 8.16.

As shown in Eq. (8.51), the average number of exceedances can be evaluated from a knowledge of the probabilities p_+ and p_-, and these can be evaluated from the joint probability distribution of two wave heights. By assuming that the wave height is twice the wave amplitude, the joint probability density of wave amplitudes $f(A_1,A_2)$ may be used for evaluating p_+ and p_-. By letting $A_*=H_*/2$, we have

$$p_+=\int_{A_*}^{\infty}\int_{A_*}^{\infty} f(A_1,A_2)\,\mathrm{d}A_1\,\mathrm{d}A_2 \Big/ \int_{0}^{\infty}\int_{A_*}^{\infty} f(A_1,A_2)\,\mathrm{d}A_1\,\mathrm{d}A_2$$

$$p_-=\int_{0}^{A_*}\int_{0}^{A_*} f(A_1,A_2)\,\mathrm{d}A_1\,\mathrm{d}A_2 \Big/ \int_{0}^{\infty}\int_{0}^{A_*} f(A_1,A_2)\,\mathrm{d}A_1\,\mathrm{d}A_2 \qquad (8.52)$$

As the joint probability density function $f(A_1,A_2)$ of two wave amplitudes, Kimura (1980) considers the two-dimensional Rayleigh probability distribution given in Eq. (3.58) in Section 3.4. The joint probability density function has a parameter k which has a functional relationship with the correlation coefficient of two random variables, wave amplitudes in the present case. Kimura's approach is to obtain the correlation coefficient between consecutive wave amplitudes from data, and then

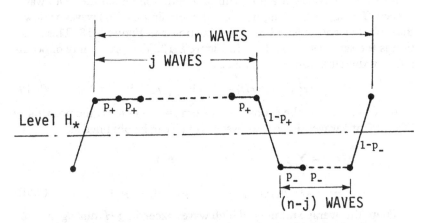

Fig. 8.18. Diagram showing a total run of length n in which the first j waves exceed H (Longuet-Higgins 1984).

evaluate the parameter k necessary to compute the joint Rayleigh probability distribution. In this procedure, the functional relationship between γ and k is given in terms of complete elliptic integrals.

On the other hand, Battjes and van Vledder (1984) and Longuet-Higgins (1984) point out that the parameter k can be evaluated for a given spectrum by the formula developed by Rice (1944). This enhances application of the Markov chain approach for evaluating group wave characteristics to a great extent. In support of this statement, let us rewrite the joint Rayleigh probability distribution applicable for wave amplitudes A_1 and A_2 developed in Section 3.4:

$$f(A_1,A_2)=\frac{A_1A_2}{(1-k^2)m_0^2}\,e^{-\frac{A_1^2+A_2^2}{2(1-k^2)m_0}}\,I_0\left(\frac{k}{1-k^2}\frac{A_1A_2}{m_0}\right) \tag{8.53}$$

where

$$k=\sqrt{\rho^2+\lambda^2}/m_0$$

$$\rho=\int_0^\infty S(\omega)\cos(\omega-\overline\omega)\tau\,d\omega$$

$$\lambda=\int_0^\infty S(\omega)\sin(\omega-\overline\omega)\tau\,d\omega$$

$$m_0=\int_0^\infty S(\omega)\,d\omega$$

$\overline\omega$ = mean frequency
I_0 = modified Bessel function of zero order.

Let two amplitudes be separated by a time interval τ. We assume that the separation equals $2\pi/\overline\omega$, then A_1 and A_2 represent, to a good approximation, the amplitudes of two successive waves. Thus, the k-value associated with two successive wave amplitudes can be evaluated for a given wave spectrum from Eq. (8.53).

Functional relationships between various parameters associated with the Markov chain approach and a narrow-band wave spectrum have been analytically studied in depth by Longuet-Higgins, whose results are summarized below.

(a) The parameter k involved in Eq. (8.53) can be approximated using moments of the spectrum by

$$k^2=1-4\pi^2\nu^2 \tag{8.54}$$

where $\nu^2=(m_2m_0/m_1^2)-1$ and m_j is the jth moment of the spectrum.

(b) A functional relationship between the correlation coefficient γ and k^2 can be obtained as shown in Figure 8.19. From the figure we have

$$\gamma \simeq \begin{cases} k^2 & \text{for } k^2 < 0.6 \\ 1-(1-k^2)/(4-\pi) & \text{for } k \text{ close to 1} \end{cases} \qquad (8.55)$$

(c) The conditional probabilities p_1 and p_2 can be evaluated as a function of k as follows:

$$1-p_+ \simeq \sqrt{1-k^2}\,(1-\exp\{-\xi^2/2\})$$

$$1-p_- \simeq \sqrt{1-k^2}\,\exp\{-\xi^2/2\} \qquad (8.56)$$

where ξ is the dimensionless specified level $A_*/\sqrt{m_0}$.

By applying the k^2-value given in Eq. (8.54), we have

$$1-p_+ = 2\pi\nu\,(1-\exp\{-\xi^2/2\})$$

$$1-p_- = 2\pi\nu\exp\{-\xi^2/2\}) \qquad (8.57)$$

Thus, the mean length of high runs, $\overline{H}$, given in Eq. (8.47) and the mean length of total runs, $\overline{G}$, given in Eq. (8.51) become

$$\overline{H} = (2\pi\nu)^{-1}\,e^{\xi^2/2}/(e^{\xi^2/2}-1)$$

$$\overline{G} = (2\pi\nu)^{-1}\,e^{\xi^2}/(e^{\xi^2/2}-1) \qquad (8.58)$$

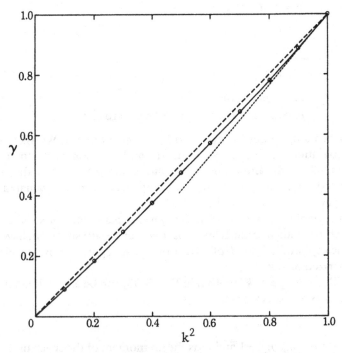

Fig. 8.19. Correlation coefficient γ as a function of the parameter k^2 (Longuet-Higgins 1984).

Although the above equations provide approximate values, only three moments (m_0, m_1 and m_2) of the spectrum are sufficient to evaluate the mean lengths.

The probabilities of n successive waves exceeding a specified level are computed by Kimura from data generated by numerical simulation techniques, and the results are compared with data shown in Figure 8.20. In these computations, the mean wave height is taken as the specified level for wave groups. The solid line in the figure shows the results computed by Kimura, while the broken line shows those computed by Longuet-Higgins' method through spectral analysis. In the latter computations, high and low frequency portions of the spectrum are filtered. As can be seen, there is almost no difference in the results computed by Kimura and Longuet-Higgins, and the computed results show good agreement with observations.

Figure 8.21 shows the probability of a total of n waves during two wave groups (total runs). The solid circles are the values obtained from Kimura's simulation data. Again, the results computed by Kimura and Longuet-Higgins (the solid line and broken line, respectively) are almost identical, and they agree very well with the observed data.

With respect to the effect of the shape of the wave spectrum on wave group properties, Medina and Hudspeth (1990) showed a functional

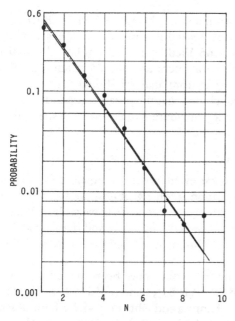

Fig. 8.20. Probability of n successive waves exceeding the mean wave height (Kimura 1980 and Longuet-Higgins 1984).

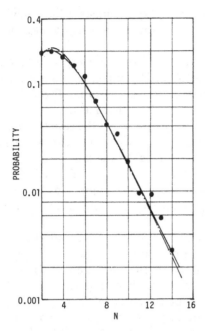

Fig. 8.21. Probability of a total of n waves during two wave groups (Kimura 1980 and Longuest-Higgins 1984).

relationship between the correlation coefficient of the joint Rayleigh distribution, γ, and the spectral peakedness parameter, Q_p given in Eq. (3.40). On the other hand, van Vledder (1992) studied the statistical properties of various shape parameters and the correlation coefficient between successive wave heights from analysis of field data and numerically simulated data. He concluded that the parameter k in Eq. (8.53) is the most significant parameter contributing to the effect of spectral shape on wave grouping.

8.3 FREAK WAVES

An unusually high single wave event observed offshore is commonly called a *freak wave*. This definition is somewhat obscure since neither the cause of the occurrence nor criteria to define freak waves have been clarified. Freak waves have been observed only rarely and these observations occurred under unexpected condition: hence, only few measured data are available.

Several studies addressing freak wave have been carried out to date. These include Klinting and Sand (1987), Sand et al. (1990), Yasuda et al. (1992), Thieke et al. (1993), Kimura and Ohta (1994). Perhaps, the most reliable existing data are those obtained at Gorm field in the North Sea where water depth is 40 m. Figure 8.22 shows an example of a recorded

Table 8.3. *Maximum wave height, significant wave height, and water depth obtained from available freak wave records (Sand et al. 1989).*

Location	Water depth (m)	Significant wave height H_s (m)	Recorded max. height H_{max} (m)	H_{max}/H_s	Registration
Hanstholm	20	3.5	7.6	2.2	Wave-rider
Gorm field	40	6.8	17.8	2.6	Radar
		7.8	16.5	2.1	
		5.0	12.0	2.4	
		5.0	11.3	2.3	
		5.0	11.0	2.2	
		4.8	13.1	2.7	
Gulf of Mexico	100	10.4	19.4	1.9	Wave staff
		10.0	23.0	2.3	

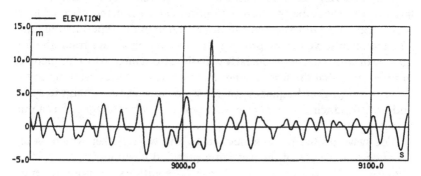

Fig. 8.22. Example of recorded freak wave (Sand *et al.* 1990).

freak wave at the Gorm field (Sand *et al.* 1990). As seen, a single isolated high wave with a sharp peak is the characteristic feature of a freak wave.

Table 8.3 is taken from Sand *et al.* (1990) which gives the water depth, significant wave height, and recorded maximum wave height obtained from available freak wave records. Visually obtained data listed in the original table are not included in Table 8.3. As seen in the table, some records show a very high ratio of maximum wave height, H_{max}, to significant wave height, H_s.

Freak waves have been defined by some as waves having a ratio of wave height to significant wave height, H/H_s, greater than 2.0. This criterion, however, may not be valid, in general. This is because the extreme wave amplitude expected to occur in a one hour observation of a storm with a risk parameter $\alpha=0.01$ is on the order of 4.3–4.7 times the rms value as shown in Figure 6.5. This value converts to 2.15–2.35 for H_{max}/H_s.

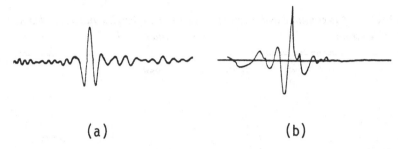

(a)   (b)

Fig. 8.23. Example of transient waves generated in the experimental tank: (a)
Davis and Zarnick (1964); (b) Kjeldsen (1982).

Examples listed in Table 8.3 have very large values of H_{max}/H_s, but many
of them are not unusually large from the viewpoint of extreme value statistics. The freak wave, however, is unique in that it occurs as a single
large wave in a group of waves of much less severity.

In considering the mechanism of the occurrence of a single isolated
large wave, it may be of interest to note that an isolated wave of large
height, called a *transient wave*, can be generated in the experimental tank.
The transient wave can be produced by generating a wave train, the frequency of which decreases linearly with time from high to low frequencies
in such a way that the fast moving (low frequency) waves catch up to the
slower ones (higher frequency waves) to coalesce at some point in space
and time producing a very large wave which may be thought of as a unit
impulse.

The wave profile to be generated by the wave maker which would
produce coalescence of the wave at a specified location was analytically
obtained by Davis and Zarnick (1964). The authors' purpose in generating a transient wave in the experimental tank was to obtain the frequency response function of a marine system to waves.

The transient wave test technique is applied for generating a single
freak wave by Kjeldsen (1982). The prime purpose of his experiments,
however, is not to clarify the mechanism of freak waves, but to generate a
freak wave for wave breaking studies.

Examples of transient waves generated in experimental tanks are
shown in Figure 8.23. The transient wave (a) is generated by Davis and
Zarnick, while the transient wave (b) is generated by Kjeldsen. If we consider some analogy exists between the transient wave and freak wave, then
wave period, in particular a sequence of periods, appears to play a significant role in the mechanism of the occurrence of freak waves observed at
sea.

9 NON-GAUSSIAN WAVES (WAVES IN FINITE WATER DEPTH)

9.1 INTRODUCTION

The statistical analysis of ocean waves discussed in previous chapters assumes that waves are a Gaussian random process; namely, waves are a steady-state, ergodic random process and displacement from the mean obeys the normal probability law. Verification that deep ocean waves are a Gaussian process was given in Section 1.1 through the central limit theorem. It has also been verified through observations at sea as well as in laboratory tests that waves can be considered a Gaussian random process even in very severe seas if the water depth is sufficiently deep.

The above statement, however, is no longer true for waves in finite water depth. Time histories of waves in shallow water show a definite excess of high crests and shallow troughs as demonstrated in the example shown in Figure 1.1(b), and thereby the histogram of wave displacement is not symmetric with respect to its mean value, as shown in Figure 1.2(b). Thus, waves in shallow water are considered to be a non-Gaussian random process. This implies that ocean waves transform from Gaussian to non-Gaussian as they propagate from deep to shallow water.

Figure 9.1 shows a portion of wave records measured simultaneously at various water depths during the ARSLOE Project carried out by the Coastal Engineering Research Center at Duck, North Carolina. In the figure, examples of records are shown at five locations where the water depth ranges from 2.23 to 24.7 m at the time of measurement. The significant wave height at the water depth 24.7 m was 4.28 m in this example. As can be seen in the figure, wave profiles transform from Gaussian to non-Gaussian with decreasing water depth.

It is noted, however, that transformation of wave characteristics from a Gaussian to a non-Gaussian random process as shown in Figure 9.1 may not always occur. If the sea severity is very mild, waves in shallow water are also a Gaussian random process. In other words, the non-Gaussian characteristics of waves depend on the water depth and sea severity. This subject will be discussed in detail in Section 9.4. In any event, the concept of a non-Gaussian random process is mandatory for statistical prediction of random waves in finite water depths.

Although considerable attention has been given to stochastic prediction of Gaussian waves, comparatively little is known of non-Gaussian waves and their associated probabilistic prediction. Probability density functions applicable to wave displacement from the mean observed in finite water depth have been developed using three different approaches. These non-Gaussian probability density functions are presented in Section 9.2, while the probability density functions applicable to peaks and troughs of non-Gaussian waves are discussed in Section 9.3.

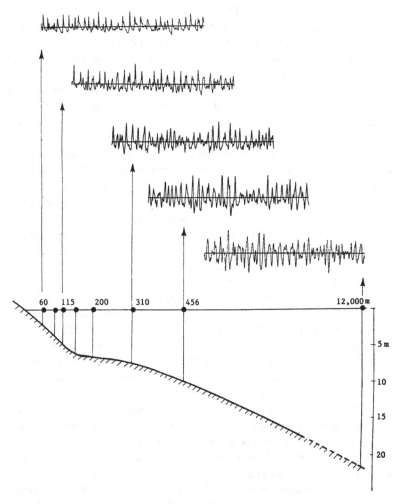

Fig. 9.1. Transformation of wave profiles from deep to shallow water.

9.2 PROBABILITY DISTRIBUTION OF NON-GAUSSIAN WAVES

The probability distribution applicable to non-Gaussian waves has been developed using three different approaches: (a) application of the Hermite polynomial which is orthogonal to the normal distribution; (b) assumption that wave profiles can be represented by Stokes second and third-order expansion; and (c) application of the stochastic solution of a nonlinear system to a random input.

9.2.1 Gram–Charlier series distribution

For analysis of non-Gaussian waves, Longuet-Higgins (1963b) derives the probability density function by applying the cumulant generating function. The density function is essentially the same as that developed by Gram and Charlier (Cramér 1966, Kendall and Stuart 1961) by applying the concept of orthogonal polynomials, and that developed by Edgeworth (1904) in connection with the probability law of error. Consequently, the probability distribution is commonly called the Gram–Charlier series probability distribution. In the following, the derivation of the Gram–Charlier series distribution will be introduced first, and then derivation of the distribution pursuing Longuet-Higgins' approach will be presented. The latter approach is concise and explicit, and thus lends itself to the derivation of the joint non-Gaussian probability density function for two random variables.

The Gram–Charlier probability density function is developed by applying the concept of polynomials orthogonal with respect to the probability density function (hereafter called the orthogonal polynomials). In particular, the Hermite polynomials are orthogonal with respect to the normal probability distribution.

Let us first give the definition of Hermite polynomials. The *Hermite polynomial* of degree n, denoted $H_n(z)$, is defined as a function which satisfies the relationship given by

$$\frac{d^n}{dz^n} e^{-z^2/2} = (-1)^n H_n(z) \cdot e^{-z^2/2} \qquad n = 0, 1, 2, \ldots \qquad (9.1)$$

From the above equation, we have the following Hermite polynomials:

$$H_0(z) = 1$$
$$H_1(z) = z$$
$$H_2(z) = z^2 - 1$$
$$H_3(z) = z^3 - 3z$$
$$H_4(z) = z^4 - 6z^2 + 3$$
$$H_5(z) = z^5 - 10z^3 + 15z$$
$$H_6(z) = z^6 - 15z^4 + 45z^2 - 15, \text{ etc.} \qquad (9.2)$$

Let $\alpha(z)$ be the standardized normal (Gaussian) probability density function given by

$$\alpha(z) = \frac{1}{\sqrt{2\pi}}\, e^{-z^2/2} \qquad (9.3)$$

It can be shown that the polynomials $(1/\sqrt{n!})\, H_n(z)$ are orthogonal with respect to the standardized normal probability density function $\alpha(z)$. In order to elaborate on this statement, let us consider integration of the product of two Hermite polynomials and the standardized normal probability density function. That is,

$$\int_{-\infty}^{\infty} H_m(z) H_n(z) \alpha(z)\,\mathrm{d}z = (-1)\int_{-\infty}^{\infty} H_m(z)\,\frac{\mathrm{d}^n}{\mathrm{d}z^n}\,\alpha(z)\,\mathrm{d}z$$

$$= (-1)^{n-1}\left[H_m(z)\,\frac{d^{n-1}}{\mathrm{d}z^{n-1}}\,\alpha(z)\right]_{-\infty}^{\infty}$$

$$+ (-1)^{n-1}\int_{-\infty}^{\infty}\frac{\mathrm{d}}{\mathrm{d}z} H_m(z)\,\frac{d^{n-1}}{\mathrm{d}z^{n-1}}\,\alpha(z)\,\mathrm{d}z \quad (9.4)$$

The first term becomes zero. For integration of the second term, we use the following property of Hermite polynomials:

$$\frac{\mathrm{d}}{\mathrm{d}z} H_n(z) = n \cdot H_{n-1}(z) \qquad (9.5)$$

By repeating the integration by parts and by repeatedly applying the property given in Eq. (9.5), we have

$$\int_{-\infty}^{\infty} H_m(z) \cdot H_n(z) \cdot \alpha(z)\,\mathrm{d}z = (-1)^{n-m} m! \int_{-\infty}^{\infty}\frac{d^{n-m}}{\mathrm{d}z^{n-m}}\alpha(z)\,\mathrm{d}z$$

$$= m! \int_{-\infty}^{\infty} H_{n-m}(z) \cdot \alpha(z)\,\mathrm{d}z = \begin{cases} n! & \text{if } m=n \\ 0 & \text{if } m \neq n \end{cases} \qquad (9.6)$$

The above equation may be written as

$$\int_{-\infty}^{\infty}\left\{\frac{1}{\sqrt{m!}} H_m(z)\right\}\left\{\frac{1}{\sqrt{n!}} H_n(z)\right\}\alpha(z)\,\mathrm{d}z = \begin{cases} 1 & \text{if } m=n \\ 0 & \text{if } m \neq n \end{cases} \qquad (9.7)$$

The relationship given in Eq. (9.7) implies that $(1/\sqrt{n!})\, H_n(z)$ is a sequence of orthogonal polynomials with respect to the standardized normal distribution $\alpha(z)$. With this in mind, let us express an arbitrarily given standardized probability density function $f(z)$ in the following form:

$$f(z) = a_0 \alpha(z) + a_1 \alpha^{(1)}(z) + a_2 \alpha^{(2)}(z) + a_3 \alpha^{(3)}(z) + \ldots \qquad (9.8)$$

where

$$a_n = \text{constant (unknown)}$$

$$\alpha^{(n)}(z) = \frac{d^n}{dz^n} \alpha(z)$$

$\alpha(z) = $ standardized normal probability density function.

From the definition of the Hermite polynomial given in Eq. (9.1), we have

$$f(z) = \alpha(z) \{a_0 H_0(z) - a_1 H_1(z) + a_2 H_2(z) - a_3 H_3(z) + \ldots$$

$$= \alpha(z) \left\{ c_0 \frac{H_0(z)}{\sqrt{0!}} - c_1 \frac{H_1(z)}{\sqrt{1!}} + c_2 \frac{H_2(z)}{\sqrt{2!}} - c_3 \frac{H_3(z)}{\sqrt{3!}} + \ldots \right\}$$

$$= \alpha(z) \sum_{n=0}^{\infty} (-1)^n c_n \frac{H_n(z)}{\sqrt{n!}} \qquad (9.9)$$

where $c_n = \sqrt{n!}\, a_n$ (unknown).

In order to determine the unknown constant, C_n, we multiply $f(z)$ by $H_n(z)/\sqrt{n!}$, and integrate over the domain $(-\infty, \infty)$. Since $H_n(z)/\sqrt{n!}$ is orthogonal with respect to $\alpha(z)$, we have

$$\int_{-\infty}^{\infty} \frac{H_n(z)}{\sqrt{n!}} f(z) dz = (-1)^n \frac{c_n}{n!} \int_{-\infty}^{\infty} H_n^2(z) \cdot \alpha(z) dz = (-1)^n c_n \qquad (9.10)$$

Thus, the constant c_n in Eq. (9.10) can be determined from

$$c_n = (-1)^n \frac{1}{\sqrt{n!}} \int_{-\infty}^{\infty} H_n(z) f(z) dz \qquad (9.11)$$

Here, the probability density function $f(z)$ is a standardized probability density function. Hence, we have

$$c_0 = \int_{-\infty}^{\infty} H_0(z) f(z) dz = 1$$

$$c_1 = -\int_{-\infty}^{\infty} H_1(z) f(z) dz = -m_1 = 0$$

$$c_2 = \frac{1}{\sqrt{2!}} \int_{-\infty}^{\infty} H_2(z) f(z) dz = \frac{1}{\sqrt{2!}} (m_2 - 1) = 0$$

$$c_3 = -\frac{1}{\sqrt{3!}} \int_{-\infty}^{\infty} H_3(z)f(z)dz = -\frac{m_3}{\sqrt{3!}}$$

$$c_4 = \frac{1}{\sqrt{4!}} \int_{-\infty}^{\infty} H_4(z)f(z)dz = \frac{1}{\sqrt{4!}}(m_4 - 3)$$

$$c_5 = \frac{1}{\sqrt{5!}} \int_{-\infty}^{\infty} H_5(z)f(z)dz = \frac{1}{\sqrt{5!}}(m_5 - 10m_3), \text{ etc.} \qquad (9.12)$$

where $m_j = j$th moment of the standardized random variable.

From Eqs. (9.10) and (9.12), the probability density function of a standardized random variable can be expressed as

$$f(z) = \frac{1}{\sqrt{2\pi}} e^{-z^2/2} \left[1 + \frac{m_3}{3!} H_3(z) + \frac{m_4 - 3}{4!} H_4(z) + \frac{m_5 - 10m_3}{5!} H_5(z) \right.$$

$$\left. + \frac{m_6 - 15m_4 + 30}{6!} H_6(z) + \dots \right] \qquad (9.13)$$

This is called the *Gram–Charlier series of Type A* probability density function.

It may be of interest to express Eq. (9.13) in terms of cumulants rather than moments so that the density function can be directly compared with that derived by Longuet-Higgins which will be presented later. The parameters m_3, m_4, m_5, etc. are the moments of the standardized random variable X. For a non-standardized random variable X which has the mean value μ and variance σ^2, the moment m_j in Eq. (9.13) can be written in the form of central moments,

$$m_j = \frac{E[(x-\mu)^j]}{\sigma^j} = \frac{E[(x-\mu)^j]}{(E[(x-\mu)^2])^{j/2}} \qquad (9.14)$$

Furthermore, the central moments can be expressed in terms of the cumulants of a random variable. That is,

$$E[(x-\mu)^2] = k_2 = \text{variance}$$
$$E[(x-\mu)^3] = k_3$$
$$E[(x-\mu)^4] = k_4 + 3k_2^2$$
$$E[(x-\mu)^5] = k_5 + 10k_3k_2$$
$$E[(x-\mu)^6] = k_6 + 15k_4k_2 + 10k_3^2 + 15k_2^3, \text{ etc.} \qquad (9.15)$$

where $k_j = $cumulants.

Thus, the probability density function of a non-standardized random variable X can be obtained from Eqs. (9.13), (9.14) and (9.15) in terms of cumulants. In particular, for a random variable with zero mean and variance σ^2, the probability density function becomes

$$f(x) = \frac{1}{\sigma\sqrt{2\pi}} e^{-x^2/2\sigma^2} \left[1 + \frac{\lambda_3}{3!} H_3\left(\frac{x}{\sigma}\right) + \frac{\lambda_4}{4!} H_4\left(\frac{x}{\sigma}\right) + \frac{\lambda_5}{5!} H_5\left(\frac{x}{\sigma}\right) \right.$$

$$\left. + \left(\frac{\lambda_6}{6!} + \frac{\lambda_3^2}{72}\right) H_6\left(\frac{x}{\sigma}\right) + \dots \right] \qquad (9.16)$$

where

$$\lambda_j = \frac{k_j}{(k_2)^{j/2}} = \frac{k_j}{\sigma^j}$$

Note that λ_3 and λ_4 are called the *skewness* and *kurtosis*, respectively (see Appendix A). Equation (9.16) is also called the *Gram–Charlier series of Type A* probability density function.

Next, the non-Gaussian probability density function developed by Longuet-Higgins based on the cumulant generating function will be presented.

In probability theory, the logarithm of the characteristic function, denoted by $\psi(t)$, is defined as the *cumulant generating function*. It can be expressed in the form of a series where the coefficient of each term is the cumulant k_j. That is,

$$\psi(t) = \ln \phi(t) = \sum_{j=1}^{\infty} \frac{(it)^j}{j!} k_j \qquad (9.17)$$

Let X be an arbitrary random variable; for the present problem X is the displacement of non-Gaussian waves. In general, the probability density function of X can be written as a function of the characteristic function $\phi(t)$ as

$$f(x) = \frac{1}{2\pi} \int_{-\infty}^{\infty} \phi(t) \, e^{-itx} dt \qquad (9.18)$$

Hence, from Eqs. (9.17) and (9.18), we can express $f(x)$ as

$$f(x) = \frac{1}{2\pi} \int_{-\infty}^{\infty} \exp\{\psi(t) - itx\} \, dt$$

$$= \frac{1}{2\pi} \int_{-\infty}^{\infty} \exp\left\{(k_1 - x)it + \frac{k_2}{2!}(it)^2 + \frac{k_3}{3!}(it)^3 + \dots\right\} dt \qquad (9.19)$$

Next, let us standardize the random variable X by letting

$$Z = \frac{X - k_1}{\sqrt{k_2}} \qquad (9.20)$$

and define

$$t = s/\sqrt{k_2}$$

$$\lambda_j = k_j/(\sqrt{k_2})^j \tag{9.21}$$

Then, the probability density function of the standardized random variable Z can be obtained from Eq. (9.19) as

$$f(z) = \frac{1}{2\pi} \int\limits_{-\infty}^{\infty} \exp\left\{ -\frac{1}{2}(s^2 + 2izs) \right\} \cdot \exp\left\{ \sum_{j=3}^{\infty} \frac{1}{j!} \lambda_j (is)^j \right\} ds \tag{9.22}$$

We may expand the second exponential term of the integrand as follows:

$$\exp\left\{ \sum_{j=3}^{\infty} \frac{1}{j!} \lambda_j (is)^j \right\}$$

$$= 1 + \sum_{j=3}^{\infty} \frac{1}{j!} \lambda_j (is)^j + \frac{1}{2!} \left(\sum_{j=3}^{\infty} \frac{1}{j!} \lambda_j (is)^j \right)^2 + \dots$$

$$= 1 + \frac{1}{3!} \lambda_3 (is)^3 + \frac{1}{4!} \lambda_4 (is)^4 + \frac{1}{5!} \lambda_5 (is)^5 + \left\{ \frac{1}{6!} \lambda_6 + \frac{1}{2!(3!)^2} \lambda_3^2 \right\} (is)^6$$

$$+ \left\{ \frac{1}{7!} \lambda_7 + \frac{1}{3!4!} \lambda_3 \lambda_4 \right\} (is)^7 + \dots \tag{9.23}$$

Hence, the standardized probability density function given in Eq. (9.22) becomes

$$f(z) = \frac{1}{2\pi} \int\limits_{-\infty}^{\infty} \exp\left\{ -\frac{1}{2}(s^2 + 2izs) \right\} \left[1 + \frac{\lambda_3}{3!}(is)^3 + \frac{\lambda_4}{4!}(is)^4 + \dots \right] ds \tag{9.24}$$

The integration involved in the above equation can be written as follows:

$$\frac{1}{\sqrt{2\pi}} \int\limits_{-\infty}^{\infty} \exp\left\{ -\frac{1}{2}(s^2 + 2izs) \right\} (is)^n ds$$

$$= \frac{(-1)^n}{\sqrt{2\pi}} \frac{d^n}{dz^n} \int\limits_{-\infty}^{\infty} \exp\left\{ -\frac{1}{2}(s^2 + 2izs) \right\} ds$$

$$= (-1)^n \frac{d^n}{dz^n} \exp\left\{ -\frac{z^2}{2} \right\} = H_n(z) \exp\left\{ -\frac{z^2}{2} \right\} \tag{9.25}$$

where $H_n(z)$ is the Hermite polynomial of degree n.

Thus, the standardized non-Gaussian random process may be expressed in the following series form:

$$f(z) = \frac{1}{\sqrt{2\pi}} e^{-z^2/2} \left[1 + \frac{\lambda_3}{3!} H_3(z) + \frac{\lambda_4}{4!} H_4(z) + \frac{\lambda_5}{5!} H_5(z) \right.$$

$$\left. + \left\{ \frac{\lambda_6}{6!} + \frac{\lambda_3^2}{2!(3!)^2} \right\} H_6(z) + \left\{ \frac{\lambda_7}{7!} + \frac{\lambda_3 \lambda_4}{3!4!} \right\} H_7(z) + \ldots \right] \qquad (9.26)$$

This is the same probability density function given in Eq. (9.16) but derived using a different approach.

If all cumulants k_j and thereby λ_j are zero, Eq. (9.26) is reduced to the standardized Gaussian probability density function.

Bitner (1980) and Ochi and Wang (1984) have demonstrated that Eqs. (9.16) and (9.26) reasonably accurately represent the probability distribution of displacement of waves obtained in shallow water areas. Unfortunately, however, the density function becomes negative for large negative displacement. As an example, Figure 9.2 shows a comparison between Eq. (9.16) and a histogram of wave displacement obtained at a water depth of 2.18 m. The variance obtained from the data is 0.26 m² which indicates a very severe sea state for this water depth. Five variations of the probability density function with various combinations of the parameters λ_3, λ_4 and λ_5 in Eq. (9.16) are compared with the histogram. As can be seen, (1) the histogram deviates substantially from the normal distribution, (2) the density function becomes negative for large negative displacements, and (3) the addition of higher-order terms of the density

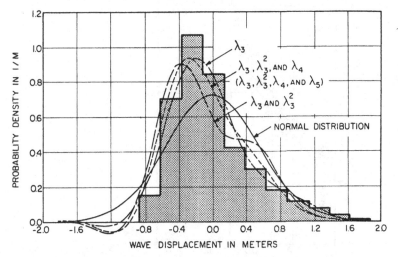

Fig. 9.2. Comparison between observed histogram, normal distribution (heavy line) and Gram–Charlier series distribution (water depth 2.18 m) (Ochi and Wang 1984).

function does not necessarily yield better agreement with the histogram. For example, the λ_5-term does not provide any improvement in the shape of the distribution.

Another drawback of applying the Gram–Charlier series density function for statistical analysis of shallow water waves is that the wave height probability density function cannot be derived from the Gram–Charlier distribution. Note that for Gaussian waves, the Rayleigh probability distribution applicable to wave height is developed based on the normal probability distribution representing the statistical properties of wave displacements.

It is of interest to note the functional relationship between the two parameters λ_3 and λ_4 shown in Figure 9.3, obtained from analysis of waves in finite water depth. As can be seen in the figure, there is a general trend that λ_4 increases with increase in λ_3, as illustrated by the average line drawn in the figure. There is considerable scatter in the value of λ_4 for λ_3 less than 0.2. These are values obtained from records taken at locations where the water depth is relatively deep; on the order of 15 to 25 m. Although the values of λ_4 vary considerably for λ_3 less than 0.2, the results of plotting the probability density functions for each combination of λ_3

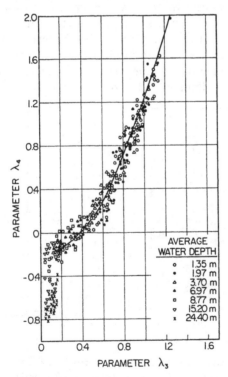

Fig. 9.3. Parameter λ_4 as a function of λ_3 (Ochi and Wang 1984).

and λ_4 show that the shape of the probability distribution is very close to that of the normal distribution.

9.2.2 Distribution based on Stokes waves

Another approach to deriving the probability distribution of non-Gaussian waves is to assume that they may be expressed as a Stokes expansion to the second and third components and consider an amplitude-modulated wave profile. This imposes a preliminary form on the wave profile for derivation of its probability distribution, in contrast to other approaches. Furthermore, the Stokes expansion is valid for waves $kh>1$, where k=wave number, h=water depth. Hence, there is some reservation in applying Stokes theory for shallow water waves unless the expansion includes higher-order terms. Irrespective of the shortcomings in this approach, the probability density function does not have any negative portion as is the case for the Gram–Charlier series distribution.

Following Stokes (1847), the wave profile may be expressed by

$$\eta = \frac{1}{2} a^2 k + a \cos \chi + \frac{1}{2} a^2 k \cos 2\chi + \frac{3}{8} a^3 k^2 \cos 3\chi + \dots$$

$$= a \cos \chi + a^2 k \cos^2 \chi + \frac{3}{8} a^3 k^2 \cos 3\chi + \dots \qquad (9.27)$$

where

$a = a(x,t) =$ amplitude
$\chi = kx - \omega t + \epsilon$
$k = \omega^2 / g =$ wave number.

The amplitude, a, and phase, χ, are both assumed to be slowly varying random variables following the Rayleigh and uniform distributions, respectively. Huang et al. (1983) derive the probability density function of the wave displacement taking terms up to the third order of Eq. (9.27). The derivation of the density function is outlined below.

The first and second moments of the displacement η can be obtained as

$$E[\eta] = \frac{1}{2} \overline{a^2} k$$

$$E[\eta^2] = \overline{a^2 \cos^2 \chi} + \overline{a^4 \cos^4 \chi} k^2 = \frac{1}{2} \overline{a^2} + \frac{3}{8} \overline{a^4} k^2$$

$$= \frac{1}{2} \overline{a^2} + \frac{3}{4} (\overline{a^2})^2 k^2 = \frac{1}{2} \overline{a^2} (1 + \frac{3}{2} \overline{a^2} k^2) \qquad (9.28)$$

where $\overline{a^2} = E[a^2] =$ mean square value of the amplitude.

Then, the variance becomes

$$\mathrm{Var}[\eta]=\sigma^2=\mathrm{E}[\eta^2]-(\mathrm{E}[\eta])^2=\frac{1}{2}\,\overline{a^2}(1+\overline{a^2}k^2) \tag{9.29}$$

It is assumed from Eq. (9.29) that the mean square value $\overline{a^2}$ may be written as $2\sigma^2$ to a first-order approximation. Next, the wave displacement η is standardized by

$$\zeta=\frac{\eta-\mathrm{E}[\eta]}{\sigma} \tag{9.30}$$

and define the random variables Z_1 and Z_2 as follows:

$$Z_1=\frac{a}{(\overline{a^2}/2)^{1/2}}\cos\chi$$

$$Z_2=\frac{a}{(\overline{a^2}/2)^{1/2}}\sin\chi \tag{9.31}$$

Then, the joint probability density function of Z_1 and Z_2 can be obtained as

$$f(z_1,z_2)=\frac{1}{2\pi}\exp\left\{-\frac{1}{2}\,(z_1^2+z_2^2)\right\} \tag{9.32}$$

Next, the wave displacement given in Eq. (9.27) may be standardized and presented in terms of Z_1 and Z_2. That is,

$$\zeta=\frac{z_1}{N}+\frac{z_1^2}{N^2}\,\sigma k+\frac{3}{8}\,\frac{z_1^3-3z_1z_2^2}{N^3}\,\sigma^2k^2-\sigma k \tag{9.33}$$

where

$$N=\frac{a}{(\overline{a^2}/2)^{1/2}}=1+\sigma^2k^2$$

By introducing an auxiliary random variable $\omega=Z$ and by assuming $\sigma k\ll1$, Z_1 and Z_2 are expressed as functions of ζ and ω using the successive approximation method, and thereby the joint probability density function $f(\zeta,\omega)$ can be obtained from the joint density function $f(z_1,z_2)$. Then, by integrating $f(\zeta,\omega)$ with respect to ω from $-\infty$ to ∞, the probability density function of the wave displacement $f(\zeta)$ can be derived as follows:

$$f(\zeta)=\frac{1}{\sqrt{2\pi}}\,e^{-H_\zeta/2}\left[\frac{\mathcal{J}_\zeta}{\sqrt{R}}+\frac{9}{8}\,\frac{(\sigma k)^2}{N}\,\frac{1}{\sqrt{R^3}}\right] \tag{9.34}$$

where

$$H_\zeta=N^2\{\zeta-\sigma k(\zeta^2-1)+(\sigma k)^2(13/8\zeta^3-2\zeta)\}^2$$

$$\mathcal{J}_\zeta=N\{1-2\sigma k\zeta+(\sigma k)^2(39/8\zeta^2-2)\}$$

$$R=1+\frac{9}{4}\,(\sigma k)^2\zeta^2.$$

Huang *et al.* write σk in the above equation as $2\pi(\sigma/\lambda)$, where λ is the wavelength corresponding to the peak of the spectrum. Figure 9.4 shows the density function given in Eq. (9.34) for various values of σ/λ ranging from 0 to 0.05 in steps of 0.01.

Tayfun (1980) develops the probability density function of the amplitude-modulated Stokes wave profile based on expansion to second order not including the constant term. For this, Tayfun writes the wave displacement in the following form:

$$\eta = \rho\cos(\overline{\chi}+\phi)+(1/2)k\rho^2\cos 2(\overline{\chi}+\phi)$$
$$= \Sigma a_n \cos(\chi_n+\epsilon_n)$$
$$+\frac{1}{2}k\{(\Sigma a_n \cos(\chi_n+\epsilon_n))^2 - (\Sigma a_n \sin(\chi_n+\epsilon_n))^2\} \qquad (9.35)$$

where

$$\chi_n = k_n x - \omega_n t$$
$$\overline{\chi} = kx - \overline{\omega}t$$
$$k_n = \omega_n^2/g$$
$$k = \overline{\omega}^2/g$$
$$\overline{\omega} = \text{mean frequency} = m_1/m_0$$
$$m_0 = \text{area under the spectrum}$$
$$m_1 = \text{first moment of the spectrum.}$$

The root-mean-square (rms) value of the wave displacement is given by

$$\eta_{\text{rms}} = \sqrt{m_0 + k^2 m_0^2} = \gamma\sqrt{m_0} \qquad (9.36)$$

where $\gamma = \sqrt{1 + k^2 m_0}$.

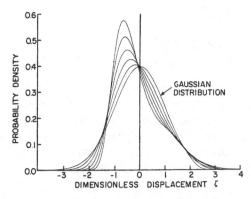

Fig. 9.4. Probability density functions of wave displacement developed based on the third-order Stokes wave model as a function of σ/λ, ranging from 0 to 0.05 in steps of 0.01 (Huang *et al.* 1983).

By letting

$$u = \frac{1}{\sqrt{m_0}} \sum a_n \cos(\chi_n + \epsilon_n)$$

$$v = \frac{1}{\sqrt{m_0}} \sum a_n \sin(\chi_n + \epsilon_n) \qquad (9.37)$$

the wave displacement can be non-dimensionalized as follows:

$$Z = \frac{\eta}{\eta_{\mathrm{rms}}} = \frac{1}{\gamma} \left\{ u + \frac{1}{2} k \sqrt{m_0} (u^2 - v^2) \right\} \qquad (9.38)$$

The cumulative distribution function of z then becomes

$$F(\zeta) = \Pr\{Z \leqslant \zeta\}$$

$$= \Pr\{u + \frac{1}{2} k \sqrt{m_0} (u^2 - v^2) \leqslant \gamma\zeta\} \qquad (9.39)$$

Since u and v are statistically independent normal variates and have zero mean, their joint probability density function is given by

$$f(u,v) = \frac{1}{2\pi} e^{-\frac{1}{2}(u^2 - v^2)} \qquad -\infty < u < \infty \qquad -\infty < v < \infty \qquad (9.40)$$

and the cumulative distribution function $F(\zeta)$ becomes

$$F(\zeta) = \frac{1}{2\pi} \int_{\alpha(\zeta)}^{\infty} e^{-\tau^2/2} [\mathrm{erf}\{A(\tau,\zeta) + \beta\} + \mathrm{erf}\{A(\tau,\zeta) - \beta\}] \mathrm{d}\tau \qquad (9.41)$$

where

$$\alpha(\zeta) = \begin{cases} 0 & \text{for} \quad \zeta > -\beta/\sqrt{2}\gamma \\ \sqrt{2}\beta\{-(1 + \sqrt{2}\gamma\zeta/\beta)\}^{1/2} & \text{for} \quad \zeta < -\beta/\sqrt{2}\gamma \end{cases}$$

$$\beta = (1/\sqrt{2}) k \sqrt{m_0}$$

$$A(\tau,\zeta) = \beta(1 + \sqrt{2}\gamma\zeta/\beta + \tau^2/\beta^2)^{1/2}$$

$$\tau = \text{error function variable.}$$

The evaluation of $F(\zeta)$ and its derivative $f(\zeta)$ requires numerical integration. As an example, Figure 9.5 shows the probability density function $f(\zeta)$ for $k\sqrt{m_0} = 0.3$ compared with results obtained from simulation. Included in the figure are the normal probability density function and Gram–Charlier series distribution with parameters λ_3 and λ_4. These parameters are given in terms of k, m_0 and γ as follows:

$$\lambda_3 = \frac{3k\sqrt{m_0}}{\gamma^3}$$

$$\lambda_4 = 3\left(\frac{1 + 6k^2 m_0 + 3k^4 m_0^2}{\gamma^4} - 1 \right) \qquad (9.42)$$

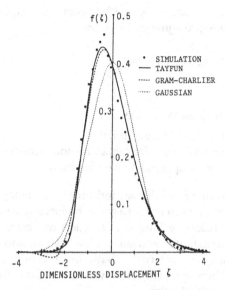

Fig. 9.5. Comparison between probability density function developed based on the second-order Stokes wave model and that obtained from simulation data (Tayfun 1980).

As can be seen in the figure, the distribution of simulation data differs significantly from the Gaussian distribution, but agrees well with the density function evaluated from Eq. (9.41) as well as that evaluated by the Gram–Charlier series distribution; the latter shows a small negative part for large negative ζ-values.

9.2.3 Distribution based on the concept of nonlinear system

As shown in Figure 9.1, wind-generated waves in deep water are considered to be a Gaussian random process, which transforms to a non-Gaussian random process as waves propagate into shallow water. This situation is likened to the response of a nonlinear system such as the heaving motion of a tension-leg platform in a seaway. The magnitude of heaving motion depends on the system's response characteristics as well as on the severity of the input.

Now, for a nonlinear system whose response can be presented in the form of two-term Volterra's stochastic series expansion, Kac and Siegert (1947) analytically show that the output of the system can be presented in terms of the standardized normal random variable as follows:

$$y(t) = \sum_{j=1}^{N} (\beta_j z_j + \lambda_j z_j^2) \qquad (9.43)$$

where $y(t)$ is the output of a nonlinear system and z_j is a standardized normal variate.

The parameters β_j and λ_j are evaluated by finding the eigenfunction and eigenvalues of the integral equation given by

$$\int K(\omega_1,\omega_2)\psi_j(\omega_2)d\omega_2 = \lambda_j\psi_j(\omega_1) \tag{9.44}$$

where

$$K(\omega_1,\omega_2) = H(\omega_1,\omega_2)\sqrt{S(\omega_1)S(\omega_2)}$$

$\psi_j(\omega)$ = orthogonal eigenfunction
$H(\omega_1,\omega_2)$ = second-order frequency response function
$S(\omega)$ = output spectral density function.

As can be seen, knowledge of the second-order frequency response function $H(\omega_1,\omega_2)$ is necessary in order to solve the integral equation Eq. (9.44). The frequency response function $H(\omega_1,\omega_2)$ has been evaluated analytically or experimentally for a given marine system, and from this the nonlinear stochastic characteristics have been estimated (Naess 1985b, for example). There is no way, however, to evaluate $H(\omega_1,\omega_2)$ for random waves in finite water depths.

In order to obtain the parameters β_j and λ_j in Eq. (9.43) from the wave record, Langley (1987) evaluates the parameters based on the concept of wave–wave interaction by applying second-order wave potential theory, and presents the functional relationship between the interaction coefficients and the parameters β_j and λ_j in a concise matrix form.

The probability distribution of wave profiles in shallow water, however, demonstrate an extremely skewed distribution associated with a strong nonlinearity. In order to address this strong nonlinearity in waves, Ochi and Ahn (1994) evaluate the nonlinear wave–wave interaction by decomposing the wave spectrum into linear and nonlinear components with the aid of bispectral analysis. Figure 9.6 shows an example of the separation of linear and nonlinear components of a spectrum. Upon clarifying the linear and nonlinear components in this manner, the parameters β_j and λ_j are evaluated by applying Langley's matrix presentation, and then the probability density function of wave displacement is numerically evaluated based on the Kac–Siegert solution given in Eq. (9.43).

Figure 9.7 shows a comparison between the probability density function thusly computed by applying the Kac–Siegert solution and the histogram constructed from data obtained in shallow water which indicates strong nonlinear characteristics. As seen, the agreement between them is excellent. Hence, it may be concluded that Eq. (9.43) can be applied for evaluating the stochastic properties of waves with strong nonlinear characteristics. Since the probability density function cannot be given in closed form, the probabilistic prediction of other wave properties, such as wave height, cannot be derived from it.

One way to derive the probability density function given in Eq. (9.43) in closed form is to present it as a function of a single random variable instead of the summation of the standardized normal distribution and its squared quantity. For this, let us present Eq. (9.43) as

$$Y = U + aU^2 \qquad (9.45)$$

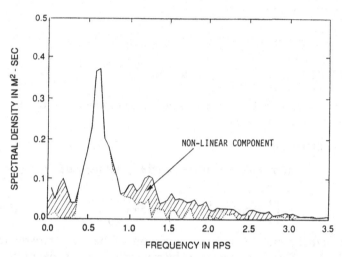

Fig. 9.6. Separation of linear and nonlinear components of wave spectrum evaluated from data obtained in shallow water (Ochi and Ahn 1994).

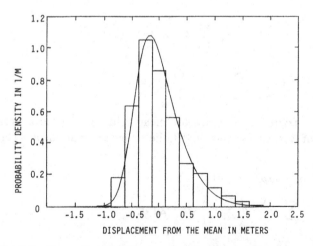

Fig. 9.7. Comparison between probability density function computed based on the Kac–Siegert solution and histogram constructed from wave data obtained in shallow water (Ochi and Ahn 1994).

where a is an unknown constant, and U is a normal variate with mean μ_* and variance σ_*^2, both of which are also unknown. The value of these unknowns will be determined by applying the cumulant generating function of Eq. (9.45) with that of the Kac–Siegert solution given in Eq. (9.43). By writing Eq. (9.45) as

$$Y=\left\{\sqrt{a}\left(U+\frac{1}{2a}\right)\right\}^2-\frac{1}{4a} \tag{9.46}$$

the characteristic function of Y can be obtained as follows:

$$\phi(t)=\frac{1}{\sqrt{1-2a\sigma_*^2 t}}\exp\left\{\left(a\mu_*^2+\mu_*+\frac{\sigma_*^2}{2}it\right)\left(\frac{it}{1-i2a\sigma_*^2 t}\right)\right\} \tag{9.47}$$

By taking the logarithm of the characteristic function and by expanding it in a series of (it), we have

$$\psi(t)=\ln\phi(t)$$
$$=(a\sigma_*^2+a\mu_*^2+\mu_*)(it)+\left\{a^2\sigma_*^4+\frac{\sigma_*^2}{2}+2a\mu_*^2(a\sigma_*^2+\mu_*)\right\}(it)^2$$
$$+\left\{\frac{4}{3}a^3\sigma_*^6+a\sigma_*^4+4a^2\sigma_*^4(a\mu_*^2+\mu_*)\right\}(it)^3+\dots \tag{9.48}$$

On the other hand, by taking the logarithm of the characteristic function of the Kac–Siegert solution, and by expanding it in a series of (it), we have

$$\psi(t)=\ln\phi(t)=-\frac{1}{2}\sum_{j=1}^{2N}\ln(1-i2\lambda_j t)-\sum_{j=1}^{2N}\frac{\beta_j t^2}{2}\frac{1}{1-i2\lambda_j t}$$
$$=\left(\sum_{j=1}^{2N}\lambda_j\right)it+\frac{1}{2!}\left(\sum_{j=1}^{2N}\beta_j^2+2\sum_{j=1}^{2N}\lambda_j^2\right)(it)^2$$
$$+\frac{1}{3!}\left(6\sum_{j=1}^{2N}\lambda_j\beta_j^2+8\sum_{j=1}^{2N}\lambda_j^3\right)(it)^3+\dots \tag{9.49}$$

Here, the first term $\Sigma\lambda_j=0$, since the mean value of Y is zero. From a comparison of Eqs. (9.48) and (9.49), the following relationships can be derived:

$$\begin{cases} a\sigma_*^2+a\mu_*^2+\mu_*=0 \\ \left(\displaystyle\sum_{j=1}^{2N}\beta_j^2\right)+2\left(\displaystyle\sum_{j=1}^{2N}\lambda_j^2\right)=\sigma_*^2-2a^2\sigma_*^4 \\ 3\left(\displaystyle\sum_{j=1}^{2N}\beta_j^2\lambda_j\right)+4\left(\displaystyle\sum_{j=1}^{2N}\lambda_j^3\right)=3a\sigma_*^4-8a^3\sigma_*^6 \end{cases} \tag{9.50}$$

It should be noted, however, that the cumulant generating function can generally be written in terms of cumulants as follows:

$$\psi(t) = \sum_{s=1}^{\infty} k_s \frac{(it)^s}{s!} = k_1(it) + \frac{1}{2!} k_2(it)^2 + \frac{1}{3!} k_3(it)^3 + \ldots \qquad (9.51)$$

where k_s are cumulants.

Hence, the three equations given in Eq. (9.50) represents the cumulants k_1, k_2 and k_3, respectively. Therefore, if we have a wave record obtained for a sufficiently long time (on the order of 20 minutes) and if we let the mean value be the zero line, we have $k_1 = 0$, and thereby k_2 and k_3 are equal to the sample moment $E[y^2]$ and $E[y^3]$, respectively. Thus, we can determine the unknown parameters a, θ_* and σ_*^2 simply by evaluating the sample moments from the wave record $y(t)$ and by applying the following relationships:

$$a\sigma_*^2 + a\mu_*^2 + \mu_* = 0$$
$$\sigma_*^2 - 2a^2\sigma_*^4 = E[y^2]$$
$$2a\sigma_*^4(3 - 8a^2\sigma_*^2) = E[y^3] \qquad (9.52)$$

Since the random variable U in Eq. (9.45) is a normal variate with known mean μ_* and variance σ_*^2, the probability density function of Y can be derived by applying the technique of change of random variables from U to Y. Unfortunately, the density function thusly derived vanishes at $y = -(1/4a)$ due to a singularity involved in the density function.

In order to overcome this difficulty, the functional relationship between Y and U given in Eq. (9.45) is inversely presented such that the random variable U is expressed approximately as a function of Y as follows:

$$U = \frac{1}{\gamma a}(1 - e^{-\gamma a Y}) \qquad (9.53)$$

where γ is a constant: 1.28 for $y \geq 0$, and 3.00 for $y < 0$. These values are determined from an analysis of the functional relationship between non-dimensionalized Y and U, and are valid for a process with very strong nonlinear characteristics.

By using the functional relationship given in Eq. (9.53), the following probability density function of Y can be derived from the probability distribution of U which is a normal probability distribution:

$$f(y) = \frac{1}{\sqrt{2\pi}\sigma_*} \exp\left\{ -\frac{1}{2(\gamma a\sigma_*)^2}(1 - \gamma a\mu_* - e^{-\gamma ay})^2 - \gamma ay \right\}$$

$$\gamma = \begin{cases} 1.28 & \text{for } y \geq 0 \\ 3.00 & \text{for } y < 0 \end{cases} \qquad (9.54)$$

It is noted that the sample space of the normal variate U defined in Eq. (9.53) is $(-\infty, 1/\gamma a)$ instead of $(-\infty, \infty)$. However, the truncation does not

(a)

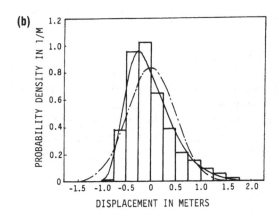

(b)

Fig. 9.8. Comparison of non-Gaussian probability density function and histogram constructed from data obtained in shallow water (Ochi and Ahn 1994).

(a)

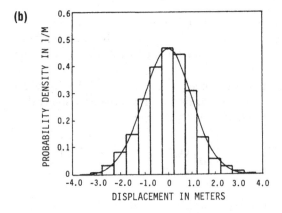

(b)

Fig. 9.9. Comparison of non-Gaussian probability density function and histogram constructed from data obtained in deep water (Ochi and Ahn 1994).

affect the probability distribution since $1/\gamma a$ is much greater than $\mu_* + 3\sigma_*$ of the normal distribution with mean μ_* and variance σ_*^2. It is also noted that Eq. (9.54) reduces to a normal distribution if $a=0$.

Figure 9.8 shows a comparison of the probability density function given in Eq. (9.54) with the histogram constructed from wave data obtained during a severe storm (variance$=0.23\,m^2$) at a water depth of 2.33 m. Included also in the figure is the normal probability density function for comparison.

Figure 9.9 shows a similar comparison for wave data obtained at a water depth of 24.7 m with the variance of 1.07 m^2. In this case, the water depth is sufficiently great for the sea severity; hence Eq. (9.54) reduces to a normal distribution and agrees well with the histogram.

9.3 PROBABILITY DISTRIBUTION OF PEAKS AND TROUGHS

For non-Gaussian waves in finite water depth, the statistical properties of positive wave displacement are different from those of negative displacement. Thus, in principle, it is highly desirable to derive the probability function applicable to peaks and troughs separately. Then, the probability function of wave height may be obtained as the sum of two independent random variables (peaks and troughs).

The probability distribution of wave peaks of non-Gaussian waves is derived by Tayfun (1980) assuming that individual waves can be expressed as a Stokes expansion to the second-order components given in Eq. (9.27). The peaks of non-Gaussian waves are expressed in dimensionless form as

$$\xi = \hat{a} + \frac{k\sqrt{m_0}}{\sqrt{2}}\,\hat{a}^2 \qquad\qquad (9.55)$$

where

$\hat{a} = a/\sqrt{2m_0} =$ dimensionless amplitude of Gaussian waves

$m_0 =$ area under the wave spectrum

$k =$ wave number.

Since the amplitude $\hat{a}$ obeys the Rayleigh probability distribution, the probability density function of ξ can be derived from Eq. (9.55) as a function of $k\sqrt{m_0}$. Figure 9.10 shows a comparison between the probability density function of ξ for $k\sqrt{m_0} = 0.14$ and 0.28 with the Rayleigh distribution. As can be seen, the density spreads toward the higher values with increase in the magnitude of $k\sqrt{m_0}$. It is noted, however, that the distribution of peaks and troughs are the same in this approach.

Arhan and Plaisted (1981), on the other hand, also developed the probability density function of peaks and troughs based on Stokes expansion to the second-order components, but expressed the peaks and troughs of non-Gaussian waves as follows:

$$\xi = a + \epsilon B a^2 \tag{9.56}$$

where

$\epsilon = +1$ for peaks, and -1 for troughs

$$B = \frac{k}{2} \coth kd \left(1 + \frac{3}{2 \sinh^2 kd} \right)$$

d = water depth

The amplitude a in Eq. (9.56) follows the Rayleigh probability distribution with the parameter m_{01}; i.e.

$$f(a) = \frac{a}{m_{01}} e^{-a^2/2m_{01}} \tag{9.57}$$

Note that the parameter m_{01} represents the portion of wave energy attributed to the first-order component, and Arhan and Plaisted theoretically derived the following relationship with the total energy m_0:

$$m_{01} = \left(-1 + \sqrt{1 + 16 m_0 B^2} \right)/8B^2 \tag{9.58}$$

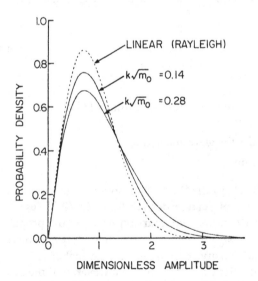

Fig. 9.10. Probability density functions of peaks developed based on the second-order Stokes wave model (Tayfun 1980).

Thus, for a given wave spectrum having an area under the density function of m_0, the probabiity density function $f(a)$ can be evaluated from Eqs. (9.57) and (9.58), and therefrom the probability density function of ξ can be derived from the relationship given in Eq. (9.56). The density function in dimensionless form becomes

$$f(\zeta)=\frac{\epsilon\alpha\left(-1+\sqrt{1+\epsilon\alpha\zeta}\right)}{\left(-1+\sqrt{1+\alpha^2}\right)\sqrt{1+\epsilon\alpha\zeta}}\cdot\exp\left\{-\frac{\epsilon\left(-1+\sqrt{1+\epsilon\alpha\zeta}\right)^2}{-1+\sqrt{1+\alpha^2}}\right\} \quad (9.59)$$

where

$$\zeta=\xi/\sqrt{m_0}$$

$$\alpha=2k\sqrt{m_0}\left(1+\frac{3}{2\sinh^2 kd}\right)\coth kd$$

Figure 9.11 shows a comparison between the cumulative distribution function computed from Eq. (9.59) with that obtained from measured data for a group of α-values ranging from 0.18 to 0.19. The data were obtained in intermediate water depth.

For representing the probability density function of wave heights observed in finite water depth, several probability distributions have been applied or developed empirically. These include the generalized gamma distribution (Ochi 1982), the truncated Rayleigh distribution (Kuo and Kuo 1975, Goda 1975), Gluhovski distribution (Shahul Hameed and Baba 1985), the Beta-Rayleigh distribution (Hughes and Borgman 1987), etc. Some of these distributions represent the histogram constructed from data very well.

For example, Figure 9.12 shows a comparison between the generalized

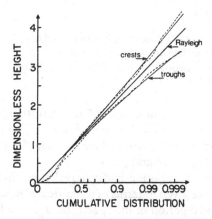

Fig. 9.11. Comparison of non-Gaussian probability distribution (solid line) with observed data (dotted line) (Arhan and Plaisted 1981).

278   NON-GAUSSIAN WAVES

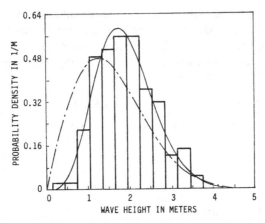

Fig. 9.12. Comparison between histogram of wave height, Rayleigh distribution (broken line) and generalized gamma distribution (solid line).

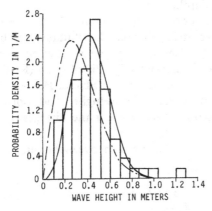

Fig. 9.13. Comparison between Beta-Rayleigh distribution (solid line), Rayleigh distribution (broken line) and histogram of wave height ($H_s=0.5$ m, water depth$=1.26$ m) (Hughes and Borgman 1987).

gamma distribution given in Eq. (5.4) in Chapter 5 and a histogram of wave height constructed from data obtained in shallow water. The parameters involved in the distribution are determined for each set of data following the procedure given in Eqs. (5.6) and (5.7).

The following Beta-Rayleigh distribution is developed such that the parameters of the distribution are given in terms of water depth, d, significant wave height, H_s, and modal period of the spectrum, T_p. The functional relationships between them are obtained from analysis of measured data:

$$f(H) = \frac{2\Gamma(\alpha+\beta)}{\Gamma(\alpha)\Gamma(\beta)} \frac{H^{2\alpha-1}}{H_b^{2\alpha}} \left(1 - \frac{H^2}{H_b^2}\right)^{\beta-1} \qquad 0 \leqslant H \leqslant H_b \qquad (9.60)$$

where

$$\alpha = \frac{k_1(k_2-k_1)}{k_1^2-k_2}$$

$$\beta = \frac{(1-k_1)(k_2-k_1)}{k_1^2-k_2}$$

$$k_1 = E[H^2]/H_b^2$$

$$k_2 = E[H^4]/H_b^4$$

H_b = breaking wave height approximated by water depth, d

$$E[H^2] = (1/2) \exp\{0.00272(d/gT_p^2)^{-0.834}\} H_s^2$$

$$E[H^4] = (1/2) \exp\{0.00046(d/gT_p^2)^{-1.208}\} H_s^2$$

T_p = modal period of wave spectrum

H_s = significant wave height.

The distribution has an upper value determined by the breaking wave height which is taken by Hughes and Borgman (1987) to be approximately equal to the water depth at a specific location. A comparison between the Beta-Rayleigh distribution, Rayleigh distribution and the histogram of wave height obtained from data is shown in Figure 9.13.

In regard to extreme values of non-Gaussian waves, it is a general trend that the extreme value estimated based on the Rayleigh distribution is greater than the measured extreme value. The results of statistical analysis of waves observed during the ARSLOE Project show that the Rayleigh distribution overpredicts the magnitude of large wave heights (Ochi 1982). The same trend is also observed in data obtained elsewhere. Forristall (1978), for example, compares the probability of exceeding a given wave height for the Rayleigh distribution and measured data as shown in Figure 9.14. In the figure, the triangles are data obtained during storms in the Gulf of Mexico, the solid line is the Rayleigh distribution, while the dashed line is the empirically fitted Weibull distribution. It is mentioned in Forristall's paper that the data show a definite excess of high crest points and a lack of low trough points. This indicates that the waves are a non-Gaussian random process for the associated water depth, and thereby the wave height distribution deviates from the Rayleigh distribution.

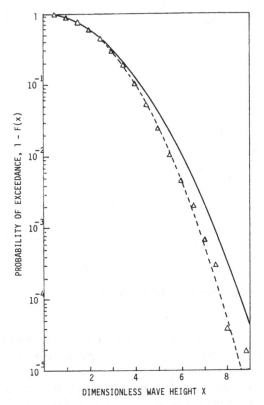

Fig. 9.14. Comparison of probabilities exceeding a specified wave height
(Forristall 1978).

9.4 TRANSFORMATION FROM GAUSSIAN TO
NON-GAUSSIAN WAVES

As demonstrated in Figure 9.1, waves which are considered to be a Gaussian
random process in deep water transform to a non-Gaussian random process
as they propagate from deep to shallow water. The question then arises as to
the location where the transition takes place in the nearshore zone. The
transition must depend on water depth as well as sea severity.

From analysis of wave data obtained during the ARSLOE Project, it is
found that $\sigma/D=0.06$ (σ=rms-value of waves at a specific location,
D=water depth at that location) is the limiting condition below which
waves are considered to be a Gaussian random process (Robillard and
Ochi 1996). Based on this limiting condition for a Gaussian random
process, Figure 9.15 shows the borderline between Gaussian and non-
Gaussian waves as a function of significant wave height, H_s, and water
depth, D. The solid line in the figure indicates $\sigma/D=0.06$ (or, assuming a
narrow-band wave, $D/H_s=4.17$), and the dotted line indicates the skew-

ness $\lambda_3=0.20$ which is the criterion established from analysis using the Gram–Charlier series probability distribution as mentioned in reference to Figure 9.3. The open circles in the figure are independently obtained from data at various water depths for which the computed parameter a of the density function given in Eq. (9.54) is zero or very close to zero, the criterion for a Gaussian wave. The results obtained from three different analyses for evaluating a non-Gaussian random process as a function of significant wave height and water depth show good agreement; therefore, it may be concluded that $D=4.17H_s$ is the limiting water depth below which waves are considered to be a non-Gaussian random process.

By using the relationship given in Figure 9.15, the transition of sea severity associated with depth variation in the nearshore zone may be estimated. It should be noted, however, that this estimation is valid only for water depths not exceeding 25 m where the data are acquired and analyzed. The basic concept of the estimation is shown in Figure 9.16. It is assumed that the sea severity is constant in the area where the water depth is greater than $4.17H_s$. This water depth is denoted by D_0 in the figure. By letting the variance of random waves at an arbitrary location in the nearshore zone be σ^2 where the water depth is D, we may obtain the value of σ/D as a function of D_0.

Figure 9.17 shows the functional relationship between σ/D and the dimensionless depth D/D_0 obtained from analysis of data. The relationship is given by

$$\sigma/D=0.06(D/D_0)^{-0.58} \qquad 0<D/D_0<1 \qquad (9.61)$$

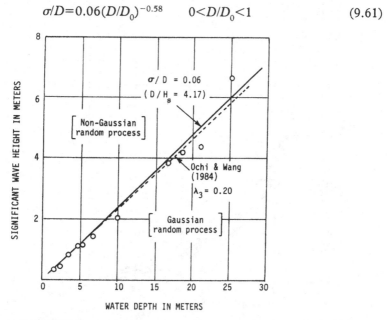

Fig. 9.15. Limiting water depth below which non-Gaussian characteristics are expected as a function of significant wave height (Robillard and Ochi 1996).

The two lines in the figure are ±5 percent deviations from the mathematical formulation which covers all data.

Thus, from knowledge of the significant wave height in deep water in the nearshore zone (but not exceeding 25 m), the limiting water depth D_0 where the transition to non-Gaussian waves takes place is estimated from Figure 9.15, and thereby the variance representing the sea severity at a specified depth, D, in the nearshore zone can be estimated based on Eq. (9.61).

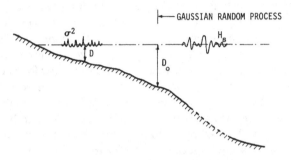

Fig. 9.16. Sketch indicating the estimated transition of sea severity due to water depth variation in the nearshore zone.

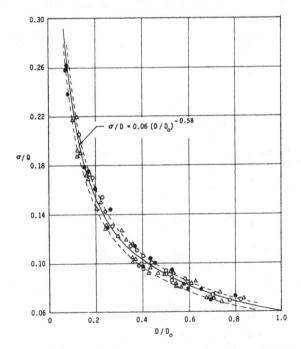

Fig. 9.17. Relationship between sea severity and water depth ratio as a function of dimensionless water depth (Robillard and Ochi 1996).

APPENDIX A   FUNDAMENTALS OF PROBABILITY THEORY

1 Probability distribution and density function

DEFINITION. The real-valued function $F(x)$ defined as

$$F(x) = \Pr\{X \leq x\} \qquad (A.1)$$

is called the *probability distribution function* or *cumulative distribution function* of the random variable X.

PROPERTIES.
   (i) $F(-\infty) = 0$, $F(\infty) = 1$, where the sample space is $(-\infty, \infty)$
   (ii) $F(x+h) \geq F(x)$ for $h > 0$
   (iii) $F(x)$ is continuous at least from the left for a discrete-type random variable.

DEFINITION. For a continuous type random variable X, if there exists a non-negative function $f(x)$ such that $F(x)$ can be expressed by

$$F(x) = \int_{-\infty}^{x} f(x)\, dx \qquad (A.2)$$

then $f(x)$ is called the *probability density function* of X.

PROPERTIES.
   (i) $f(x) \geq 0$ and is integrable over every real value in sample space

   (ii) $\displaystyle\int_{-\infty}^{\infty} f(x)\, dx = 1.$ \qquad (A.3)

NOTES. For a discrete-type random variable X, the *probability mass function* $p(x_i)$ is defined as

$$F(x) = \sum_{x < x_i} p(x_i)$$

283

where $p(x_i) = \Pr\{X = x_i\}$ and

$$\sum_{i=1}^{\infty} p(x_i) = 1$$

2. Joint probability distribution and density function of two random variables

DEFINITION. The real-valued function $F(x,y)$ defined as

$$F(x,y) = \Pr\{X \leq x, Y \leq y\} \tag{A.4}$$

is called the *joint probability distribution function* of X and Y. A real-valued function $f(x,y)$ is called the *joint probability density function* of X and Y if it satisfies the condition

$$F(x,y) = \int_{-\infty}^{y} \int_{-\infty}^{x} f(x,y) \, dx \, dy \tag{A.5}$$

PROPERTIES.

(i) $f(x,y) > 0$

(ii) Integrable over all real values in the sample space

(iii) $\int_{-\infty}^{\infty} \int_{-\infty}^{\infty} f(x,y) \, dx \, dy = 1.$ \hfill (A.6)

DEFINITION. For a joint distribution function $f(x,y)$ of two random variables X and Y, the *marginal distribution function* of X and Y is defined as

$$F_x(x) = F(x,\infty) = \Pr\{X \leq x, Y < \infty\}$$
$$F_y(y) = F(\infty,y) = \Pr\{X < \infty, Y \leq y\} \tag{A.7}$$

For continuous-type random variables, the *marginal density function* of X and Y can be obtained as

$$f(x) = \int_{-\infty}^{\infty} f(x,y) \, dy$$

$$f(y) = \int_{-\infty}^{\infty} f(x,y) \, dx \tag{A.8}$$

DEFINITION. Two random variables X and Y are said to be *statistically independent* if the joint probability distribution is equal to the product of the individual distribution functions. That is,

$$F(x,y)=F(x)\,F(y) \tag{A.9}$$

Hence, for continuous-type random variables, we have

$$f(x,y)=f(x)\,f(y) \tag{A.10}$$

3. Conditional probability distribution and density function

DEFINITION. The *conditional probability distribution function* of a random variable Y given that a random variable X has the value x is defined by

$$F(y|x)=\Pr\{Y\leqslant y|X=x\}=\frac{\Pr\{Y\leqslant y \text{ and } X=x\}}{\Pr\{X=x\}} \tag{A.11}$$

The *conditional probability density function* of a random variable Y given that X has the value x can be written as

$$f(y|x)=\frac{f(x,y)}{f(x)}=\frac{f(x,y)}{\displaystyle\int_{-\infty}^{\infty}f(x,y)\mathrm{d}y} \tag{A.12}$$

The conditional probability density function of three random variables X, Y and Z may be defined as

$$f(z|x,y)=\frac{f(x,y,z)}{\displaystyle\int_{-\infty}^{\infty}f(x,y,z)\mathrm{d}z} \tag{A.13}$$

$$f(y,z|x)=\frac{f(x,y,z)}{\displaystyle\int_{-\infty}^{\infty}\int_{-\infty}^{\infty}f(x,y,z)\mathrm{d}y\,\mathrm{d}z} \tag{A.14}$$

DEFINITION. Let S_* be a subset of the sample space of a continuous random variable X with a probability density function $f(x)$. The following conditional distribution of X in sample space S_* is called the *truncated probability distribution*:

$$F_*(x)=\Pr\{X\leqslant x|X\epsilon S_*\}=\frac{\displaystyle\int_{-\infty\cap S_*}^{x}f(x)\,\mathrm{d}x}{\displaystyle\int_{S_*}f(x)\,\mathrm{d}x} \tag{A.15}$$

The *truncated probability density function*, denoted by $f_*(x)$ is defined as

$$f_*(x) = \frac{f(x)}{\displaystyle\int_{S_*} f(x)\,dx} \qquad \text{where} \quad x \in S_* \qquad (A.16)$$

NOTES. For a discrete-type random variable X with the probability mass function $p(x_i)$, the *truncated probability mass function* is defined as follows

$$p_*(x) = \frac{p(x_i)}{\displaystyle\sum_{x_i \in S_*} p(x_i)} \qquad \text{where } x_i \in S_* \qquad (A.17)$$

4. Expected value

DEFINITION. The *expected value* (or *expectation*) of a function $u(x)$ of a random variable X is defined as follows:

$$E[u(x)] = \begin{cases} \displaystyle\sum_{i \in S} u(x_i)p(x_i) \text{ for a discrete-type random variable} \\ \displaystyle\int_S u(x)f(x)\,dx \text{ for a continuous-type random variable} \end{cases} \qquad (A.18)$$

where S stands for the sample space.

DEFINITION. The expected value of $u(x) = x^k$ is called the *kth moment* of the random variable X. In particular, for $k=1$, $E[x]$ is called the *mean value*.

DEFINITION. The expected value of $u(x) = (x-\mu)^k$, where $\mu =$ mean value is called the *kth central moment* of the random variable X. In particular, for $k=2$, $E[(x-\mu)^2]$ is called the *variance*, denoted by σ^2. The square-root of the variance, σ, is called the *standard deviation*.

PROPERTIES.
    (i) $\text{Var}[x] = E[x^2] - (E[x])^2$
    (ii) $\text{Var}[ax] = a^2\text{Var}[x]$ where $a =$ constant
    (iii) $\text{Var}[x+a] = \text{Var}[x]$. $\qquad\qquad\qquad\qquad\qquad$ (A.19)

DEFINITION. A random variable X that has a zero mean, $E[x]=0$, and unit variance, $Var[x]=1$, is called the *standardized random variable*. For a random variable X with a mean μ and variance σ^2,

$$Z=\frac{X-\mu}{\sigma} \qquad (A.20)$$

is called the *standardized random variable* of X. The random variable Z has zero mean and unit variance.

DEFINITION. Let X be a random variable with mean μ and variance σ^2. The following parameters are often considered for representing the shape of a probability density function:

$Coefficient\ of\ variation = \sigma/\mu$
$Skewness = E[(x-\mu)^3]/\sigma^3$
$Kurtosis = E[(x-\mu)^4]/\sigma^4 - 3$ $\qquad$ (A.21)

NOTES. Some define the kurtosis as $E[(x-\mu)^4]/\sigma^4$ (Johnson and Leone 1964, for example). See also the definition in terms of cumulants.

DEFINITION. The *covariance* of two random variables X and Y, denoted $Cov[x,y]$ or σ_{xy}, is defined by

$$Cov[x,y]=E[(x-\mu_x)(y-\mu_y)]=E[xy]-E[x]E[y] \qquad (A.22)$$

If $Cov[x,y]=0$, then the two random variables are called *uncorrelated*.

DEFINITION. The *correlation coefficient* of two random variables X and Y, denoted ρ_{xy}, is defined by

$$\rho_{xy}=\frac{Cov[x,y]}{\sqrt{Var[x]\ Var[y]}}=\frac{\sigma_{xy}}{\sigma_x\sigma_y} \qquad (A.23)$$

PROPERTIES.
   (i) $-1\leqslant\rho_{xy}\leqslant 1$
   (ii) If two random variables are statistically independent, then they are uncorrelated. However, the reverse is not always true.

5. Characteristic function and moment generating function

DEFINITION. The function $\phi(t)=E[e^{itx}]$, where t is a real number is called the *characteristic function* of the random variable X. That is,

$$\phi(t) = \begin{cases} \sum_{j\epsilon s} e^{itx_j} p(x_j) \text{ for a discrete-type random variable} \\ \int_S e^{itx} f(x) \, dx \text{ for a continuous-type random variable} \end{cases}$$

$$(A.24)$$

where S stands for the sample space.

PROPERTIES.
  (i) $\phi(t)$ always exists
  (ii) $\phi(0) = 1$
  (iii) $|\phi(t)| \leqslant 1$
  (iv) $\phi(-t) = \phi^*(t)$, the complex conjugate of $\phi(t)$
  (v) $E[x^r] = (1/i^r) |d^r/dt^r)|_{t=0}$
  (iv) Let $\phi_x(t)$ be the characteristic function of a random variable X. Then, the characteristic function of $Y = aX + b$, where a and b are constants is given by $\exp\{ibt\} \phi_x(at)$.
  (vii) Let the characteristic functions of two statistically independent random variables X and Y be $\phi_x(t)$ and $\phi_y(t)$, respectively. Then, the characteristic function of the sum of two random variables $Z = X + Y$ is given by $\phi_z(t) = \phi_x(t) \phi_y(t)$.

DEFINITION. The function $M(t) = E[e^{tx}]$, where t is a real number is called the *moment generating function* of the random variable X. That is,

$$m(t) = \begin{cases} \sum_{j\epsilon s} e^{tx_j} p(x_j) \text{ for a discrete-type random variable} \\ \int_S e^{tx} f(x) \, dx \text{ for a continuous-type random variable} \end{cases}$$

$$(A.25)$$

where S stands for the sample space.

PROPERTIES.
  (i) $m(t)$ does not always exist
  (ii) $E[x^r] = \left| \dfrac{d^r}{dt^r} m(t) \right|_{t=0}$

DEFINITION. The logarithm of the characteristic function of the random variable X is defined as the *cumulant generating function*, denoted by $\psi(t)$. That is,

$$\psi(t) = \ln \phi(t) = \sum_{s=1}^{\infty} k_s \frac{(it)^s}{s!} \qquad (A.26)$$

The coefficients k_s are called *cumulants* or *semi-invariants* of the random variable X.

PROPERTIES.

(i) $k_1 = E[x]$

(ii) Skewness $= k_3/k_2^{1.5}$

Kurtosis $= k_4/k_2^2$

6. Transformation of random variables

THEOREM. Let X be a continuous random variable with the probability density function $f_x(x)$. Let Y be a new random variable having a single-valued functional relationship with X given by $y = g(x)$. The probability density function of Y is then given by

$$f(y) = f_x[h(y)] \left| \frac{\mathrm{d}}{\mathrm{d}y} h(y) \right| \qquad (A.27)$$

where $h(y)$ is the inverse function of $y = g(x)$.

For a discrete random variable X, we have

$$p(y) = p_x[h(y)] \qquad (A.28)$$

NOTES. If $Y = aX^2$ where $0 < X < \infty$, and a is a positive constant, we have $f(y) = f_x(\sqrt{y/a}) |1/2\sqrt{ay}|$, $0 < y < \infty$.

THEOREM. Let $f_{xy}(x,y)$ be the joint probability density function of the random variables X and Y. Consider new random variables $U = g_1(X,Y)$ and $V = g_2(X,Y)$, where the functions g_1 and g_2 have continuous partial derivatives with respect to x and y. Then, the joint probability density function of U and V is given by

$$f(u,v) = f_{xy}[h_1(u,v), h_2(u,v)] \times \begin{vmatrix} \dfrac{\partial h_1}{\partial u} & \dfrac{\partial h_1}{\partial v} \\[2mm] \dfrac{\partial h_2}{\partial u} & \dfrac{\partial h_2}{\partial v} \end{vmatrix} \qquad (A.29)$$

where $x = h_1(u,v)$ and $y = h_2(x,y)$ are the inverse functions of $u = g_1(x,y)$ and $v = g_2(x,y)$.

THEOREM. Let the random variable Z be the sum of two continuous-type random variables X and Y with the joint probability density function $f_{xy}(x,y)$. The probability density function of Z becomes

$$f(z)=\int_{s_x} f_{xy}(x,z-x)\mathrm{d}x=\int_{s_y} f_{xy}(z-y,y)\,\mathrm{d}y \qquad (A.30)$$

where s_x and s_y are the sample spaces of the random variables X and Y, respectively. If X and Y are statistically independent, then we have

$$f(z)=\int_{s_x} f_x(x)f_y(z-x)\mathrm{d}x=\int_{s_y} f_x(z-y)f_y(y)\,\mathrm{d}y \qquad (A.31)$$

Care must be taken in the integration domain of s_x and s_y, since it is often necessary to divide the sample space into two domains for integration.

THEOREM. Let the random variable Z be the difference of two random variables X and Y ($Z=X-Y$), with the joint probability density function $f_{xy}(x,y)$. Then, the probability density function of Z is given by

$$f(z)=\int_{s_x} f_{xy}(x,x-z)\mathrm{d}x=\int_{s_y} f_{xy}(z+y,y)\,\mathrm{d}y \qquad (A.32)$$

If X and Y are statistically independent,

$$f(z)=\int_{s_x} f_x(x)f_y(x-z)\,\mathrm{d}x=\int_{s_y} f_x(z+y)f_y(y)\,\mathrm{d}y \qquad (A.33)$$

THEOREM. Let the random variable Z be the product of two random variables X and Y with the joint probability density function $f_{xy}(x,y)$. Then, the probability density function of Z is given by

$$f(z)=\int_{s_x} \frac{1}{|x|} f_{xy}(x,z/x)\,\mathrm{d}x=\int_{s_y} \frac{1}{|y|} f_{xy}(z/y,y)\,\mathrm{d}y \qquad (A.34)$$

If X and Y are statistically independent,

$$f(z)=\int_{s_x} \frac{1}{|x|} f_x(x)f_y(z/x)\,\mathrm{d}x=\int_{s_y} \frac{1}{|y|} f_x(z/y)f_y(y)\,\mathrm{d}y \qquad (A.35)$$

THEOREM. Let the random variable Z be the ratio of two random variables X and Y ($Z=X/Y$) with the joint probability density function $f_{xy}(x,y)$. Then, the probability density function of Z is given by

$$f(z) = \int_{S_y} |y| f_{xy}(zy,y) \, dy \tag{A.36}$$

If X and Y are statistically independent,

$$f(z) = \int_{S_y} |y| f_x(zy) f_y(y) \, dy \tag{A.37}$$

7. Some discrete-type probability distributions

DEFINITION. *Binomial distribution*

$$p(x) = \binom{n}{x} p^x q^{n-x}$$

where

$$p + q = 1 \text{ and } n = 0, 1, 2, \ldots, n \tag{A.38}$$

(i) $E[x] = np$, $\text{Var}[x] = npq$
(ii) Binomial distribution with small p and with large n can be approximated by the Poisson distribution with $\mu = np$.

DEFINITION. *Poisson distribution*

$$p(x) = \frac{\mu^x}{x!} e^{-\mu}$$

where

$$x = 0, 1, 2, \ldots, \infty \tag{A.39}$$

(i) $E[x] = \mu$, $\text{Var}[x] = \mu$
(ii) The parameter μ is often not constant; instead, it is a random variable having a gamma probability distribution. In this case, the probability distribution becomes the negative binomial distribution.

8. Some continuous-type distributions

DEFINITION. *Normal distribution*

$$f(x) = \frac{1}{\sqrt{2\pi}\sigma} \exp\left\{ -\frac{1}{2} \left(\frac{x-\mu}{\sigma} \right)^2 \right\} \qquad -\infty < x < \infty \tag{A.40}$$

(i) $E[x] = \mu$, $\text{Var}[x] = \sigma^2$

(ii) $\Pr\{a<x<b\} = \Phi\left(\dfrac{b-\mu}{\sigma}\right) - \Phi\left(\dfrac{a-\mu}{\sigma}\right)$

where

$$\Phi(z) = \frac{1}{\sqrt{2\pi}} \int\limits_{-\infty}^{z} \exp\{-z^2/2\}\,dz$$

THEOREM. Central limit theorem. Let $X_1, X_2, \ldots, X_n$ be n independent random variables having unspecified but identical distributions with mean μ and variance σ^2. Then, the average value $(1/n)\Sigma_{i=1}^{n} x_i$ has a normal distribution with mean μ and variance σ^2/n if n is large.

DEFINITION. *Log-normal distribution*

$$f(x) = \frac{1}{\sqrt{2\pi}\sigma x} \exp\left\{-\frac{1}{2}\left(\frac{\ln x - \mu}{\sigma}\right)^2\right\} \qquad 0<x<\infty$$

$E[x] = \exp\{\mu + \sigma^2/2\} \qquad \mathrm{Var}[x] = \exp\{2\mu + \sigma^2\}(\exp\{\sigma^2\} - 1)$

$$(\text{A.41})$$

DEFINITION. *Chi-square (χ^2) distribution with r degrees of freedom*

$$f(x) = \frac{1}{\Gamma(r/2)} \frac{1}{2^{r/2}} x^{(r/2)-1} \exp\{-x/2\} \qquad 0<x<\infty$$

$E[x] = r \qquad \mathrm{Var}[x] = 2r$

$$(\text{A.42})$$

DEFINITION. *Gamma distribution*

$$f(x) = \frac{1}{\Gamma(m)} \lambda^m x^{m-1} \exp\{-\lambda x\} \qquad 0<x<\infty \qquad (\text{A.43})$$

(i) $E[x] = m/\lambda$, $\mathrm{Var}[x] = m/\lambda^2$
(ii) The gamma distribution with $m=1$ is called the *exponential distribution*
(iii) *Generalized-gamma distribution*

$$f(x) = \frac{c}{\Gamma(m)} \lambda^{cm} x^{cm-1} \exp\{-(\lambda x)^c\} \qquad 0<x<\infty$$

DEFINITION. *Weibull distribution*

$$f(x) = c\lambda^c x^{c-1} \exp\{-(\lambda x)^c\} \qquad 0<x<\infty$$

$$E[x] = \frac{1}{\lambda}\, \Gamma\left(1+\frac{1}{c}\right)$$

$$\text{Var}[x] = \frac{1}{\lambda^2}\left[\Gamma\left(1+\frac{2}{c}\right) - \left\{\Gamma\left(1+\frac{1}{c}\right)\right\}^2 \right] \tag{A.44}$$

DEFINITION. *Rayleigh distribution*

$$f(x) = (2x/R)\exp\{-x^2/R\} \qquad 0<x<\infty$$

$$\text{E}[x] = \sqrt{\pi R}/2 \qquad \text{Var}[x] = \left(1-\frac{\pi}{4}\right)R \tag{A.45}$$

THEOREM. Let $A_1, A_2, ..., A_n$ be mutually exclusive events and let B be any other event. We have then

$$\Pr\{A_i|B\} = \frac{\Pr\{A_i\}\ \Pr\{B|A_i\}}{\Pr\{B\}} \tag{A.46}$$

The above formula is called *Bayes' theorem*. $\Pr\{A_i|B\}$ may be interpreted as the probability of event A_i after B has occurred; hence it is called *posterior probability*. On the other hand, $\Pr\{A_i\}$ is the probability of the event A_i before the information about B is available, therefore it is called *a priori probability*.

APPENDIX B FUNDAMENTALS OF STOCHASTIC PROCESS THEORY

In general, there are four different types of stochastic processes depending on whether the state and time of a process are continuous or discrete. The following presentation is limited to continuous stochastic processes both in time and space except for the Markov chain process.

1. Stochastic process

DEFINITION. A set (family) of random variables consisting of n elements of records taken simultaneously at a specified time t, $\{^1x(t), {}^2x(t)..., {}^nx(t)\}$, is called an *ensemble* of a stochastic process (or random process).

DEFINITION. A stochastic process $x(t)$ is said to be an *ergodic process* if all statistical properties of the ensemble are equal to those for a single record $x(t)$ taken for a sufficiently long period of time. If this restricted condition is limited to the mean and covariance, then the stochastic process is called a *weakly ergodic random process*. In this case, we have

$$\frac{1}{n}\sum_{k=1}^{n} {}^kx(t) = \lim_{T\to\infty} \frac{1}{T}\int_0^T x(t)\,\mathrm{d}t = \mu$$

$$\frac{1}{n}\sum_{k=1}^{n} \{^kx(t)-\mu\}\,\{^kx(t+\tau)-\mu\}$$

$$= \lim_{T\to\infty} \frac{1}{T}\int_0^T \{x(t)-\mu\}\,\{x(t+\tau)-\mu\}\,\mathrm{d}t \qquad (\text{B.1})$$

DEFINITION. A stochastic process $x(t)$ is said to be a *weakly stationary process* if its mean value is constant and its covariance function depends on a time shift τ for all t. That is,

$$\lim_{T\to\infty} \frac{1}{T}\int_0^T f(t)\,\mathrm{d}t = \text{constant}$$

$$\lim_{T\to\infty} \frac{1}{T}\int_0^T \{x(t)-\mu\}\{x(t+\tau)-\mu\}\,dt$$

$$=\lim_{T\to\infty} \frac{1}{T}\int_0^T x(t)x(t+\tau)\,dt - \mu^2 = R(\tau)-\mu^2 \qquad (B.2)$$

where $R(\tau)$ is called the *auto-correlation* function.

Ocean waves are considered to be a weakly steady state (stationary) ergodic random process.

DEFINITION. A steady-state random process $x(t)$ is called a *narrow-band random process* if the spectral density function (defined later) is sharply concentrated in the neighbourhood of a specified frequency ω_0. The narrow-band random process $x(t)$ is generally presented as

$$x(t)=A(t)\cos\{\omega_0 t+\epsilon(t)\} \qquad (B.3)$$

Note that the amplitude $A(t)$ and phase $\epsilon(t)$ are random variables, but the frequency ω_0 is a constant.

DEFINITION. A steady state ergodic random process $x(t)$ is said to be a *Gaussian* (or *normal*) *random process* if a set of randomly sampled displacements from the mean value is normally distributed.

NOTES. The profile of ocean waves in deep water is known to be normally distributed with zero mean and a certain variance depending on sea severity.

DEFINITION. A stochastic process $x(t)$ is said to be a *Markov process* if it satisfies the following conditional probability:

$$\Pr\{x(t_n)<x_n|x(t_1)=x_1,x(t_2)=x_2,\ldots,x(t_{n-1})=x_{n-1}\}$$
$$=\Pr\{x(t_n)<x_n|x(t_{n-1})=x_{n-1}\} \qquad \text{where } t_1<t_2<\ldots<t_n \qquad (B.4)$$

The Markov process implies that the conditional probability of a random process $x(t)=x_n$ at time t_n, given that its value at some earlier time is known, depends only on the immediate past value at t_{n-1} and is independent of its history prior to t_{n-1}. The state and time of a Markov process can be discrete as well as continuous. In particular, the Markov process with a discrete state is called a *Markov chain*, while that with a continuous state is called a *diffusion process*.

2. Spectral density function

DEFINITION. The *spectral density function (spectrum)* of a random process $x(t)$ is defined as

$$S_{xx}(\omega)=\lim_{T\to\infty}\frac{1}{2\pi T}|X(\omega)|^2 \tag{B.5a}$$

where $X(\omega)=$ Fourier transform of $x(t)$.
In terms of frequency f, we have

$$S_{xx}(f)=\lim_{T\to\infty}\frac{1}{T}|X(f)|^2 \tag{B.5b}$$

THEOREM. The time average of energy of a random process $x(t)$, denoted as $\bar{P}_x$, can be presented with the aid of Parseval's theorem as follows:

$$\bar{P}_x=\int_0^\infty S_{xx}(\omega)\,d\omega=\int_0^\infty S_{xx}(f)\,df \tag{B.6}$$

THEOREM. For a steady state ergodic random process $x(t)$, its auto-correlation function $R_x(\tau)$ and spectral density function $S_x(\omega)$ or $S_x(f)$ are related by the Fourier transform. This is called the *Wiener–Khintchine theorem*. That is,

$$S_{xx}(\omega)=\frac{1}{\pi}\int_{-\infty}^\infty R_{xx}(\tau)\,e^{-i\omega\tau}d\tau=\frac{2}{\pi}\int_0^\infty R_{xx}(\tau)\cos\omega\tau\,d\tau$$

$$R_{xx}(\tau)=\frac{1}{2}\int_{-\infty}^\infty S_{xx}(\omega)\,e^{i\omega\tau}d\omega=\int_0^\infty S_{xx}(\omega)\cos\omega\tau\,d\omega \tag{B.7a}$$

$$S_{xx}(f)=2\int_{-\infty}^\infty R_{xx}(\tau)\,e^{-i2\pi f\tau}d\tau=4\int_0^\infty R_{xx}(\tau)\cos 2\pi f\tau\,d\tau$$

$$R_{xx}(\tau)=\frac{1}{2}\int_{-\infty}^\infty S_{xx}(f)\,e^{i2\pi f\tau}df=\int_0^\infty S_{xx}(f)\cos 2\pi f\tau\,df \tag{B.7b}$$

NOTES. The auto-correlation function and spectral density function of derived random processes are as follows:

(i) Velocity process $\dot{x}(t)$

$$R_{\dot{x}\dot{x}}(\tau)=-\frac{d^2}{d\tau^2}R_{xx}(\tau)$$

$$S_{\dot{x}\dot{x}}(\omega)=\omega^2 S_{xx}(\omega)\qquad S_{\dot{x}\dot{x}}(f)=(2\pi f)^2 S_{xx}(f) \tag{B.8}$$

(ii) Acceleration process $\ddot{x}(t)$

$$R_{\ddot{x}\ddot{x}}(\tau) = \frac{d^4}{d\tau^4} R_{xx}(\tau)$$

$$S_{\ddot{x}\ddot{x}}(\omega) = \omega^4 S_{xx}(\omega) \qquad S_{\ddot{x}\ddot{x}}(f) = (2\pi f)^4 S_{xx}(f) \qquad \text{(B.9)}$$

DEFINITION. The cross-spectral density function (*cross-spectrum*) of two random processes $x(t)$ and $y(t)$ is defined as

$$S_{xy}(\omega) = \lim_{T \to \infty} \frac{1}{2\pi T} X^*(\omega) Y(\omega)$$

$$S_{xy}(f) = \lim_{T \to \infty} \frac{1}{T} X^*(f) Y(f) \qquad \text{(B.10)}$$

where $X^*(\omega)$, $X^*(f)$ are the conjugate functions of $X(\omega)$ and $X(f)$, respectively.

PROPERTIES.
  (i) The Wiener–Khintchine theorem is also applicable to the cross-spectrum.
  (ii) The cross-spectrum is a complex function in contrast to the auto-spectrum. The real and imaginary parts of the cross-spectrum are referred to as the *co-spectrum*, $C_{xy}(\omega)$, and the *quadrature spectrum*, $Q_{xy}(\omega)$, respectively. That is,

$$S_{xy}(\omega) = C_{xy}(\omega) + iQ_{xy}(\omega) \qquad \text{(B.11)}$$

  (iii) The co-spectrum is an even function, and we have

$$C_{xy}(-\omega) = C_{xy}(\omega) = C_{yx}(\omega) = C_{yx}(-\omega) \qquad \text{(B.12)}$$

  (iv) The quadrature spectrum is an odd function, and we have

$$Q_{xy}(-\omega) = -Q_{xy}(\omega) = Q_{yx}(\omega)$$
$$Q_{yx}(-\omega) = -Q_{yx}(\omega) = Q_{xy}(\omega) \qquad \text{(B.13)}$$

  (v) By using the properties given in the above equations, we can derive the following relationship:

$$S_{xy}(-\omega) = S_{xy}^*(\omega) = S_{yx}(\omega) \qquad \text{(B.14)}$$

where $S_{xy}^*(\omega)$ is the complex conjugate of $S_{xy}(\omega)$.

DEFINITION. The two-dimensional auto-correlation function of a random variable $x(t)$, denoted by $M_x(\tau_1, \tau_2)$, is defined by

$$M_x(\tau_1, \tau_2) = E[x(t)x(t+\tau_1)x(t+\tau_2)]$$

$$=\lim_{T\to\infty}\frac{1}{2T}\int_{-T}^{T} x(t)x(t+\tau_1)x(t+\tau_2)\,dt \qquad (B.15)$$

PROPERTIES.

$$M_x(\tau_1,\tau_2)=M_x(\tau_2,\tau_1)=M_x(-\tau_2,\tau_1-\tau_2)$$
$$=M_x(\tau_1-\tau_2,-\tau_2)=M_x(-\tau_1,\tau_2-\tau_1)=M_x(\tau_2-\tau_1,-\tau_1) \qquad (B.16)$$

DEFINITION. The two-dimensional spectral density function of a random variable $x(t)$ is called the *bispectrum*, and is defined as

$$B(\omega_1,\omega_2)=\lim_{T\to\infty}\frac{1}{2\pi T}X(\omega_1)X(\omega_2)X^*(\omega_1+\omega_2)$$

$$=\lim_{T\to\infty}\frac{1}{2\pi T}X(\omega_1)X(\omega_2)X(\omega_3) \qquad (B.17)$$

where $\omega_1+\omega_2+\omega_3=0$ and $X^*(\omega)$ is the conjugate of $X(\omega)$.

PROPERTIES.

(i) The Wiener–Khintchine theorem can also be applied to $M(\tau_1,\tau_2)$ and $B(\omega_1,\omega_2)$. That is,

$$B(\omega_1,\omega_2)=\frac{1}{\pi^2}\int_{-\infty}^{\infty}\int_{-\infty}^{\infty} M(\tau_1,\tau_2)\,e^{-i(\omega_1\tau_1+\omega_2\tau_2)}d\tau_1\,d\tau_2$$

$$M(\tau_1,\tau_2)=\frac{1}{2^2}\int_{-\infty}^{\infty}\int_{-\infty}^{\infty} B(\omega_1,\omega_2)\,e^{i(\omega_1\tau_1+\omega_2\tau_2)}d\omega_1\,d\omega_2 \qquad (B.18)$$

(ii) Since there are six two-dimensional auto-correlation function for a given τ_1 and τ_2, there exist six bispectra for a given ω_1 and ω_2 having the same value. These are

$$B(\omega_1,\omega_2)=B(\omega_2,\omega_1)=B(\omega_1,-\omega_1-\omega_2)$$
$$=B(-\omega_1-\omega_2,\omega_1)=B(\omega_2,-\omega_1-\omega_2)=B(-\omega_1-\omega_2,\omega_2) \qquad (B.19)$$

(iii) Bispectrum is a complex function; hence, there exists a conjugate function for each bispectrum. That is,

$$B(\omega_1,\omega_2)=B^*(-\omega_1,-\omega_2)$$

Following this property, there are six conjugate bispectra. The absolute value of each of the twelve bispectra is the same. Hence, it is sufficient to evaluate the bispectrum only in the positive octant domain given by $0\leqslant\omega_2<\omega_1\leqslant|\omega_1+\omega_2|$

(iv) From Eqs. (B.15) and (B.18), we have

$$M(0,0) = E[x^3(t)] = \int\limits_{0}^{\infty} \int\limits_{0}^{\infty} B(\omega_1,\omega_2) d\omega_1 \, d\omega_2 \qquad (B.20)$$

The above equation implies that the volume of the bispectrum $B(\omega_1,\omega_2)$ yields the third moment of the random process $x(t)$. Hence, six times the integrated volume of $B(\omega_1,\omega_2)$ in the positive octant domain yields the third moment.

APPENDIX C FOURIER TRANSFORM AND HILBERT TRANSFORM

1. Fourier transform

DEFINITION. The following transformation between the time and frequency domain is called the *Fourier transform pair*, and denoted by $x(t) \leftrightarrow X(\omega)$.

$$X(\omega) = \int_{-\infty}^{\infty} x(t)\, e^{-i\omega t} dt$$

$$X(t) = \frac{1}{2\pi} \int_{-\infty}^{\infty} X(\omega)\, e^{i\omega t} d\omega \qquad (C.1)$$

For the frequency f in Hz, the Fourier transform pair is given by

$$X(f) = \int_{-\infty}^{\infty} x(t)\, e^{-i2\pi ft} dt$$

$$x(t) = \int_{-\infty}^{\infty} X(f)\, e^{i2\pi ft} df \qquad (C.2)$$

NOTES.

(i) If $x(t)$ is a real-valued even function, then the Fourier transform pair is given by

$$X(\omega) = 2 \int_{0}^{\infty} x(t)\, \cos \omega t\, dt$$

$$x(t) = \frac{1}{\pi} \int_{0}^{\infty} X(\omega)\, \cos \omega t\, d\omega \qquad (C.3)$$

(ii) If $x(t)$ is a real-valued odd function, we have

$$X(\omega) = -2 \int x(t) \sin \omega t \, dt$$

$$x(t) \;\; = -\frac{1}{\pi} \int X(\omega) \sin \omega t \, d\omega \qquad\qquad\qquad (C.4)$$

PROPERTIES. Let $x(t) \leftrightarrow X(\omega)$. Then, we have the following properties.

(i) Linearity

$$\sum_j a_j x_j(t) \leftrightarrow \sum_j a_j X_j(\omega) \qquad\qquad\qquad (C.5)$$

(ii) Change of sign

$$x(-t) \leftrightarrow X(-\omega) = X^*(\omega) \qquad\qquad\qquad (C.6)$$

(iii) Time shifting

$$x(t-t_0) \leftrightarrow X(\omega) \, e^{-i\omega t_0} \qquad\qquad\qquad (C.7)$$

(iv) Frequency shifting

$$x(t) \, e^{i\omega_0 t} \leftrightarrow X(\omega - \omega_0) \qquad\qquad\qquad (C.8)$$

(v) Time scaling

$$x(at) \leftrightarrow \frac{1}{|a|} \, X\!\left(\frac{\omega}{a}\right) \qquad\qquad\qquad (C.9)$$

(vi) Symmetry

$$X(t) \leftrightarrow 2\pi x(-\omega) \qquad\qquad\qquad (C.10)$$

(vii) Frequency convolution

$$x_1(t) x_2(t) \leftrightarrow \frac{1}{2\pi} \, X_1(\omega) \star X_2(\omega)$$

$$\text{where } X_1(\omega) \star X_2(\omega) = \int X_1(\omega - \lambda) X_2(\lambda) d\lambda \qquad\qquad (C.11)$$

(viii) Time convolution

$$x_1(t) \star x_2(t) \leftrightarrow X_1(\omega) X_2(\omega)$$

$$\text{where } x_1(t) \star x_2(t) = \int x_1(t-\tau) x_2(\tau) d\tau \qquad\qquad (C.12)$$

(ix) Time differentiation

$$\frac{\mathrm{d}^n}{\mathrm{d}t^n}x(t) \leftrightarrow (\mathrm{i}\omega)^n X(\omega) \qquad (C.13)$$

(x) Frequency differentiation

$$(-\mathrm{i}t)^n x(t) \leftrightarrow \frac{\mathrm{d}^n}{\mathrm{d}t^n}X(\omega) \qquad (C.14)$$

2. Hilbert transform

DEFINITION. Let $x(t)$ be a real-valued function in the interval $-\infty < x < \infty$. The *Hilbert transform* of $x(t)$, denoted $\tilde{x}(t)$, is defined as

$$\tilde{x}(t) = \frac{1}{\pi}\int\limits_{-\infty}^{\infty}\frac{x(\tau)}{t-\tau}\,\mathrm{d}\tau \qquad (C.15)$$

NOTES. The right side of Eq. (C.15) is a convolution integral of $x(t)$ and $1/\pi t$.

PROPERTIES.

(i) Linearity

$$\text{Hilbert T.}\left\{\sum_j a_j x_j(t)\right\} = \sum_j a_j \tilde{x}_j(t) \qquad (C.16)$$

(ii) Successive Hilbert transform

$$\text{Hilbert T.}\{\tilde{x}(t)\} = -x(t) \qquad (C.17)$$

(iii) Orthogonality

$$\int\limits_{-\infty}^{\infty} x(t)\tilde{x}(t)\,\mathrm{d}t = 0 \qquad (C.18)$$

(iv) Convolution

$$\begin{aligned}\text{Hilbert T.}\{x_1(t)\star x_2(t)\} &= \tilde{x}_1(t)\star x_2(t)\\ &= x_1(t)\star \tilde{x}_2(t)\end{aligned} \qquad (C.19)$$

(v) Fourier transform of $\tilde{x}(t)$

$$\text{Fourier T.}\{\tilde{x}(t)\} = \begin{cases} -\mathrm{i}X(\omega) & \text{for } \omega<0 \\ 0 & \omega=0 \\ \mathrm{i}X(\omega) & \omega>0 \end{cases} \qquad (C.20)$$

(vi) Inverse Hilbert transform

$$x(t) = -\int_{-\infty}^{\infty} \frac{\tilde{x}(u)}{\pi(t-u)}\, du \qquad\qquad (C.21)$$

APPLICATION. Let $z(t)$ be a random process defined by

$$z(t) = x(t) + i\tilde{x}(t)$$

$z(t)$ may be written as

$$z(t) = \rho(t) \exp\{i\phi(t)\}$$

Then, the envelope process $\rho(t)$ is given by

$$\rho(t) = \{x^2(t) + \tilde{x}^2(t)\}^{1/2} \qquad\qquad (C.22)$$

REFERENCES

Allender, J. *et al.* (1989). The WADIC Project: a comprehensive field evaluation of directional wave instrumentation. *Ocean Engng.*, **16**, 505–36.

Arhan, M.K., Cavanié, A. and Ezraty, R. (1976). Etude théorique et expérimentale de la relation hauteur-période des vagues de tempete. I.F.P. 24191. Centre National pour l'Exploitation des Oceans.

Arhan, M. and Ezraty, R. (1978). Statistical relationship between successive wave heights. *Oceanol. Acta*, **1**, 151–8.

Arhan, M. and Plaisted, R.O. (1981). Non-linear deformation of seawave profiles in intermediate and shallow water. *Oceanol. Acta*, **4**, 2, 107–15.

Athanassoulis, G.A., Skarsoulis, E.K. and Belibassakis, K.A. (1994). Bivariate distribution with given marginals with an application to wave climate description. *J. Appl. Ocean Res.*, **16**, 1–17.

Athanassoulis, G. A. and Soukissian, T.H. (1993). A new model for long-term stochastic prediction of cumulative quantities. In *Proc. 12th Conf. on Offshore Mech. & Arctic Engng.*, **2**, 417–24.

Athanassoulis, G.A. and Stefanakos, Ch. N. (1995). A nonstationary stochastic model for long-term time series of significant wave height. *J. Geophys. Res.*, **100**, C8, 14149–62.

Athanassoulis, G.A., Vranas, P.B. and Soukissian, T.H. (1992). A new model for long-term stochastic analysis and prediction – part 1: theoretical background. *J. Ship Res.*, **36**, 1–16.

Banner, M.L. and Phillips, O.M. (1974). On the incipient breaking of small scale waves. *J. Fluid Mech.*, **65**, 4, 647–56.

Battjes, J.A. (1972). Long-term wave height distribution at seven stations around the British Isles. *Dtsch. Hydrogr. Z.*, **25**, 4, 180–9.

Battjes, J.A. and van Vledder, G.P. (1984). Verification of Kimura's theory for wave group statistics. In *Proc. 19th Coastal Engng. Conf.*, **1**, 642–8, Houston, Texas.

Battjes, J.A., Zitman, T.J. and Holthuijsen, L.H. (1987). A reanalysis of the spectra observed in JONSWAP. *J. Phys. Oceanogr.*, **17**, 8, 1288–95.

Benoit, M. (1992). Practical comparative performance survey of methods used for estimating directional wave spectra from heave–pitch–roll data. In *Proc. 23rd Conf. Coastal Engng.*, **1**, 62–75, Venice.

Benoit, M. (1993). Extensive comparison of directional wave analysis methods from gauge array data. In *Proc. 2nd Symp. Ocean Wave Measurement and Analysis*, 740–54, New Orleans, Louisiana.

Benoit, M. and Teisson, C. (1994). Laboratory comparison of directional wave measurement systems and analysis techniques. In *Proc. 24th Int. Conf. Coastal Engng.*, 1, 42–56, Kobe, Japan.

Bitner, E.M. (1980). Nonlinear effects of the statistical model of shallow-water wind waves. *Appl. Ocean Res.*, 2, 2, 63–73.

Borgman, L.E. (1969). Directional spectra models for design use. In *Offshore Tech. Conf. OTC 1069.*, 1, 721–36, Houston, Texas.

Borgman, L.E. (1973). Probabilities for highest wave in hurricane. *J. Waterways, Harbours & Coastal Engng.*, 99, WW2, 185–207.

Borgman, L.E. and Panicker, N.N. (1970). Design study for a suggested wave gage array off Point Mugu, California. Technical Report HEL. 1–14. Hydraulic Engng. Lab., Univ. of California, Berkeley.

Bouws, E. (1978). Wind and wave climate in the Netherlands sectors of the North Sea between 53° and 54° north latitude. Report W.R. 78–9, Koninklijk Netherlands Meterologisch Inst.

Bouws, E., Günther, H., Rosenthal, W. and Vincent, C.L. (1985). Similarity of the wind wave spectrum in finite depth water : 1. Spectral form. *J. Geophys. Res.*, 90, C1, 975–86.

Bretschneider, C.L. (1959). Wave variability and wave spectra for wind-generated gravity waves. Tech. Memo. No. 118. Beach Erosion Board, US Army Corps of Eng., Washington, DC.

Bretschneider, C.L. and Tamaye, E.E. (1976). Hurricane wind and wave forecasting techniques. In *Proc. 15th Coastal Engng. Conf.*, 1, 202–37, Honolulu, Hawaii.

Briggs, M.J. (1984). Calculation of directional wave spectra by the Maximum Entropy Method of spectra analysis. In *Proc. 19 Int. Conf. on Coastal Engng.*, 1, 484–500, Houston, Texas.

Brissette, F.P. and Tsanis, I.K. (1994). Estimation of wave directional spectrum from pitch–roll buoy data. *J. Waterway, Port, Coastal and Ocean Engng.*, 120, 1, 93–115.

Burg, J.P. (1967). Maximum entropy spectral analysis. In *Proc. 37th Annual Meeting, Soc. Explor. Geophysics.*

Burling, R.W. (1959). The spectrum of waves at short fetches. *Dtsch. Hydrogr. Z.*, 12, 2, 45–117.

Burrows, R. and Salih, B.A. (1986). Statistical modelling of long-term wave climates. In *Proc. 20th Int. Conf. Coastal Engng.*, 1, 42–56, Taipei, Taiwan.

Capon, J. (1969). High-resolution frequency-wave number spectrum analysis. *Proc. IEEE*, 57, 8, 1408–18.

Capon J., Greenfield, R.J. and Kolker, R.J. (1967). Multidimensional maximum-likelihood processing of a large aperture seismic array. *Proc. IEEE*, 55, 192–211.

Cardone, V.J., Pierson, W.J. and Ward, E.G. (1976). Hindcasting the directional spectra of hurricane-generated seas. *J. Petroleum Technol.*, 28, 385–95.

Carter, D.J.T. and Challenor, P.G. (1983). Methods of fitting the Fisher–Tippett Type I extreme value distribution. *Ocean Engng*, 10, 3, 191–9.

Cartwright, D.E. (1961). The use of directional spectra in studying the output of a wave recorder on a moving ship. In *Proc. Conf. Ocean Wave Spectra*, 203–18, Easton, Maryland.

Cartwright, D.E. and Longuet-Higgins, M.S. (1956). The statistical distribution of the maxima of a random function. *Proc. Roy. Soc. London, Ser. A*, **237**, 212–32.

Cartwright, D.E. and Smith, N.D. (1964). Buoy techniques for obtaining directional wave spectra. In *Proc. Buoy Technol. Symp.*, 173–82, Washington, DC: Mar. Tech. Soc.

Cavanié, A., Arhan, M.K. and Ezraty, E. (1976). A statistical relationship between individual heights and periods of storm waves. In *Proc. Conf. Behaviour Offshore Structure*, **2**, 354–60, Trondhheim.

Chakrabarti, S.K., Snider, R.H. and Feldhausen, P.H. (1974). Mean length of runs of ocean waves. *J. Geophys. Res.*, **79**, 36, 5665–7.

Charnock, H.A. (1958). A note on empirical wind-wave formulae. *Quart. J. Roy. Meteorol. Soc.*, **84**, 362.

Cokelet, E.D. (1977). Breaking waves. *Nature*, **267**, 5614, 769–74.

Cramér, H. (1966). *Mathematical Methods of Statistics*. Princeton Univ. Press, Princeton, NJ.

Davenport, W.B. and Root, W.L. (1958). *An Introduction to the Theory of Random Signals and Noise*. McGraw-Hill, New York.

Davis, M.C. and Zarnick, E.E. (1964). Testing ship models in transient waves. In *Proc. 5th Symp. on Naval Hydrogr.*, Office of Naval Research, ACR-112, 507–43.

Davis, R.E. and Regier, L.A. (1977). Methods for estimating directional wave spectra from multi-element arrays. *J. Marine Res.*, **35**, 3, 453–77.

Dawson, T.H., Kriebel, D.L. and Wallendorf, L.A. (1991). Experimental study of wave groups in deep-water random waves. *Appl. Ocean Res.*, **13**, 3, 116–31.

Dawson, T.H., Kriebel, D.L. and Wallendorf, L.A. (1993). Breaking waves in laboratory-generated JONSWAP seas. *Appl. Ocean Res.*, **15**, 85–93.

Dean, R.G. (1968). Breaking wave criteria: A study employing a numerical wave theory. In *Proc. 11th Int. Conf. Coastal Engng.*, **1**, 108–123.

Donelan, M. (1990). Air–sea interaction. In *The Sea, Ocean Engineering Science*, **9**, Part A, 239–92, John Wiley & Sons, New York.

Donelan, M.A., Hamilton, J. and Hui, W.H. (1985). Directional spectra for wind-generated waves. *Phil. Trans. Roy. London*, **A 315**, 509–62.

Donelan, M., Longuet-Higgins, M.S. and Turner, J.S. (1972). Periodicity in whitecaps. *Nature*, **239**, 449–50.

Donoso, M.C.I., LeMehaute, B. and Long, R.B. (1987). Data base of maximum sea states during hurricanes. *J. Waterway, Port, Coastal and Ocean Engng*, **113**, 4, 311–27.

Edgeworth, F.Y. (1904). The law of error. *Trans. Camb. Phil. Soc.* 20, 36–65.

Ellison, T.H. (1956). *Atmospheric Turbulence, Surveys in Mechanics*. Cambridge Univ. Press, London.

Epstein, B. (1949). The distribution of extreme values in sample whose members are subject to a Markov chain condition. *Annals Math. Stats*, 20, 590–4.

Ewing, J.A. (1973). Mean length of runs of high waves. *J. Geophys., Res.*, **78**, 12, 1933–6.

Fang, Z.S. and Hogben, N. (1982). Analysis and prediction of long term probability distribution of wave heights and periods. National Marine Institute Report R416, Feltham, England.

Fisher, R.A. (1922). On the mathematical foundations of theoretical statistics. *Phil. Trans. Roy. Soc. Series A*, **222**, 309–68.

Fisher, R.A. and Tippett, L.H.C. (1928). Limiting forms of the frequency distributions of the largest or smallest number of a sample. *Proc. Cambridge Philos. Soc.*, **24**, Part 2, 180–90.

Forristall, G.Z. (1978). On the statistical distribution of wave heights in a storm. *J. Geophys. Res.*, **83**, 2353–8.

Forristall, G.Z. (1981). Measurements of a saturated range in ocean wave spectra. *J. Geophys. Res.*, **86**, C9, 8075–84.

Forristall, G.Z. (1984). The distance of measured and simulated wave heights as a function of spectral shape. *J. Geophys. Res.*, **89**, C6, 10547–52.

Foster, E.R. (1982). JONSWAP spectral formulation applied to hurricane-generated seas. University of Florida, Report UFL/COEL-82/004.

Fréchet, M. (1927). Sur la loi de probabilité de l'écart maximum. *Ann. de la Soc. Polonaise de Math.*, **6**, 93–116.

Garrett, J. (1969). Some new observations on the equilibrium region of the wind-wave spectrum. *J. Marine Res.*, **27**, 273–7.

Glad, I.K. and Krogstad, H.E. (1992). The maximum-likelihood property of estimators of wave parameters from heave, pitch and roll buoys. *J. Atmospheric and Oceanic Technol.*, **9**, 2, 169–73.

Gnedenko, B.V. (1943). Sur la distribution limite du terme maximum d'une série aléatoire. *Ann. Math.*, **44**, 423–53.

Goda, Y. (1970). Numerical experiments on wave statistics with spectral simulation. Report Port Harbour Res. Inst. **9**, 3, 3–57.

Goda, Y. (1975). Irregular wave deformation in the surf zone. *Coastal Engng.*, Japan, **18**, 13–26.

Goda, Y. (1976). On wave groups. In *Proc. Conf. Behaviour Offshore Structures*, **1**, 1–14, Trondheim.

Gumbel, E.J. (1958). *Statistics of Extremes*. Columbia Univ. Press, New York.

Gumbel, E.J., Greenwood, J. A. and Durand, D. (1953). The circular normal distribution; theory and tables. *J. Am. Stat. Assoc.*, **48**, 131–152.

Haring, R.E., Osborne, A.R. and Spencer, L.P. (1976). Extreme wave parameters based on continental shelf wave records. In *Proc. 15th Coastal Engng. Conf.*, **1**, 151–70, Honolulu, Hawaii.

Hashimoto, N. and Kobune, K. (1988). Directional spectrum estimation from a Bayesian approach. In *Proc. 21st Coastal Engng. Conf.*, **1**, 62–76, Costa del Sol-Malaga, Spain.

Hashimoto, N., Nagai, T. and Asai, T. (1994). Extension of the maximum entropy principle method for directional wave spectrum estimation. In *Proc. 24th Conf. on Coastal Engng.*, **1**, 232–46.

Hasselmann, D.E., Dunckel, M. and Ewing, J.A. (1980). Directional wave spectra observed during JONSWAP 1973. *J. Phys. Oceanogr.*, **10**, 1264–80.

Hasselmann, K. (1962). On the nonlinear energy transfer in a gravity wave spectrum. 1; General theory. *J. Fluid Mech.*, **12**, 481–500.

Hasselmann, K. *et al.* (1973). Measurements of wind-wave growth and swell decay during the Joint North Sea Wave Project (JONSWAP). Dtsch. Hydrogr. Inst., Hamburg.

Hasselmann, K., Munk, W. and MacDonald, G. (1962). Bispectra of ocean waves. In *Proc. Symp. Time Ser. Anal.*, Chapter 8, 125–39.

Haver, S. (1985). Wave climate off Northern Norway. *Appl. Ocean Res.*, **7**, 2, 85–92.

Hess, G.D., Hidy, G.M. and Plate, E.J. (1969). Comparison between wind waves at sea and in the laboratory. *J. Marine Res.*, **27**, 2, 216–25.

Holthuijsen, L.H. (1981). Wave directionality inferred from stereo photos. In *Proc. Dir. Wave Spectra Application*, 42–9, Berkeley: Univ. of California.

Holthuijsen, L.H. and Herbers, T.H.C. (1986). Statistics of breaking waves observed as whitecaps in the open sea. *J. Phys. Oceanogr.*, **16**, 290–7.

Huang, N.E., Long, S.R., Tung, G.G. and Yuan, Y. (1983). A non-Gaussian statistical model for surface elevation of nonlinear random waves. *J. Geophysical Res.*, **88**, C12, 7597–606.

Hudspeth, R.T. and Chen, M-C. (1979). Digitial simulation of nonlinear random waves. *J. Waterway, Port, Coastal & Ocean Engng.*, **105**, WW1, 67–85.

Hughes, S.A. and Borgman, L.E. (1987). Beta-Rayleigh distribution for shallow water wave heights. In *Proc. Conf. Coastal Hydrodynamics*, 17–31, Delaware: University of Delaware.

Isobe, M. (1988). On joint distribution of wave heights and directions. In *Proc. 21st Conf. Coastal Engng.*, **1**, 524–38.

Isobe, M., Kondo, K. and Horikawa, K. (1984). Extension of MLM for estimating directional wave spectrum. In *Proc. Sympos. on Description and Modelling of Directional Seas*, Danish Hydraulic Inst.

Isaacson, M. and Mackensie, N.G. (1981). Long-term distribution of ocean waves; A review. *J. Waterways, Port, Coastal & Ocean Engng*, **107**, 93–109.

Jaynes, E.T. (1957). Information theory and statistical mechanics. *Phys. Rev.*, **106**, 620–30.

Jeffreys, H. (1924). On the formation of waves by wind. *Proc. Roy. Soc. London Ser. A*, **107**, 189–206.

Jeffreys, H. (1925). On the formation of waves by wind, II. *Proc. Roy. Soc. London Ser. A*, **110**, 341–7.

Johnson, N.L. and Leone, F.C. (1964). *Statistics and Experimental Design in Engineering and the Physical Sciences*. John Wiley & Sons, New York.

Kac, M. and Siegert, A.J.F. (1947). On the theory of noise in radio receivers with square law detectors. *J. Appl. Phys.*, **8**, 383–97.

Kendall, M. G. and Stuart, A. (1961). *The Advanced Theory of Statistics*, Vol. 1. Hafner, New York.

Kahma, K.K. (1981). A study of the growth of the wave spectrum with fetch. *J. Phys. Oceanogr.*, **11**, 1503–15.

Kerstens, J.G.M., Pacheco, L.A. and Edwards, G. (1988). A Bayesian method for the estimation of return values of wave heights. *Ocean Engng*, **15**, 2, 153–70.

Kimeldorf, G. and Sampson, A. (1975). One-parameter families of bivariate distribution with fixed marginals. *Commun. Stat.*, **4**, 293–301.

Kimura, A. (1980). Statistical properties of random wave groups. In *Proc. 17th Int. Conf. Coastal Engng.*, **3**, 2955–73, Sydney.

Kimura, A. and Ohta, T. (1994). Probability of the freak wave appearance in a 3-dimensional sea condition. In *Proc. 24th Int. Conf. on Coastal Engng.*, **1**, 356–69.

Kinsman, B. (1965). *Wind Waves*. Prentice Hall, Englewood Cliffs, NJ.

Kitaigorodskii, S. A. (1961). Application of the theory of similarity to the analysis of wind-generated wave motion as a stochastic process. *Bull. Akad. Nauk SSSR, Ser. Geophys.*, **105–117**, 73–80 (translated from Russian).

Kitaigorodskii, S.A. (1973). *The Physics of Air–Sea Interaction.* (translated from Russian), TT72-50062, Nat. Tech. Inf. Cent., Springfield, Virginia.

Kitaigorodskii, S.A. (1983). On the theory of the equilibrium range in the spectrum of wind-generated gravity waves. *J. Phys. Oceanogr.,* 13, 5, 816–27.

Kitaigorodskii, S.A., Krasitskii, V.P. and Zaslavskii, M.M. (1975). On Phillips' theory of equilibrium range in the spectra of wind-generated gravity waves. *J. Phys. Oceanogr.,* 5, 3, 410–20.

Kjeldsen, S.P. (1982). The two and three dimensional deterministic freak waves. In *Proc. 18th Conf. on Coastal Engng.,* 1, 677–94, Cape Town.

Kjeldsen, S.P. and Myrhaug, D. (1978). Kinematics and dynamics of breaking waves. Report STT 60A78100, Ships in Rough Seas, Part 4, Norwegian Hydro. Lab.

Klinting, P. and Sand, S.E. (1987). Analysis of prototype freak waves. In *Proc. Coastal Hydrodynamics,* Ame. Soc. Civil Eng., 618–32.

Kobune, K. and Hashimoto, N. (1986). Estimation of directional spectra from the maximum entropy principle. In *Proc. 5th Symp. Offshore Mech. and Arch. Eng.,* 80–5.

Krogstad, H. (1985). Height and period distributions of extreme waves. *J. Appl. Ocean Res.,* 7, 3, 158–65.

Krogstad, H.E., Gordon, R.L. and Millar, M.C. (1988). High-resolution directional wave spectra from horizontally mounted acoustic Doppler current meters. *J. Atmos. Oceanic Technol.,* 5, 340–52.

Kuo, C.T. and Kuo, S.J. (1975). Effect of wave breaking on statistical distribution of wave heights. In *Proc. Civil Eng. in the Oceans III,* Ame. Soc. Civil Eng., 1211–31.

Kwon, J.G. and Deguchi, I. (1994). On the joint distribution of wave height, period and direction of individual waves in a three-dimensional random seas. In *Proc. 24th Int. Conf. Coastal Engng.,* 1, 370–83, Kobe, Japan.

Labeyrie, J. (1990). Stationary and transient states of random seas. *Marine Structures,* 3, 43–58.

Langley, R.S. (1987). A statistical analysis of non-linear random waves. *Ocean Engng,* 14, 5, 389–407.

Lettenmaier, D.P. and Burges, S.J. (1982). Gumbel's extreme value I distribution; A new look. *J. Hydraul. Div. ASCE,* 108, HY4, 502–14.

Lindgren, G. (1970). Some properties of a normal process near a local maximum. *Ann. Math. Stat.,* 41, 1870–83.

Lindgren, G. (1972). Wave-length and amplitude in Gaussian noise. *Adv. Appl. Probability,* 4, 81–108.

Lindgren, G. and Rychlik, I. (1982). Wave characteristic distributions for Gaussian waves – wave-length, amplitude and steepness. *Ocean Engng.,* 9, 5, 411–32.

Long, R.B. (1980). The statistical evaluation of directional spectrum estimates derived from pitch/roll buoy data. *J. Phys. Oceanogr.,* 10, 944–52.

Long, R.B. and Hasselmann, K. (1979). A variational technique for extracting directional spectra from multicomponent wave data. *J. Phys. Oceanogr.,* 9, 373–81.

Longuet-Higgins, M.S. (1952). On the statistical distribution of the heights of sea waves. *J. Marine Res.,* 11, 3, 245–66.

Longuet-Higgins, M.S. (1957). The statistical analysis of a random moving surface. *Philos. Trans. Roy. Soc. London, Ser. A*, **249**, 321–87.

Longuet-Higgins, M.S. (1962). The distribution of intervals between a stationary random function. *Philos. Trans. Roy. Soc. London, Ser. A*, **254**, 557–99.

Longuet-Higgins, M.S. (1963a). The generation of capillary waves by steep gravity waves. *J. Fluid Mech.*, **16**, 138–59.

Longuet-Higgins, M.S. (1963b). The effect of non-linearities on statistical distributions in the theory of sea waves. *J. Fluid Mech.*, **17**, Part 3, 459–80.

Longuet-Higgins, M.S. (1969). On wave breaking and equilibrium spectrum of wind-generated waves. *Proc. Roy. Soc. London, Ser. A*, **310**, 151–9.

Longuet-Higgins, M.S. (1974). Breaking waves – In deep or shallow water. In *Proc. 10th Symp. on Naval Hydrogr.*, pp. 597–605.

Longuet-Higgins, M.S. (1975). On the joint distribution of the periods and amplitudes of sea waves. *J. Geophys. Res.*, **80**, 18, 2688–94.

Longuet-Higgins, M.S. (1976). Recent development in the study of breaking waves. In *Proc. 15th Conf. Coastal Engng.*, **1**, 441–60, Honolulu, Hawaii.

Longuet-Higgins, M.S. (1978a). The instabilities of gravity waves of finite amplitude in deep water. I: Superharmonics. *Proc. Roy. Soc. London, Ser. A*, **360**, 471–88.

Longuet-Higgins, M.S. (1978b). The instabilities of gravity waves of finite amplitude in deep water. II: Subharmonics. *Proc. Roy. Soc. London, Ser. A*, **360**, 489–505.

Longuet-Higgins, M.S. (1979). The almost-highest wave: A simple approximation. *J. Fluid Mech.*, **94**, 269–273.

Longuet-Higgins, M.S. (1983). On the joint distribution of wave periods and amplitudes in a random wave field. *Proc. Roy. Soc. London, Ser. A*, **389**, 241–58.

Longuet-Higgins, M.S. (1984). Statistical properties of wave groups in a random sea-state. *Philos. Trans. R. Soc. Lond., Ser. A*, **312**, 219–50.

Longuet-Higgins, M.S. (1985). Accelerations in steep gravity waves. *J. Phys. Oceanogr.*, **15**, 1570–9.

Longuet-Higgins, M.S. Cartwright, D.E. and Smith N.D. (1961). Observations of the directional spectrum of sea waves using the motions of a floating buoy. In *Proc. Conf. Ocean Wave Spectra*, pp. 111–32.

Longuet-Higgins, M.S. and Cokelet, E.D. (1978). The deformation of steep surface waves on water. II: Growth of normal-mode instabilities. *Proc. Roy. Soc. London, Ser. A*, **364**, 1–28.

Longuet-Higgins, M.S. and Fox, M.J.H. (1977). Theory of the almost-highest wave: The inner solution. *J. Fluid Mech.*, **80**, 721–41.

Lygre, A. and Krogstad, H.E. (1986). Maximum entropy estimation of the directional distribution in ocean wave spectra. *J. Phys. Oceanogr.*, **16**, 2052–60.

Mardia, K.V. (1967). Some contributions to contingency-type bivariate distributions. *Biometrika*, **54**, 235–49.

Marsden, R.F. and Juszko, B.A. (1987). An eigenvector method for the calculation of directional spectra from heave, pitch and roll buoy data. *J. Phys. Oceanogr.*, **17**, 2157–67.

Mathiesen, J. and Bitner-Gregersen, E. (1990). Joint distribution for significant wave height and zero-crossing period. *Appl. Ocean Res.*, **12**, 2, 93–103.

Medina, J.R., Giménez, M.H. and Hudspeth, R.T. (1991). A wave climate simulator. In *Proc. 24th Int. Assoc. Hydrogr. Res. Congress*, B521-8.

Medina, J.R. and Hudspeth, R.T. (1990). A review of the analyses of ocean wave groups. *Coastal Engng.*, 14, 515-42.

Melville, W.K. (1982). The instability and breaking of deep-water waves. *J. Fluid Mech.*, 115, 165-85.

Mesa, D. (1985). Statistical estimation of extreme sea severity. University of Florida Report, UFL/COEL-85/004.

Michel, J.H. (1893). The highest waves in water. *Philos. Mag.*, 36, 430-7.

Middleton, D. (1960). *An Introduction to Statistical Communication Theory*. McGraw-Hill, New York.

Miles, J.W. (1957). On the generation of surface waves by shear flows. *J. Fluid Mech.*, 3, 185-204.

Miles, J.W. (1959a). On the generation of surface waves by shear flows, Part 2. *J. Fluid Mech.*, 6, 568-82.

Miles, J.W. (1959b). On the generation of surface waves by shear flows, Part 3. *J. Fluid Mech.*, 6, 583-98.

Miles, J.W. (1960). On the generation of surface waves by turbulent shear flows. *J. Fluid Mech.*, 7, 469-78.

Miles, J.W. (1962). On the generation of surface waves by shear flows, Part 4. *J. Fluid Mech.*, 13, 433-48.

Mitsuyasu, H. *et al.* (1975). Observations of the directional spectrum of ocean waves using a cloverleaf buoy. *J. Phys. Oceanogr.*, 5, 750-60.

Mitsuyasu, *et al.* (1980). Observation of the power spectrum of ocean waves using a clover-leaf buoy. *J. Phys. Oceanogra.*, 10, 2, 286-96.

Mollo-Christensen, E. and Ramamonjiarisoa, A. (1978). Modelling the presence of wave groups in a random wave field. *J. Geophys. Res.*, 83, C8, 4117-22.

Morgenstern, D. (1956). Einfache beispiele zweidimensionaler verteilungen. *Mitteilungsblatt für Math. Statistik*, 8, 234-5.

Moskowitz, L. Pierson, W.J. and Mehr, E. (1963). Wave spectra estimated from wave records obtained by the OWS WEATHER EXPLORER and the OWS WEATHER REPORTER. New York University, College of Engineering, Department of Meteorology and Oceanography.

Moyal, J.E. (1949). Stochastic processes and statistical physics. *J. Roy. Stat. Soc.*, B,, 11, 150-210.

Muir, L.R. and El-Shaarawi, A.H. (1986). On the calculation of extreme wave heights: A review. *Ocean Engng*, 1, 13, 93-118.

Myrhaug, D. and Rue, H. (1993). Note on a joint distribution of successive wave periods. *J. Ship Res.*, 37, 3, 208-12.

Naess, A. (1982). Extreme value estimates based on the envelope concept. *J. Appl. Ocean Res.*, 4, 3, 181-7.

Naess, A. (1985a). On the distribution of crest to trough wave heights. *Ocean Engng.*, 12, 3, 221-34.

Naess, A. (1985b). Statistical analysis of second-order response of marine structures. *J. Ship Res.*, 29, 4, 270-84.

Nath, J.H. and Ramsey, F.L. (1976). Probability distributions of breaking wave height emphasizing the utilization of the JONSWAP spectrum. *J. Phys. Oceanogr.*, 6, 316-23.

Neumann, G. (1953). On ocean wave spectra and a new method of forecasting wind-generated sea. Tech. Memo. No. 43, Beach Erosion Board, U.S. Army Corps of Engineers, Washington, DC.

NOAA Buoy Office (1975). Buoy observation during Hurricane ELOISE.

NOAA Buoy Office (1978). Buoy observations during Hurricane BELLE, August 1976.

NOAA Data Buoy Office (1981). Buoy observations during Hurricane FREDERIC.

NOAA, National Data Buoy Center (1986). Buoy and C-man observations during the 1985 Atlantic hurricane season.

NOAA, National Data Buoy Center (1990). Climatic summaries for NDBC buoys and stations.

Nolte, K.G. and Hsu, F.H. (1972). Statistics of ocean wave groups. In *Proc. 4th Offshore Technol. Conf.*, 2, 637–44.

Nwogu, O. (1989). Maximum entropy estimation of directional wave spectra from an array of wave probes. *Appl. Ocean Res.*, 11, 4, 176–82.

Ochi, M.K. (1973). On prediction of extreme values. *J. Ship Res.*, 17, 1, 29–37.

Ochi, M.K. (1978a). Wave statistics for the design of ships and ocean structures. *Trans. Soc. Nav. Arch. Mar. Eng.*, 86, 47–76.

Ochi, M.K. (1978b). On long-term statistics for ocean and coastal waves. In *Proc. 16th Conf. Coastal Engng.*, 1, 59–75, Hamburg.

Ochi, M.K. (1979a). A series of JONSWAP wave spectra for offshore structure design. In *Proc. Int. Conf. Behav. Offshore Struct.*, pp. 75–86.

Ochi, M.K. (1979b). Extreme values of waves and ship responses subject to the Markov chain condition. *J. Ship Res.*, 23, 3, 188–97.

Ochi, M.K. (1981). Principles of extreme value statistics and their application. In *Proc. Extreme Loads Response Symposium, Soc. Naval Arch. Mar. Eng.*, pp. 15–30.

Ochi, M.K. (1985). Application of half-cycle analysis for estimation of fatigue loads on offshore structures. In *Proc. 4th Conf. on Structural Safety and Reliability*, 2, 361–70, Kobe.

Ochi, M.K. (1987). Design extreme value based on long-term and short-term approaches. In *Proc. 3rd Int. Symp. on Practical Design of Ships and Mobile Units*, 2, 808–17, Trondheim.

Ochi, M.K. (1990a). *Applied Probability and Stochastic Processes.* John Wiley & Sons, New York.

Ochi, M.K. (1990b). Stochastic description of offshore environment, In *Water Waves Kinematics*, Kluwer Academic, pp. 31–56.

Ochi, M.K. (1992). New approach for estimating the severest sea state from statistical data. In *Proc. 23rd Int. Conf. on Coastal Engng.*, 1, 512–25, Venice.

Ochi, M.K. (1993). On hurricane-generated seas. In *Proc. 2nd Int. Conf. on Ocean Wave Measurement and Analysis*, pp. 374–87, New Orleans.

Ochi, M.K. and Ahn, K. (1994). Non-Gaussian probability distribution of coastal waves. In *Proc. 24th Conf. Coastal Engng.*, 1, 482–96, Kobe.

Ochi, M.K. and Bales, S.L. (1977). Effect of various spectral formulations in predicting responses of marine vehicles and ocean structures. In *Proc. Offshore Technol. Conf., OTC 2743*, 1, 133–48, Houston.

Ochi, M.K. and Eckhoff, M.A. (1984). Prediction of wave heights based on half-cycle excursion analysis. *Ocean Engng*, 11, 6, 581–91.

Ochi, M.K. and Hubble, E.N. (1976). On six-parameter wave spectra. In *Proc. 15th Conf. Coastal Engng.*, 1, 301–28.

Ochi, M.K., Malakar, S.B. and Wang. W.C. (1982). Statistical analysis of coastal waves observed during the ARSLOE Project. Univ. of Florida Report, UFL/COEL/TR-45.

Ochi, M.K., Mesa D. and Liu, D.F. (1986). Estimation of extreme sea severity from measured daily maxima. In *Proc. 20th Int. Conf. Coastal Engng.*, 1, 647–59, Taipei.

Ochi, M.K., Passiliao, E.L. and Malakar, S.B. (1996). Joint probability distribution of significant wave height and average period. Univ. of Florida Rep. UFL/COEL/TR-110.

Ochi, M.K. and Sahinoglou, I.I. (1989a). Stochastic characteristics of wave group in random seas; Part 1, Time duration of and number of waves in a wave group. *J. Appl. Ocean Res.*, 11, 1, 39–50.

Ochi, M.K. and Sahinoglou, I.I. (1989b). Stochastic characteristics of wave group in random seas; Part 2, Frequency of occurrence of wave group. *J. Appl. Ocean Res.*, 11, 2, 89–99.

Ochi, M.K. and Tsai, C.H. (1983). Prediction of occurrence of breaking waves in deep water. *J. Phys. Oceanogr.*, 13, 11, 2008–19.

Ochi, M.K. and Wang, W.C. (1984). Non-Gaussian characteristics of coastal waves. In *Proc. 19th Coastal Engng. Conf.*, 1, 516–31, Houston.

Ochi, M.K. and Whalen, J.E. (1980). Prediction of the severest significant wave height. In *Proc. 17th. Conf. Coastal Engng.*, 1, 587–99, Houston.

Oltman-Shay, J. and Guza, R.T. (1984). A data adaptive ocean wave directional spectrum estimator for pitch/roll type measurements, *J. Phys. Oceanogr.*, 14, 1800–10.

Panicker, N.N. and Borgman, L.E. (1970). Directional spectra from wave-gage arrays. In *Proc. 12th Coastal Engng. Conf.*, 1, 117–36.

Patterson, M.M. (1974). Oceanographic data from hurricane CAMILLE. In *Proc. Offshore Technol. Conf. OTC 2109*, 2, 781–90, Houston.

Pawka, S.S. (1983). Island shadows in wave directional spectra. *J. Geophys. Res.*, 88, 2579–91.

Phillips, O.M. (1958). The equilibrium range in the spectrum of wind-generated waves. *J. Fluid Mech.*, 4, 785–90.

Phillips, O.M. (1966). *The Dynamics of the Upper Ocean*. Cambridge Univ. Press, London.

Phillips, O.M. (1967). The theory of wind-generated waves. In *Adv. Hydrosci.*, 4, 119–49.

Phillips, O.M. (1985). Spectral and statistical properties of the equilibrium range in wind-generated gravity waves. *J. Fluid Mech.*, 156, 505–31.

Pierce, R.D. (1985). Extreme value estimates for arbitrary band width Gaussian processes using the analytic envelope. *Ocean Engng*, 12, 6, 493–529.

Pierson, W.J. (1952). A unified mathematical theory for the analysis propagation and refraction of storm generated ocean surface waves – Parts I and II. Res. Division, College of Engineering, New York Univ.

Pierson, W.J. (1955). Wind generated gravity waves. In *Advances in Geophysics*, Chapter 2, pp. 93–178.

Pierson, W.J. and Moskowitz, L. (1964). A proposed spectral form for fully developed wind seas based on the similarity theory of S.A. Kitaigorodskii. *J. Geophys. Res.*, **69**, 24, 5181–90.

Pierson, W.J., Neumann, G. and James, R.W. (1958). Practical methods for observing and forecasting ocean waves by means of wave spectra and statistics. Hydrogr. Off. Publ., No. 603, U.S. Navy Hydrogr. Off. Washington, DC.

Plackett, R.L. (1965). A class of bivariate distributions. *J. Am. Stat. Assoc.* **60**, 516–22.

Ramamonjiarisoa, A. (1974). Contribution à l'Etude de la structure statistique et des mécanismes de géneration des vagues de vent. Thèse de doctorat détat. Univ. de Provence, Provence, France.

Ramberg, S.E. and Griffin, M. (1987). Laboratory study of steep and breaking deep water waves. *J. Waterway, Port, Coastal and Ocean Engng. ASCE*, **113**, 5, 493–506.

Rice, S.O. (1944). Mathematical analysis of random noise. *Bell Syst. Tech. J.*, **23**, 282–332.

Rice, S.O. (1945). Mathematical analysis of random noise. *Bell Syst. Tech. J.*, **24**, 46–156.

Rice, S.O. (958). Distribution of the duration of random noise. *Bell Syst. Tech.J.*, **37**, 581–635.

Robillard, D.J. and Ochi, M.K. (1997). Transition of stochastic characteristics of waves in the nearshore zone. In *Proc. 25th Coastal Engng. Conf.*, **1**, 878–88, Orlando.

Ross, D.B. (1979). Observing and predicting hurricane wind and wave conditions. Atlantic Oceanographic Meteoro. Lab. NOAA, Collected Reprints, 309–21.

Ross, D. and Cardone, V. (1978). A comparison of parametric and spectral hurricane wave prediction products. NATO Conf. on Turbulent Flax., NOAA Collected Reprint, 549–567.

Rudnick, P. (1951). Correlograms for Pacific Ocean waves. In *Proc. 2nd Berkeley Symp. on Math. Stat. & Probability*, 627–38.

Rye, H. (1974). Wave group formation among storm waves. In *Proc. 14th Int. Conf. on Coastal Engng.*, **1**, 164–83, Copenhagen.

Rye, H. (1979). Wave parameter studies and wave groups. In *Proc. Conf. on Sea Climatology*, pp. 89–123.

St. Denis, M. and Pierson, W.J. (1953). On the motions of ships in confused seas. *Trans. Soc. Nav. Archit. Mar. Eng.*, **61**, 280–357.

Sand, S.E., et al. (1990). Freak wave kinematics. *Water Wave Kinematics NATO AS1 Series E.*, **178**, 535–50, Kluwer Academic, Dordrecht.

Sawhney, M.D. (1962). A study of ocean wave amplitudes in terms of the theory of runs and a Markhov chain process. Tech. Report, Dept. Meteor. & Oceanography, New York University.

Scheffner, N.W. and Borgman, L.E. (1992). Stochastic time-series representation of wave data. *J. Waterway, Port, Coastal Ocean Engng*, **118**, 4, 337–51.

Shahul Hameed, T.S. and Baba, M. (1985). Wave height distribution in shallow water. *Ocean Engng.*, **12**, 4, 309–19.

Shum, K.T. and Melville, W.K. (1984). Estimates of the joint statistics of amplitudes and periods of ocean waves using an integral transform technique. *J. Geophy. Res.*, **89**, C4, 6467–76.

Skarsoulis, E.K. and Athanassoulis, G.A. (1993). The use of classical maximum likelihood method for estimating directional wave spectra from heave-pitch-roll time series. In *Proc. 3rd Int. Offshore and Polar Eng. Conf.*, Singapore, **3**, 1–8.

Sneider, R.H. and Chakrabarti, S.K. (1973). High wave conditions observed over the North Atlantic in March 1968. *J. Geophys. Res.*, **78**, 36, 8793–807.

Snyder, R.L. and Kennedy, R.M. (1983). On the formation of whitecaps by a threshold mechanism. Part 1: Basic formalism. *J. Phys. Oceangr.*, **13**, 1482–92.

Sobey, R.J. (1992). The distribution of zero-crossing wave heights and periods in a stationary sea state. *Ocean Engng*, **19**, 2, 101–18.

Srokosz, M.A. (1985). On the probability of wave breaking in deep water. *J. Phys. Oceanogr.*, **16**, 382–5.

Srokosz, M.A. and Challenor, P.G. (1987). Joint distributions of wave height and period: A critical comparison. *Ocean Engng.*, **14**, 4, 295–311.

Steele, K.E., Teng, C.C. and Wang, D.W.C. (1992). Wave direction measurements using pitch–roll buoys. *J. Ocean Engng.*, **19**, 4, 349–75.

Stokes, G.G. (1847). On the theory of oscillatory waves. *Trans. Camb. Philos. Soc.*, **8**, 441–55.

Sue, M.Y., Bergin, M., Marler, P. and Myrick, R. (1982). Experiments on nonlinear instabilities and evolution of steep gravity-wave trains. *J. Fluid Mech.*, **124**, 45–72.

Sverdrup, H.U. and Munk, W.H. (1947). Wind, sea and swell: Theory of relations for forecasting. Hydrogr. Off. Publ., No. 601. US Navy Hydrographic Office, Washington, DC.

Tayfun, M.A. (1980). Narrow-band nonlinear sea waves. *J. Geophys. Res.*, **85**, C3, 1548–52.

Tayfun, M.A. (1981). Distribution of crest-to-trough wave heights. *Proc. Am. Soc. Civ. Eng.*, **107**, WW3, 149–58.

Tayfun, M.Z. (1983). Frequency analysis of wave heights based on wave envelope. *J. Geophy. Res.*, **88**, C12, 7573–87.

Teng, C.C., Timpe, G.L., Palao, I.M. and Brown, D.A. (1993). Design waves and wave spectra for engineering application. In *Proc. Ocean Wave Measuremt and Analysis*, pp. 993–1007.

Thieke, R.J., Dean, R.G. and Garcia, A.W. (1993). The Daytona beach "large wave" event of 3 July 1992. In *Proc. Ocean Wave Measuremt. Analysis*, pp. 45–60, New Orleans.

Thompson, E.F. and Harris, D.L. (1972). A wave climatology for U.S. coastal waters. In *Proc. Offshore Technol. Conf., OTC 1693*, **2**, 675–88.

Toba, Y. (1973). Local balance in the air–sea boundary processes III; On the spectrum of wind waves. *J. Oceanogr. Soc. Jpn*, **29**, 209–20.

Toba, Y. (1978). Stochastic form of the growth of wind waves in a single-parameter presentation with physical implications. *J. Phys. Oceanogr.*, **8**, 3, 494–507.

Tung, C.C. (1974). Peak distribution of random wave-current force. *J. Eng. Mech. EM5*, 873–84.

Tung, C.C. and Huang, N.E. (1987). The effect of wave breaking on the wave energy spectrum. *J. Phys. Oceanogr.*, 17, 1156–62.

Uhlenbeck, G.E. (1943). Theory of random processes. Mass. Inst. Technol., Radiation Lab. Report 454.

Van Dorn, W.G. and Pazan, S.E. (1975). Laboratory investigation of wave breaking; Part 2, Deep water waves. Adv. Ocean Engng. Lab. Rep. 71, Scripps Inst. Oceanogr.

Van Vledder, G.P. (1992). Statistics of wave group parameters. In *Proc. 23rd Coastal Engng. Conf.*, 1, 946–59, Venice.

Weissman, M.A., Atakturk, S.S. and Katsaros, K.B. (1984). Detection of breaking events in a wind-generated wave field. *J. Phys. Oceanogr.* 14, 1608–19.

Wu, J. (1969). Wind stress and surface roughness at air–sea interface. *J. Geophys. Res.*, 74, 2, 444–55.

Wu, J. (1980). Wind-stress coefficients over sea surface near neutral conditions – A revisit. *J. Phys. Oceanogr.*, 10, 5, 727–40.

Wu, J. (1982). Wind-stress coefficients over sea surface from breeze to hurricane. *J. Geophys. Res.*, 87, C12, 9704–6.

Xu, D., Hwang, P.A. and Wu, J. (1986). Breaking of wind-generated waves. *J. Phys. Oceanogr.*, 16, 2172–8.

Yasuda, M. Mori, N. and Ito, K. (1992). Freak waves in unidirectional wave trains and their properties. In *Proc. 23rd Coastal Engng. Conf.*, 1, 751–64.

Yim, S.C.S., Burton, R.M. and Goulet, M.R. (1992). Practical methods of extreme value estimation based on measured time series for ocean systems. *Ocean Engng.*, 19, 3, 219–38.

Young, I.R. (1988). Parametric hurricane wave prediction model. *J. Waterway, Port, Coastal & Ocean Engng.*, 114, 5, 637–52.

Young, I.R. (1994). On the measurement of directional wave spectra. *J. Appl. Ocean Res.*, 16, 283–94.

Young, I.R. and Sobey, R.J. (1981). The numerical prediction of tropical cyclone wind-waves. Res. Bulletin No. CS20, University of North Queensland.

Yuen, H.C. and Lake, B.M. (1975). Nonlinear deep water waves: Theory and experiments. *Phys. Fluids.*, 18, 8, 956–60.

INDEX